Carbon Dioxide as a Source of Carbon

Biochemical and Chemical Uses

NATO ASI Series

Advanced Science Institutes Series

A series presenting the results of activities sponsored by the NATO Science Committee, which aims at the dissemination of advanced scientific and technological knowledge, with a view to strengthening links between scientific communities.

The series is published by an international board of publishers in conjunction with the NATO Scientific Affairs Division

A Life Sciences B Physics	Plenum Publishing Corporation London and New York
C Mathematical and Physical Sciences	D Reidel Publishing Company Dordrecht, Boston, Lancaster and Tokyo
D Behavioural and Social Sciences E Engineering and Materials Sciences	Martinus Nijhoff Publishers Dordrecht, Boston and Lancaster
F Computer and Systems Sciences G Ecological Sciences H Cell Biology	Springer-Verlag Berlin, Heidelberg, New York, London, Paris, and Tokyo

Series C: Mathematical and Physical Sciences Vol. 206

Carbon Dioxide as a Source of Carbon

Biochemical and Chemical Uses

edited by

M. Aresta

Department of Chemistry,
University of Bari, Italy

and

G. Forti

Department of Biology,
University of Milan, Italy

D. Reidel Publishing Company

Dordrecht / Boston / Lancaster / Tokyo

Published in cooperation with NATO Scientific Affairs Division

Proceedings of the NATO Advanced Study Institute on
Carbon Dioxide: Biochemical and Chemical Uses as a Source of Carbon
Pugnochiuso, Italy
June 22-July 3, 1986

Library of Congress Cataloging in Publication Data

NATO Advanced Study Institute on Carbon Dioxide: Biochemical and Chemical Uses as a Source of Carbon (1986: Pugnochiuso, Italy)
Carbon dioxide as a source of carbon.

(NATO ASI series. Series C, Mathematical and physical sciences; vol. 206)
"Published in cooperation with NATO Scientific Affairs Division."
"Proceedings of the NATO Advanced Study Institute on Carbon Dioxide: Biochemical and Chemical Uses as a Source of Carbon, Pugnochiuso, Italy, June 22–July 3, 1986"—T.p. verso.
Includes index.
1. Carbon dioxide—Metabolism—Congresses. 2. Carbon dioxide—Congresses. I. Aresta, M. (Michele), 1940– . II. Forti, Giorgio. III. Title. IV. Series: NATO ASI series. Series C, Mathematical and physical sciences; vol. 206.
QP535.C1N38 1986 661'.0681 87–12886
ISBN 90–277–2544–6

Published by D. Reidel Publishing Company
P.O. Box 17, 3300 AA Dordrecht, Holland

Sold and distributed in the U.S.A. and Canada
by Kluwer Academic Publishers,
101 Philip Drive, Assinippi Park, Norwell, MA 02061, U.S.A.

In all other countries, sold and distributed
by Kluwer Academic Publishers Group,
P.O Box 322, 3300 AH Dordrecht, Holland

D. Reidel Publishing Company is a member of the Kluwer Academic Publishers Group

Printed in The Netherlands

To

Lella and Anna Lucia

TABLE OF CONTENTS

ACKNOWLEDGEMENTS

We wish to thank the NATO Science Committee for having supported the ASI Summer School on "Carbon Dioxide : Chemical and Biochemical Uses as a Source of Carbon".
This was a great opportunity for Biochemists and Chemists for discussing in detail the carbon dioxide fixation processes.
The School was attended by several Students involved in this Research and the comparison of the different aspects of the problem resulted very helpful for a better understanding of the fundamentals and for designing a strategy for overcoming some of the difficulties usually encountered.
The importance of the subject and the great number of questions that stay still opened make this School a starting point and confirm that the interdisciplinary approach to the problem is of fundamental importance.
We wish to thank also : The University of Bari, the Dipartimento di Chimica - University of Bari, the Consiglio Nazionale delle Ricerche (CNR)-Rome, ENEA, ENIChem Sintesi, the Regione Puglia, the Provincia di Foggia-Assessorato alla Pubblica Istruzione, the Società Chimica Italiana-Divisione di Chimica Inorganica, the Associazione Italiana di Chimica Inorganica, the Consorzio per la Università di Capitanata for their interest and help.

The Carbon Dioxide Problem

Mankind Energy Needs and Environmental Pollution

Michele Aresta
Dipartimento di Chimica
173 via Amendola
70126 Bari, Italy

The world energy demand is covered essentially by the following primary sources today : fossil fuel (coal,oil,natural gas), nuclear fission,solar, geothermal, wind- and water-power. Several studies ascribe a different index of development to the above sources and a different commercial value to the so produced energy .

We expect that by 1990' , 84% of the world wide utilized energy will come from fossil fuels and that coal itself will nearly supply one fifth of the total energy (1). As carbon dioxide is formed in the combustion of these fuels, its concentration in the atmosphere will steadly continue to increase,as it happens since about 1850'.

It is a fact that the concentration of carbon dioxide in the atmosphere is increased from 270 ppm of the preindustrial age to 338.6 ppm , the estimated 1980 level .

Besides fuel burning, cement manufacture is an important "industrial" source of carbon dioxide (2). The total antropogenic carbon dioxide emission in the atmosphere is evaluated close to 5 Gt of carbon today, while the biotic source of carbon ranges around 20 Gt (3). These figures are of wide concern as the accumulation of carbon dioxide in the atmosphere can cause an important climate change .

On the other hand,carbon dioxide is a cheap "source of carbon" and its use as a "raw material" in synthetic chemistry might be quite interesting from several points of view. Its recovery and utilization seems to be, thus, a fundamental task from both an ecological and an economic point of view.

M. Aresta and G. Forti (eds.), Carbon Dioxide as a Source of Carbon, 1–22.

The effects of the increased concentration of carbon dioxide into the atmosphere will be felt by future generations,but for the present one it is a must to make decisions in order to avoid such a global,adverse environmental impact.

Carbon Dioxide emission in the atmosphere.

Using different procedures, several authors have estimated the amount of carbon dioxide released into the atmosphere The accuracy of the published figures concerning the contribution from fuels depends on the reliability of four sets of data : i) the amount of extracted fuels; ii) the fraction of carbon in fuels; iii) the efficacy of the combustion process; iv) the fraction of fuel diverted to non-fuel uses (utilization as a row material in the chemical industry for the synthesis of plastics, fibers, proteins etc.).

As these figures may change every year, for a correct evaluation of the amount of carbon dioxide discharged into the atmosphere through the flue gases a continuous updating is required

Keeling (4) based his analysis mainly on fuel data published by the United Nations Statistical Office,Energy Section, and estimated the fraction of each type of fossil fuel that is not oxidized Marland (5) reviewed the methods to estimate the industrial carbon dioxide sources. It is known that more reliable figures can be obtained when the amount of released carbon dioxide is related to the real average caloric value of different fuels and when the producing country and the year of production are considered too (this is true for coal, especially). The relevance of these factors has been clearly demonstrated : for example,the annual average caloric value for US bituminous coal and lignite has changed from 7279 cal g^{-1} prior to 1955,to 7129 for 1960,6829 for 1970,6432 for 1976 (6). Obviously ,the coal caloric value is strictly related to its carbon content. Rotty has revised the methods to evaluate the carbon dioxide emission and has based the calculation of industrial production of carbon dioxide more on fuel production than on fuel consumption (7-9) . He has also regularly updated the data since 1973. Rotty (7b) and Baes (10) have calculated for the carbon dioxide industrial production an exponential growth at a rate of 4.3 - 4.5 % per year from 1950 to 1973. Since1973, the

annual growth rate is around 2.25 %.
The change is due to the fact that in 1973 the OPEC (Organization of Petroleum Exporting Countries) increased the oil price so that consumers were obliged to revalue their oil needs and utilization Since that time the search for alternative energy sources (not based on fossil fuels) has been continuous and nuclear and solar energy have found several practical applications. Then, being the oil price presently reduced the fluid fuels consumption will probably grow again.
We have to point out that coal combustion is growing at the same rate (1.9 %) since the fifties,while after 1973 the oil and natural gas burning is significantly slowing down (from 7.08 % per year to 1.74 and from 8.03 to 2.79 %,respectively).
In 1980, despite this reduction,5.170 Gt of carbon (as carbon dioxide) have been discharged into the atmosphere,compared to the 3.126 Gt of 1965 and to the 1.618 Gt of 1950.
The carbon dioxide emission is not homogeneous all over the world : the developed countries are contributing with a considerable higher amount than developing countries. The amount of carbon dioxide emitted into the atmosphere in Northern Hemisphere is higher than in the Southern one,the largest contribution coming from the 45° - 60° N latitude countries However,the pattern of global carbon dioxide emission is changing rapidly and it should be emphasized that the growing rate of industrial carbon dioxide production did not show any abrupt change for the Developing Countries and Eastern European Countries in 1973 (8). The exponential growth was around 6.3 % through the 1970s and the first 1980s. In 1980 North America contributed the 26.7 % of the global carbon dioxide emission compared to 24.2 % of USSR and East Europe; 16.5 % of West Europe; 12.2 % of Developing Countries; 8.5 % of some Asian Countries (having a Central Planned Economy) and 5 8% of Japan and Australia . The 1965 figures are 32.1 %for N.America; 24.0 % for USSR and E.Europe Countries; 20.6 % for W.Europe; 8.0 % for Developing Countries ; 5.7% for CPE Asian Countries and 4.4% for Japan and Australia. These figures can also give an idea of the development of alternative energy sources of some Countries.
At the present time, the world average annual carbon emission in the atmosphere is evaluated around 1.2 t per person. If the growth

of the population is such to double the number of persons living on our Planet within next forty years,the carbon dioxide emission in the atmosphere will double at that time too This means that nearly 10.5 Gt of carbon per year will be released into the atmosphere. This value might be much higher if the improvement of the standard of living is not accompanied by the utilization of alternative sources of energy.

It is not easy to find out correct parameters to state the population increase,the standard of living improvement,the incidence of non-fossil fuels energy on the total amount of used energy. This can create some difficulty for a correct projection of 2000s.

Obviously,the limit to fossil fuels burning comes from their availability.

Oil and natural gas will be available at the current rate and at a suitable extraction cost until 2030s. On the contrary ,coal will constitute a large reserve of energy at least until 2300s.

As Chemistry provides ways to convert coal into gas and petrol, carbon based fuels will probably represent the main source of energy for mankind for the coming hundred years.

As the effects of carbon dioxide can be serious, it is of great interest, thus,to make correct predictions about its concentration in the atmosphere by the years 2000s - 2030s.

The Carbon Cycle and Accumulation of Carbon Dioxide in the Atmosphere

As part of the carbon dioxide is removed through the plant photosynthesis or is dissolved in the oceans , not all of it accumulates into the atmosphere. It is considered as retained by the atmosphere, an average value of 57% of the emitted carbon dioxide (11).A typical distribution of carbon between atmosphere and Earth is given below (12).

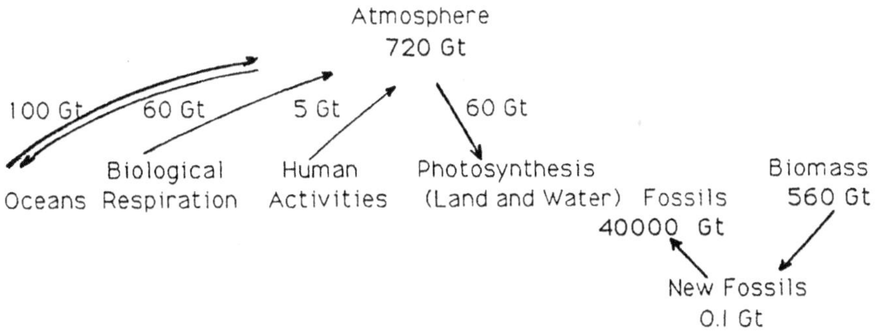

Bold characters indicate reserve. Plane characters indicate fluxes per year.

Several observatories around the Planet measure the atmospheric carbon dioxide concentration using two techniques : NDIR (Non Dispersive Infrared) for local continuous measurement and flask sampling of air for analysis at quasi regular intervals in a central laboratory (13). Since 1958 the Manua Loa and Antarctic observatories represent two of the most important sources of data for carbon dioxide concentration and most "carbon cycles" are tuned on their records.

The global average carbon dioxide increase depends on factors such as deforestation and changes in land management too : the exact evaluation of this contribution is not straight. The carbon dioxide input due to human interaction with the biosphere cannot be exactly defined (14) .

Oceans play a key role in reducing the effect of discharged carbon dioxide. Their sink effect has probably reduced to a half the atmospheric carbon dioxide increasing . A great effort is currently made to understand if they can be considered as "infinite buffers".

It has been proposed that the traditional model uptake by oceans,based on the air-sea partial pressure of carbon dioxide should be modified for the wind effect that can pump gas into water (15). Thus this effect takes into account the change of climatic conditions and can not be exactly evaluated in a general way . It has been proposed that the rate of exchange air-water for carbon dioxide can be accelerated by enzymatic

catalysis,namely by Carbonic Anhydrase (16). This suggestion is certainly based on the CA role in living organisms and on the consideration that some CA can be present in the oceans. Actually, it does not seem to apply unambiguously to this system (17).
Most of the uncertainties on future carbon dioxide level in the atmosphere derive thus from our indecision about

i) Standardization of the recording techniques of carbon dioxide level;
ii) Comparability and representativeness of data;
iii) Availability of data all over the Planet for a long period;
iv) Trend of input in the atmosphere from fossil fuels and cement;
v) Other human impact on the biosphere;
vi) Fate of non airborne excess carbon dioxide;
vii) Disturbances of carbon dioxide concentration in the different large reservoirs;
viii) Dependence of the model on the long-term prognoses.

However,the carbon cycle is regarded as a linear system and the carbon dioxide concentration into the atmosphere can be expressed as a function of the amount formed in the fossil combustion and of the fraction related to the global biomass.
According to one of these "models" (18),based on the 1940-1980 trend,in 2030s the atmospheric carbon dioxide concentration will range around 450 ppm. Other scenarios yield projections of 660 ppm in the 2030s and 1600 ppm in the 2060s. These values are subject to a considerable debate and pose a problem that is essentially ethyc today ,that is :the responsability of the present generation towards the future ones.
The problem of the carbon dioxide pollution of the atmosphere is often compared with the environment pollution by nuclear wastes. We have to underline that even if the emission of carbon dioxide into the atmosphere is,as we have previously pointed out, for developed Countries much more significant today,in a relatively short time (roughly one month) the carbon dioxide is dispersed within the troposphere and one hemisphere and finally it equilibrates all over the Planet within one year . The carbon dioxide emission in the atmosphere is a global problem.

Effects of Carbon Dioxide Accumulation in the Atmosphere.

The increase of atmospheric carbon dioxide causes a perturbation of the atmosphere thermal structure as a consequence of the trapping of earth's surface and troposphere's emitted infrared radiation which will generate a warming of surface and troposphere.
Theoretical studies (19,20) have shown that the increase of atmospheric carbon dioxide would have caused tropospheric temperature increase and stratospheric temperature decrease occuring because of the "green house" effect and infrared radiative cooling.
These studies have also demonstrated that the thermal balance of the earth-atmosphere system depends on gases other than carbon dioxide : i.e. methane,fluorinated hydrocarbons,NOx,ozone. It has been suggested that increasing carbon dioxide causes the increasing of stratospheric ozone equilibrium levels.This influences the upper stratosphere temperature in a negative way (the temperature lowering is estimated (21) around 8-10 K for a 16% increase of ozone).
A more recent study (22),based on the analysis of the atmospheric chemistry and photochemistry of species such as Ox,HOx,NOx,in relation to the concentration perturbation of CH_4,CF_2Cl_2,$CFCl_3$, H_2O,CO,CO_2,has discussed the basic chemical-climate mechanism in one dimension.The vertical transport was considered from the ground to 50 km,but latitudinal variations were not taken into account. The conclusion was that carbon dioxide increase would have had effects on ozone,methane,water and dinitrogen oxide concentrations and the thermal structure of the troposphere and stratosphere would depend on these coupled interactions that can also provide for compensating effects. The net influence of increasing carbon dioxide level into the atmosphere on the planet climat can not be straightly evaluated. The observed temperature increase of 0.5 K in the last hundred years can not be attributed only to carbon dioxide.
The signal due to carbon dioxide will probably emerge from the natural variability at the end of this century. The warming has been evaluated around 2.6 K for 2 x CO_2.

The carbon dioxide increase will cause the change of either water vapor content in the troposphere or in the proximity of the surface and this would influence the agricultural productivity. The carbon dioxide level increase would favour the primary productivity of most cultivated species, C_3 species essentially , and decrease the transpiration, in C_4 plants especially.

Although C_3 and C_4 plants present different response to evapotranspiration (23),the photosynthetic production per mol of water consumed by plants through transpiration might increase in both C_3 and C_4.

Some authors (24-26) uphold that today the water supply and nutrient availability have a greater influence on the global productivity than the carbon dioxide shortage. However,if this is true,it means that as the carbon dioxide level increases into the atmosphere,the agricultural productivity decreases. Events such as root growth and branching, flowering time,maturation,might be influenced by carbon dioxide in a negative way Studies have been carried out in order to set a precise relation between carbon dioxide concentration and photosynthetic productivity. It seems that only at a 5000 ppm level a definite negative response is obtained.

The major question is if the indications obtained from systems operating under controlled conditions will hold in the field. It is not possible now to draw any definite, positive or negative, conclusion about the influence of carbon dioxide concentration on the agricultural productivity.

Carbon Dioxide as a Source of Carbon.

The future concentration of carbon dioxide in the atmosphere and its effects on climate and agricultural primary productivity are quite unknown. The general feeling that the consequences of the carbon dioxide accumulation might not be positive persuaded several Scientists to make efforts in order to put light on the various aspects of carbon dioxide problem. The most relevant one is that carbon dioxide is a cheap and abundant raw material to be used in synthetic chemistry as a source of carbon.

Two major uses are possible:

i) utilization as a synthon for organic compounds containing the carboxylic group -C(O)O-(acids,esters,lactones,carbonates, carbamates and derivatives) or the carboxamidic group -C(O)NR- (ureas,amides and derivatives);

ii) utilization in reduction processes yielding other C1 energy richer molecules ($HCOOH$,CO,CH_2O,CH_3OH,CH_4),

based on chemical processes which require a quite different energy input.
Moreover,(i) yields products characterized by a high added value ("fine chemicals"), while (ii) affords products that can be used as a source of energy in combustion processes or as raw materials in the bulk chemical industry (low "added value" ,but large amount).
The two uses of carbon dioxide are reminiscent of natural processes such as carboxylation,carbamatation and carbon dioxide reduction by enzymes. Since a long time, Chemists are actually trying to use carbon dioxide in the synthesis of chemicals,but today only very few industrial processes are operating (see the Keim contribution in this Book). In the last decade, a "carbon dioxide chemistry" has been developed.
As the search for new energy sources and materials is perdurable , the utilization of carbon dioxide in synthetic chemistry or in other industrial applications might become quite worthy in a short time. This new chemistry can give, then, a substantial contribute to the safety either in industrial laboratories or in plants as carbon dioxide can substitute,for example, carbonyl chloride in the synthesis of carbamates and chloroformates in the synthesis of carbonates.
Our actual knowledge about this utilization doesn't allow us to take into consideration this one as "the way" to undo the carbon dioxide unexpected polluting effect .This, probably will not be the way as the amount of carbon dioxide employed is not very large. Way (i),in fact, can be only considered as a process to recover some carbon and to secure more contained production prices for the synthesis of a few chemicals. A cheap material is utilized and new operative conditions characterized by a low risk and a lower energetic cost are adopted. The efforts currently made are intended to find out the correct reaction conditions and catalysts to

accomplish the synthesis with high yields and good selectivity. Considerations about pathway (ii) may be different.
The amount of carbon monoxide,methanol,methane currently used is very large as they are used in the synthesis of fuels (see, for example,the carbon monoxide use in the Fisher-Tropsch synthesis of fuels) or are directly used as fuels (methane). A remarkable point is that reduction requires a high energy input,an "oxygen sink" and a source of hydrogen. Nature uses visible light and water in the carbon dioxide reduction and Chemists are trying to mimic the natural photosynthetic process. Effective methods to reduce large masses of carbon dioxide at a high rate and low cost are not accessible today.
It would be of great interest to use energy transfer systems able to capture solar energy and to put it in the carbon dioxide reduction. If we look at the normal yields of solar energy utilization in the conversion of inorganic carbon into organic carbon by green plants, we find out that it is quite low (usually 1.2%,in some special cases it can be just a bit higher than 2%). This could wrongly convince us that it might be easy to reach higher yields in laboratories and/or in industrial plants.
Man-made systems able to capture solar energy are well known but, when semiconductors are used as energy transfer systems in carbon dioxide reduction processes, the solar- to chemical-energy conversion yield appears to be at least of two orders of magnitude lower than that observed in naturally occurring photosynthetic processes (26). This peculiar aspect of the carbon dioxide reduction is of obvious interest from a practical point of view as light has been used to convert carbon dioxide and water into formic acid, formaldehyde, methanol and methane,mimicking the natural process. Moreover,these studies can bring some support to the suggested hypothesis that metal oxides and ternary compounds present on the earth surface might have been responsible for the carbon dioxide conversion and dioxygen enrichment of the atmosphere in prebiotic conditions. The oxides used up to now did not show high selectivity and ultraviolet light has often been employed. When solar light is used,the efficiency of the conversion is very low. Obviously, the use of u.v.-light is not so interesting as the use of visible light.
New more active and selective catalysts must be synthetized and

more effective energy transfer systems must be studied to be employed in the carbon dioxide reduction through this procedure. At least two other major processes can be considered for the carbon dioxide reduction :

- the utilization of transition metal complexes for solar energy transfer ;
- the utilization of enzymes isolated from bacteria.

The two subjects are discussed in this book.
A number of systems able to capture visible light are known .
The isolation of the enzymes responsible for the carbon dioxide utilization as a source of carbon in natural fundamental processes (plants and bacteria) and the exact knowledge of how they work can help to solve several practical problems and can make possible the development of new synthetic technologies as the application of Biotechnologies in the synthesis of chemicals.
A lot of work must be done in this field.
Another interesting aspect related to this point must be considered .
Most of the enzymes involved in the CO_2 reduction by plants and by bacteria contain metal atoms as the active center (27-29).
The exact knowledge of their mode of action can be very helpful to Chemists as it might adress to the synthesis of similar and simpler metal systems to be employed in synthetic chemistry.
Biochemists and Chemists can give a great contribution to highlight some aspects of the carbon dioxide problem.
The state of the art on carbon dioxide utilization is presented in this book. One of the outstanding conclusions is that the collaboration between Biochemists and Chemists must be encouraged in order to gain an insight into the reaction mechanism of those enzymes that catalyses any C-1 molecule conversion in a biosystem. These studies can bring fundamental contribution to the general knowledge and suggest operative conditions to reach the carbon dioxide utilization goal in laboratory .

Other Industrial Applications of Carbon Dioxide.

As well as in the chemical industry, carbon dioxide finds out an important application in other industries and specialist sectors requiring an advanced technology.

a. Extractive industry;
b. Growing of plants and flowers in "greenhouses".

This aspect can be of primary importance whereas there is a carbon dioxide source with a high concentration. Moreover,(a) and (b) demand different purity requirements. The carbon dioxide sources can be identified in:

1. Vulcanoes discharging into the atmosphere carbon dioxide with a purity around 95-98%;
2. Coke-, natural gas-, oil-feeded stations power and cement plants.

In these cases,the flue gases usually present a carbon dioxide concentration in the range around 10-16 % and 23-30 % in volume, respectively (30).
The utilization as a solvent concerning the application of supercritical carbon dioxide in the extraction from a biological material is reviewed in this book (extraction from coffee beans and similar products).
Supercritical carbon dioxide can find also an important utilization to increase oil production through miscible flooding via injection into oil reservoirs (31-32). This new application demands carbon dioxide recovery and processes based on permeable membrane (33) are under study. These membranes can also be of great utility to recover carbon dioxide from gas mixtures. The utilization of metal complexes to separate carbon dioxide from other small molecules has been studied too (34).
Obviously, the carbon dioxide utilization in these industrial applications must not be considered as one of its possibilities to be recycled, but only as a procedure to develop new technologies. Conversely, the carbon dioxide utilization in greenhouses is interesting from an economical and ecological point of view, as growing of flowers and plants is speeded up and as plants can fix and eliminate it from the environment. In this case the purity of

the used gas is the most important factor and,thus,the study of systems able to separate carbon dioxide from other gases (pollutants or not) is one of the outstanding research fields.

The Carbon Dioxide Electronic Structure,Spectroscopic and Thermodynamic Properties : the Starting Point to Design any Carbon Dioxide Reduction Process .

The natural carbon dioxide reduction process is quite complex and requires an energetic input : a cheap energy source is used, the sun. This point is most relevant for any utilization man attempts. Carbon dioxide utilization in the laboratory is accomplished using electron rich species or energy rich compounds or furnishing energy to the reaction system.
The analysis of the carbon dioxide bonding molecule, of its spectroscopic and thermodynamic properties is necessary to understand the aspects of its reactivity . The bonding and the spectroscopic data are presented in this Chapter , while the thermodynamic properties will be discussed in detail in other Chapters of this Book.

Carbon dioxide has 16 valence electrons and in its ground state is a linear molecule belonging to the $D_{\infty h}$ point group. The C-O distance is 0.116 nm.

Both the electronic spectrum (35) and the infrared spectrum (36-38,43) have been considered by several authors. The infrared spectrum is of particular interest as infrared spectroscopy is used for a continuous recording of carbon dioxide level in the atmosphere and for the analysis of remote sensing data. The infrared absorption data are reported in Table 2.

TABLE 2

	O=C=O Sym C=O	$\overset{-}{O}=\overset{+}{C}=\overset{-}{O}$ O=C=O Bending	O=C=O Asym C=O
Gaseous	(1285-1388) Raman	667	2349
Aqueous Solution			2342
Solid		660 653	2344

The u.v.-visible spectrum shows three maxima at 147.5 , 133.2 , and 112.1 nm. All of them show a superimposed vibrational structure. The bands are respectively assigned to $^1\Sigma g \rightarrow {}^1Au$, $^1\Sigma u \rightarrow {}^1\pi g$, $^1\Sigma g \rightarrow {}^1\Sigma u$ transitions. The band at 112.1 nm is the strongest and is relative to a permitted transition.
The molecular orbitals diagram (Table 3) for carbon dioxide indicates that the C=O bond order is two. The energy varies from the linear molecule to a bent configuration.
The energy difference between the HOMO ($1\pi_g$) and the LUMO ($2\pi_u$) degenerated orbitals ranges around 31 e V in the linear configuration. As we have indicated in Table 3 the energy of the orbitals varies with the OCO angle and the ΔE ($E_{LUMO} - E_{HOMO}$) varies with the configuration of the molecule. When the OCO angle is 133°, the ΔE is around 4 e V for the two orbitals ($4b_2 - 6a_1$). The value of the OCO angle found in several metal complexes containing coordinated carbon dioxide and in organic carboxylic acids is very close to 133° (see the Chapter on Carbon Dioxide Metal Complexes in this Book).
The mode of variation of the energy of the $4b_2$ orbital is responsible for the linearity of the molecule in the ground state and the decrease of the ΔE between the $4b_2$ and the $6a_1$ orbitals as the OCO angle grows smaller explains why the excited states of carbon dioxide and the CO_2^- species do not present a linear geometry (40,41).
It is worth to note that the LUMO are located essentially on the carbon atom,while the HOMO are located on the oxygen ones.
From an Acid-Base point of view,carbon dioxide can be defined as

an anphoteric oxide,even if its Lewis acid character predominates. In fact, carbon dioxide reacts with electron rich species (Lewis bases, electron rich metal centers) that bind to the carbon atom, but it reacts also with electron deficient compounds (protons,metal ions) that can interact with the oxygen atoms.
As a result of this tendency,a simultaneous acid-base catalysis is required for carbon dioxide reactions (for the fixation on an organic substrate or for the simple hydration reaction) as discussed in this book.
A basic center and an acid one are often required for carbon dioxide coordination to metal centers. Carbon dioxide is in a bent configuration in all the isolated complexes.
Carbon dioxide might be, presumably, in a linear geometry in complexes in which it would interact with metal centers through the oxygen atoms alone. Any time the carbon atom is involved ,the molecule assumes a bent configuration.
Carbon dioxide seems to loose its linearity also when it gives a weak interaction with a metal surface (42).

In Table 3 the character of the molecular orbitals obtained from the combination of the 2s and 2p atomic orbitals is presented (simplified Walsh diagram).

The molecule is considered in a y-z plane that contains also the orbitals.

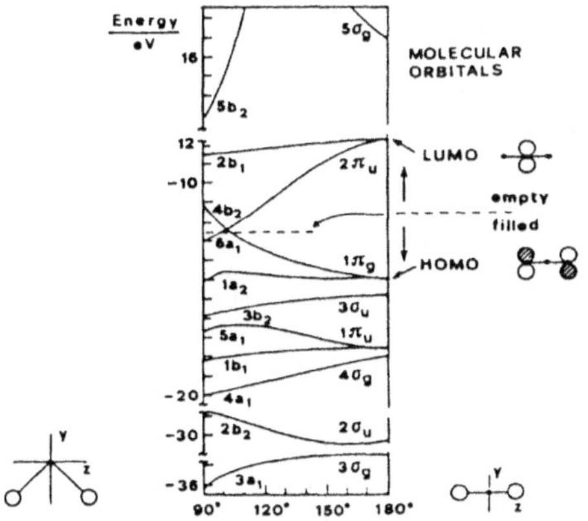

C_{2v} O-C-O angle $D_{\infty h}$

Table 3.

Molecular Orbitals for Carbon Dioxide.

$$(3\sigma_g)^2\ (2\sigma_u)^{*2}\ (4\sigma_g)^2\ (1\pi_u)^4\ (3\sigma_u)^2\ (1\pi_g,nb)^4\ (2\pi_u)\ \ (5\sigma_g).$$

$3a_1\ (\sigma_g)^2$	From $2s_O$ and $2s_C$ atomic orbitals. Strongly bonding in the ground state; bending stabilizes the orbital.
$2b_2\ (\sigma_u)^{*2}$	From $2s_O$ and $2p_C$ atomic orbitals. Antibonding with a node on C. Bending causes an increase of energy.
$4a_1\ (\sigma_g)^2$	From $2s_O$,$2p_O$,$2s_C$ atomic orbitals. Bonding,with nodes that intersect the O-C bonds;bending causes decrease of energy.
$1b_1$,$5a_1$ $(\pi_u)^4$	From 2p from the three atoms. Degenerate π orbitals.Strongly bonding;upon bending up to 100°,$1b_1$ decreases its energy,while $5a_1$ increases its energy.
$3b_2\ (\sigma_u)^2$	From $2p_O$,$2p_C$,$2s_O$ atomic orbitals. Strongly bonding; bending causes a decrease of the energy.
$1a_2$,$4b_2$ $(\pi g)^4$	From 2p atomic orbitals of the three atoms. Degenerate π orbitals with nodes intersecting the two C-O bonds.This orbital has a non-bonding character.
$6a_1$,$2b_1$ $(\pi_u)^*$	From $2p_C$ and $2p_O$ atomic orbitals. It has an antibonding character. Bending causes a net decrease of the $6a_1$ energy, while the energy of the $2b_1$ is nearly constant.
$7a_1$ $(\sigma_g)^*$	From $2s_C$ and $2s_O$ atomic orbitals. It has a strongly antibonding character.The molecule bending causes a noticeable increase of energy.

It is worth to compare the spectroscopic data and bonding geometries and distances for CO_2, CO_2^+, and CO_2^- (Table 4).

Table 4.

Comparison of Infrared, U.V.-Visible data, bond length and angle for CO_2, CO_2^+, CO_2^-.

Species	Infrared bands cm^{-1}			U.V.-Visible peaks nm			D(C-O) nm	OCO angle	ΔE^a eV
	ν_3	ν_1	ν_2						
CO_2	2349^b	1331^b	667^b	147.5^c	133.2^c	112.1^c	0.1162^d	180	
CO_2^+	1469^e	1280^e	498^e				0.1177^f	180	
CO_2^-	1665^g						0.1282	137^h	2.6

a) Energy variation respect to CO_2; b) From Ref. 44; c) From Ref. 35; d) From Ref. 39; e) From Ref. 44; f) From Ref. 45; g) From Ref. 46-47; h) The OCO angle value is in the interval 127° - 138°.

The radical anion CO_2^- has a great importance as it can be considered as the first step in the carbon dioxide reduction. It has been isolated in a solid matrix at room temperature (46-48) ($t_{1/2}$> 1 year) or it has been trapped in an aqueous medium (49). In the gas phase it is quite reactive and it half-life is very short (40).
The radical anion is very reactive towards several agents such as water ($2\ CO_2^- + H_2O = HCO_3^- + HCO_2^-$) and carbon dioxide itself. This aspect is discussed in detail in the chapters concerning the carbon dioxide electrochemical reduction.
The radical anion is formed in the reaction of carbon dioxide with alkaly metal atoms in rare gas matrices (50). It has been demonstrated that several species are formed depending on the metal, the gas matrix, the temperature.
The carbon dioxide interaction with transition metal atoms at low temperature (51) has been studied too. Some very useful

informations have been obtained on the carbon dioxide way of bonding to the metal center and on the reactivity of the coordinated carbon dioxide. These data can be very informative for catalytic reactions involving carbon dioxide on heterogeneous catalysts.

References and Notes.

1) *Administration's Annual Report to Congress* ,1977 Washington D.C.,Department of Energy ,1978.
2) United Nations , *Monthly Bulletin of Statistics* , N.Y. 1981b.
3) Lemons,J., 'Carbon Dioxide and Environment : a Problem of Uncertainty',*J. of Environmental Sciences* ,1985, 60.
4) Keeling,C.D.,'Industrial Production of Carbon Dioxide from Fossil Fuel and Limestone' , *Tellus* ,1973 , **25**,174.
5a) Marland,G., 'Dynamic Models of Environment : CO_2 Cycle'. Annual Report for 1980. *Annual Report to the Gas Research Institute,Oak Ridge* ,Associated Univ.,Institute for Energy Analysis,Oak Ridge,Tenn.,1980.
b) Marland,G.,'Carbon Dioxide Discharged from the Combustion of Coal',*CO_2 Workshop* , Paper Contrib. 81-24,Oak Ridge Associated University.Inst. for Energy Analysis,Oak Ridge,Tenn.,1981.
6) U.S. Bureau of Mines,*Minerals Yearbook, Vol. 1,Metals,Minerals and Fuels* ,U.S. Government Printing Office,Washington D.C., 1976.
7a) Rotty,R.M.,'Commentary on and Extension of Calculative Proce dure for CO_2 Production, *Tellus* ,1973,**25**,508.
b) Rotty,R.M.,'Global CO_2 Production from Fossil Fuels and Cement,A.D.1950 - A.D. 2000', *The Fate of Fossil Fuel Carbon Dioxide in the Oceans*,R.N.Anderson and A.Malahoff Ed.,Plenum New York,1977.
c) Rotty,R.M.,'Past and Future Emission of Carbon Dioxide',*Experientia*,1980,**36**,781.
d) Rotty,R.M.,'Distribution and Change in Industrial Carbon Dioxide Production',*J.of Geophys.Res.* ,1983,**88,**1301.
8) Rotty,R.M.,G.Marland and N.Treat,'The Changing Pattern of Fossil Fuel Carbon Dioxide Emissions',Oak Ridge Ass.Univ.,Inst. of Energy Analysis

Report for Dept. of Energy, 1984.

9) Marland, G., and R.M. Rotty, 'Carbon Dioxide Emission from Fossil Fuels : a Procedure for Estimation and Results for 1950 - 1982', *Tellus*, 1984, **36 B**, 232.

10) Baes, C.F., Jr., H.E. Goeller, J.S. Olson, and R.M. Rotty, ' The Global Carbon Dioxide Problem', ORNL - 5194, Oak Ridge National Laboratory, Oak Ridge, Tenn., 1976.

11) Fraser, P.J., G.J., Pearman, and P. Hyson, ' The Global Distribution of Atmospheric Carbon Dioxide. 2. A Review of Provisional Background Observations, 1978-1980.', *J. Geophys. Res.*, 1983, **88,** 3591.

12) Elliott, W.P., L. Machta and C.D. Keeling, 'An Estimate of Biotic Contribution to the Atmospheric Carbon Dioxide Increase Based on Direct Measurements at Mauna Loa Observatory', *J. of Geophys. Res.*, 1985, **90,** 3471.

13) Several Institutions collect data on the atmospheric carbon dioxide concentration, but not all data are referred to a common calibration scale. Four major sources of calibrated data are : i) The University of California's Scripps Institution of Oceanography (SIO); ii) The National Oceanic and Atmospheric Administration's Program, Geophysical Monitoring for Climatic Change (NOAA-GMCC); iii) The Australian Common wealth Scientific and Industrial Research Organization, Division of Atmospheric Physics (ACSIRO); iv) The University of Stockholm's Department of Meteorology (SWE).

14) Hartwig, S., W. Heudorfer and G. Schnatz, 'Possible Climatic Change by Antropogenic Carbon Dioxide Emission', *I. Chem. E. Symposium, Series 78*, 1982.

15) Smith, S.D., and E.P. Jones, 'Evidence for Wind-Pumping of Air-Sea Gas Exchange Based on Direct Measurements of Carbon Dioxide Fluxes', *J. of Geophys. Res.*, 1985, **90**, 869.

16) Berger, R., and W.F. Libby, *Science*, 1969, **164**, 1395.

17) Goldman, J.C., and M.R. Dennett, 'Carbon Dioxide Exchange Between Air and Seawater : no Evidence for Rate Catalysis', *Science*, 1983, **220**, 199.

18) Oeschger, H., and M. Heimann, 'Uncertainties of Predictions of Future Atmospheric Carbon Dioxide Concentrations', *J. of Geophys. Res.*, 1983, **88,** 1258.

19) Manabe, S., and R.T. Wetherald, 'Thermal Equilibrium of the

Atmosphere with a given Distribution of Relative Humidity',*J.Atmos.Sci.*, 1967,**24**,241,
20) Boughner,R.E.,V.Ramanathan,'Climatic Consequences of Increasing Carbon Dioxide. Second Conference on Atmospheric Radiation',*Am.Meteorol. Soc.*, Arlington,Virginia,1975.
21) Boughner,R.E.,'The Effect of Increased Carbon Dioxide Concentrations on Stratospheric Ozone',*J.Geophys.Res.*,1978, **83**,1326.
22) Callis,L.B.,M.Natarajan,R.E.Boughner,'On the Relationship between the Greenhouse Effect,Atmospheric Photochemistry and Species Distribution', *J.Geophys.Res.*,1983, **88**,1401.
23) Rosenberg,N.J.,'The Increasing Carbon Dioxide Concentration in the Atmosphere and its Implication on Agricultural Productivity', *Climate Change*,1981, **3**,265.
24) Lemon,E.R.,'The Lands Response to more Carbon Dioxide',in *The Fate of Fossil Fuels in the Oceans* ,N.J.Anderson and A.Malahoff Eds.,Plenum Publ.Co.,NY,1976.
25) Goudrian,J., and G.L.Ajtay,'The Possible Effects of Increased Carbon Dioxide on Photosynthesis',in *The Global Carbon Cycle*, B.Bolin, E.T.Degens,S.Kempe,and P.Ketner ,Eds.,J.Wiley and Sons. Publ.,NY,1979.
26) 'Photoelectrochemical Reduction of Carbon Dioxide, Report to the Commission of European Communities',EUR 9529,EN,1984.
27) Thauer ,R.K.,G.Fuchs,K.Jungermann,'Role of Iron Sulfur Proteins in Formate Metabolism',in *Iron Sulfur Proteins* ,1977, III,121.
28) Diekert,G.,G.Fuchs,R.K.Thauer,'Properties and Function of Carbon Monoxide Dehydrogenase from Anaerobic Bacteria', in *Microbial Gas Metabolism : Mechanistic,Metabolic and Biotechnological Aspects'*, 1985, 115.
29) Wolfe,R.S.,'Unusual Coenzymes of Methanogenesis',*TIBS*,1985, 396.
30) Anada,H.R.,M.D.Fraser,D.F.King,A.P.Seskus,J.T.Sears,' Feasibility and Preliminary Economics of by-product Carbon Dioxide Recovery and Transportation for EOR',*American Institute of Chemical Engineers*1983 Spring National Meeting and Petro Expo1983,Astrohall, Houston Texas, March,1983, 27.
31) Holm,L.,W.,'Annual Production Report',*Oil and Gas J.*,1982,139.
32) Holm,L.W.,'Carbon Dioxide Flooding:Its Time has Come',*J.of Petroleum Technology*,1982, 2739.

33) Coady,A.B.,Davis,J.A.,'A new Gas Permeation Process for Recovery of Carbon Dioxide',*Delta Project Limited* ,P.O.Box 5244 Station A,Calgary, Alberta,Canada.
34) Nelson,D.A.,R.T.Hallen,M.A.Lilga,'Separation of Gas Mixtures by Supported Complexes',Report to the U.S.Department of Energy,Pacific Northwest Laboratory,Battelle Memorial Institute,1985.
35) See for example : Rabalais,J.W.,J.M.McDonald,V.Scherr and S.P.McGlynn,'Elecronic Spectroscopy of Isoelectronic Molecules',*Chem.Rev.*,1971,**III**,71, and references there in.
36) Rinsland,C.P.,D.C.Benner,V.Malathy Devi,P.S.Ferey,C.H.Sutton, and D.J. Richardson,'Atlas of high Resolution Infrared Spectra of Carbon Dioxide', NASA Tech.Memo,1984, 85764.
37) Toth,R.A.,'Line Position and Strenghts of Carbon Dioxide in the 1200-1430 cm^{-1} Region',*Applied Optics*,1985, **24**,261.
38) Rinsland,C.P.,D.C.Benner and V.Malathy Devi,'Absolute Line Intensities in Carbon Dioxide Bands near 4.8 μm',*Applied Optics*,1986,**25**,1204,and references there in.
39) Sutton,L.E.,*Tables of Interatomic Distances and Configurations in Molecules and Ions*,The Chemical Society,London,1958.
40) Claydon,R.C.,G.A.Segal,H.S.Taylor,'Theoretical Interpretation of the Electron Scattering Spectrum of CO_2',*J.Chem.Phys.*,1970, **52**,3387
41) Jacox,M.E.,and D.E.Milligan,'Vibrational Spectrum of CO_2^-', *Chem.Phys.Lett.*,1974, **28**,163.
42) Ozin, G.A,H.Huber,D.McIntosh,'Metal Atom Chemistry and Surface Chemistry : (Carbon Dioxide)Silver,Ag(CO_2).A Localized Bonding Model for Weakly Chemisorbed CO_2',*Inorg.Chem.*,1978, **17**,1472.
43) Pugh,L.A.,and K.N.Rao, *Molecular Spectroscopy : Modern Research.*, Vol.II,K.N.Rao,Ed.,Academic Press,New York,1976.
44) Bondybey,V.E.,J.H.English,'Spectroscopy and Relaxation of CS_2^+',*J.Chem.Phys.* ,1980, **73**,3098.
45) Chin,S,W.B.Person,'Quantitative Prediction of Infrared Spectra of Isolated Molecular Ions and Radicals',*J.Phys.Chem.*,1984, **88**,553.
46) Hisatsune,I.C.,T.Adl,E.C.Beahm,R.J.Kempf,'Matrix Isolation and

Decay of Carbon Dioxide and Carbonate Anion Free Radicals',*J.Phys.Chem.*,1970,**74**,3225.
47) Hartman,K.O.,and I.C.Hisatsune,'Infrared Spectrum of Carbon Dioxide Anion Radical',*J.Chem.Phys.*,1966, **44**,1913.
48) Lunsford,J.H.,and J.P.Jayne,'Formation of CO_2^- Radical Ions when Carbon Dioxide is Adsorbed on Irradiated MgO', *J.Phys.Chem.*,1965, **69**,2182.
49) Harbour,J.R., and W.L.Hair,' Spin Trapping of CO_2^- Radical in Aqueous Medium',*Can.J.Chem.*,1979, **57**,1150.
50) Kafafi,Z.H.,R.H.Hange,W.E.Billups and J.L.Margrave, 'Carbon Dioxide Activation by Lithyum Metal', *J.Am.Chem.Soc.*,1983, **105**, 3883, and references there in.
51) Mascetti,J.,and M.Tranquille,'I.R. Evidence for the Formation of Carbon Dioxide Transition-Metal Atom Complexes in Low Temperature Matrixes',*Surf.Sci.*,1985,**156**,201.

INDUSTRIAL USES OF CARBON DIOXIDE

Willi Keim
Institute of Technical Chemistry and Petrochemistry
Aachen University of Technology, Worringer Weg 1,
D-5100 Aachen
Federal Republik of Germany

ABSTRACT. This chapter deals with industrial utilizations of carbon dioxide. The preparation and the various applications making use of CO_2's properties in cooling, inertnes, pressure and solubility characteristics and reactivity pattern will be discussed. Major emphasis will be paid to chemical reactions covering present and potential future uses.

1. INTRODUCTION

Carbon dioxide is one of the most abundant chemicals in nature. Nearly unlimited amounts are found in the atmosphere, dissolved in water, in gas wells and in solids as carbonates. Due to the widely practised combustion of the various carbonic resources the amount of CO_2 is steadily increasing. Based on its availability carbon dioxide represents an ideal compound for industrial applications.

2. MANUFACTURE OF CARBON DIOXIDE

The major amounts of industrially used carbon dioxide stem from gas wells, chemical plants, fermentation, and separation in combustion processes. Sometimes carbonates (for instance limestone) are heated to produce CO_2. Worldwide numerous natural wells produce CO_2 sometimes in purities of > 99 %. Also the off-gas in grain fermentation consists of CO_2 in 99 % purity. Merchant productions often stem from chemical plants providing hydrogen, which is used in substantial amounts to synthesize fertilizers such as ammonia. Here the CO_2 is derived via the shift reaction exemplified in Figure 1 for methane conversion. Methane is converted by a combination of combustion and shift reaction to a mixture of CO_2 and H_2 from which CO_2 is separated by absorption (srubbing).
Carbon dioxide is also separated in the many gas treating plants practised to remove CO_2 and/or H_2S from natural gases. A great variety of solvents is available to chemically, physically or chemically/physically absorb CO_2. The principal of this separation methode is shown in Figure 2.

M. Aresta and G. Forti (eds.), Carbon Dioxide as a Source of Carbon, 23–31.

$$CH_4 + 2\,H_2O \longrightarrow 4\,H_2 + CO_2$$

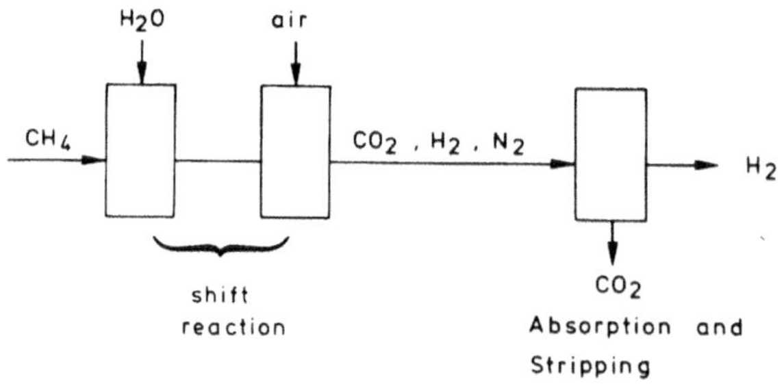

Fig. 1 CO manufacture from hydrogen synthesis

Raw gas containing CO enters the absorber. Loaded gas flows to the desorption where CO is flashed out and the absorber liquid is recyled. An alternative to conventional gas treating via absorption is separation by membranes.

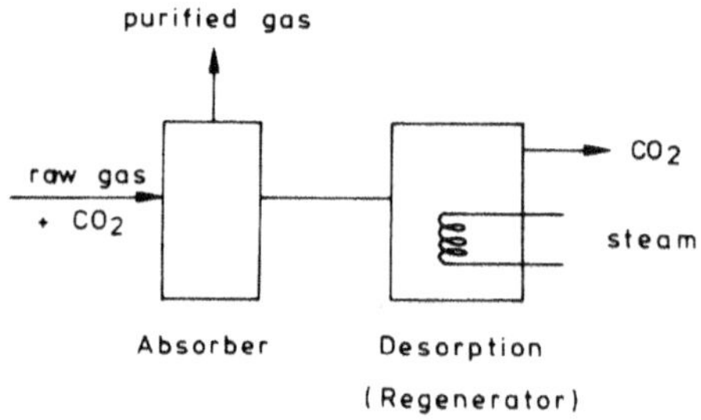

Fig. 2 CO_2 removal by absorption

Due to environmental concern various processes to remove acids from flue gas are under development or have been introduced. Here, CO_2 will also be separated and could be isolated.
Table 1 lists a variety of solvents and processes to remove CO_2

Table 1 Separation of CO_2

Physical Absorption	Chemical Absorption
CH_3OH (Rectisol)	
Molecular sieve (Union Carbide)	Monoethanol amin
Membrane (Enstar Eng. Co.)	Diethanol amin
Dimethylether of polyethylene glycol (Selectol)	Triethanol amin
Gas/Spec (Dow Chem.)	$CaO + CO_2 \longrightarrow CaCO_3$
Propylene carbonate	(CO_2 acceptor)
NMP (Purisol/Lurgi)	
Sulfinol (Shell)	

3. USE PATTERN OF CO_2

Carbon dioxide is non-toxic, a solid at -78,9°C, it sublimes (r_S = 573,02 KJ/Kg) and it frees 645 KJ/Kg upon heating from -78,9°C to 0°C. In addition, CO_2 is very stable and dissociates only to 0,032 at 1200°C.

$$CO_2 + 283\ KJ \rightleftharpoons CO + 1/2\ O_2$$

All these properties open many routes to utilize CO_2 in numerous industrial applications. Major uses of CO_2 in USA are the following:

- Refrigeration (40 %)
- Beverage carbonation (20 %)
- Enhanced oil recovery (5-10 %)
- Metal fabrication (10 %)
- Manufacture of chemicals (10 %)

All these applications of CO_2 utilize the physical and chemical behaviour elucidated in Figure 3.

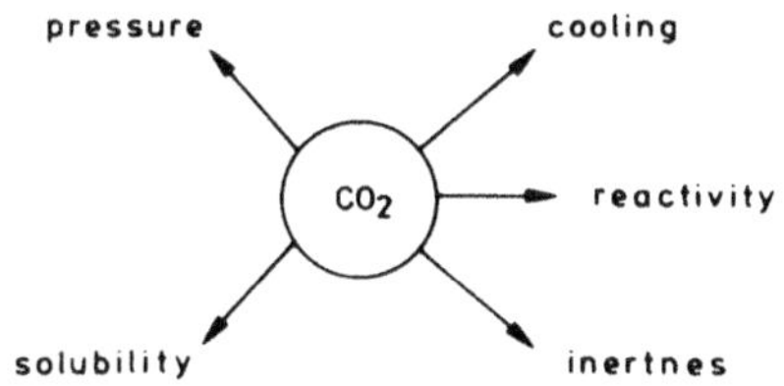

Fig. 3 Applications of CO_2 based on its chemical/physical behaviour

3.1 USE OF COOLING CHARACTERISTICS

Based on its chemical and physical properties (see P,T-diagram) CO_2 is an ideal cooling medium in the food industry. A large use for carbon dioxide is in the form of dry ice as a refrigerant. Frozen foods, dairy/meat products and food packing industry consume substantial amounts. Without CO_2, storage and transportation of perishable food is unthinkable. The metal industry uses frozen CO_2 in metal forming. The chemical industry needs CO_2 to clean and to pulverize rubber parts and elastomeric materials.

3.2 USE OF INERTNES

Beeing inert and non-toxic gives CO_2 a broad spectrum of applications in the food industry. The durability of food is extended when packed or stored under CO_2. Besides in the food industry many uses for CO_2 exist utilizing its inertnes. Metal welding is carried out under the protection of CO_2. Fire exstinguishers built on CO_2 are very common. Explosion prevention is achieved using a CO_2 blanket.

3.3 USE OF CO_2's SOLUBILITY CHARACTERISTICS

As the p,T-diagram of CO_2 exhibits, carbon dioxide can be used to foam food (dough in a bakery) and to extract unwanted components from food (see chapter by Eisenbach in this book "Supercritical Carbon dioxide: Properties and uses").
At 1 bar only 0,1 % of CO_2 is dissociated in water giving a pH-value of 3,7. On this behaviour rests CO_2's use in carbonation of beverages. A promissing area of utilizing CO_2 is in enhanced oil recovery. When carbon dioxide contacts crude oil under reservoir conditions, it dissolves in the oil reducing the viscosity thus enhancing oil migration. As the amount of CO_2 added to crude oil is increased, a two-phase system (see Figure 4) of a CO_2 and an oil rich phase is formed. The carbon-dioxide-rich phase may be either vapor or liquid depending on the conditions in the reservoir, and it consists of a mixture of CO_2 and hydrocarbons extracted from the crude oil.

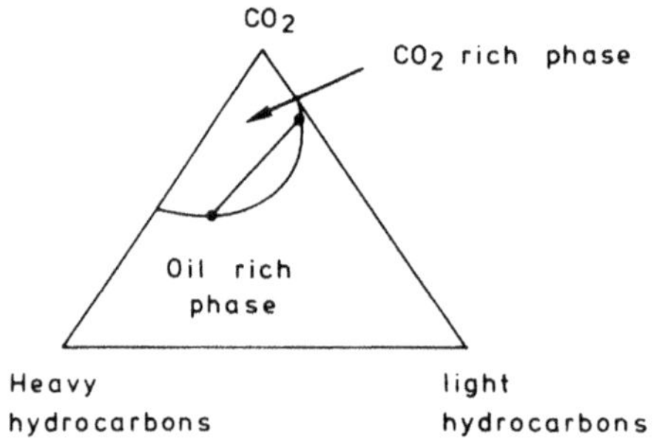

Fig. 4 CO_2 in enhanced oil recovery

3.4 USE OF CO_2's PRESSURE CHARACTERISTICS

In the food industry the pressure behaviour of CO_2 is used to transport liquids from one container/tank into other ones. Aerosols are obtained using CO_2. Cells are cracked with CO_2. Applications in injection molding, for instance, exhibit the broad applicability of CO_2 based on its pressure behaviour.

3.5 USE OF CO_2's REACTIVITY

At 1 bar the pH of CO_2 in H_2O is 3,7. Increase of pressure lowers this value to 3,3. The acidity of CO_2 is used to control the pH in food and drinking water.
Recently CO_2 is increasingly used to neutralize waste water. Advantages come from two points: a) using CO_2 yields carbonates thus circiumventing the conventional use of H_2SO_4 or HCl leading to sulfates and chlorides; b) applying a strong acid gives a steep pH-change between pH 2-11. With CO_2 a maximal pH of 6,5 prevails in a broad zone thus enabling easier pH control.

It is well known that plants grow better at an increased level of CO_2. Therefore, CO_2 is often applied for plant growth in green houses.

3.5.1 USE OF CO_2 FOR CHEMICALS

The chemical industry uses CO_2 for the synthesis of a variety of organic and inorganic chemicals. Many inorganic carbonates are manufactured via the general route

$$\text{Metal salt} + CO_2 \longrightarrow \text{Metal carbonate}$$

In this way carbonates such as Na_2CO_3, K_2CO_3, $BaCO_3$ and pigments are synthesized. The major organic chemicals made from CO_2 as building block are shown in Figure 5.

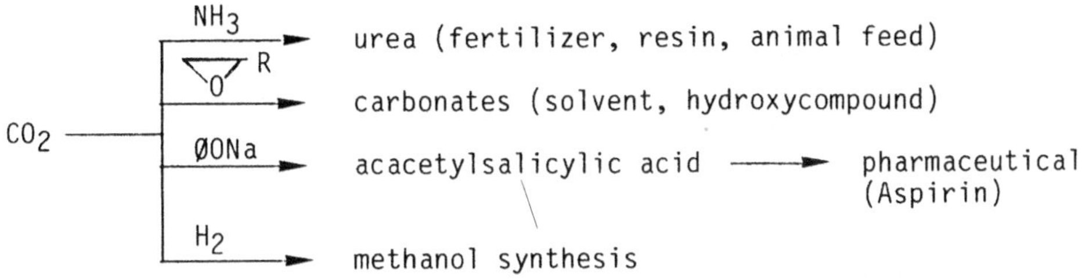

Fig. 5 Major organic chemicals made from CO_2

Urea and its derivatives can be made by reacting CO_2 with NH_3 (or amins). Considering that 42 mio t of urea are manufactured worldwide and keeping in mind that 1,3 tons of CO_2 are needed for 1 t of urea, makes it obvious that the manufacture of urea is the biggest application of CO_2 in the chemical industry.
The second largest consumption of CO_2 is provided in the methanol synthesis. Here CO_2 is used to balance the $CO:H_2$ ratio and to control the heat of the CO hydrogenation. Up to 5 % of CO_2 are added to the raw material CO/H_2 thus emphasizing at a world CH_3OH capacity of 25 mio tons the size of its consumption.
Recently mechanistic evidence has been quoted claiming that the methanol synthesis proceeds via formate formation on the surface. This can be considered as a hint that the direct methanol synthesis from CO_2 and hydrogen may be feasible in the future.
A synthesis practised for a long time is shown in Figure 6.

Alone in the USA 25.000 tons of Aspirin were made in 1980.
Organic carbonates are manufactured by reacting ethylene oxide and propylene oxide with CO_2. Based on the carbonates obtained a broad chemistry can be built as is exhibited in Figure 7.

An interesting approach using carbon dioxide is the Calcor process (Hydrocarbon processing May 1985). In this process CO_2 and hydrocarbons are heated over a catalyst yielding CO of > 99 % purity.

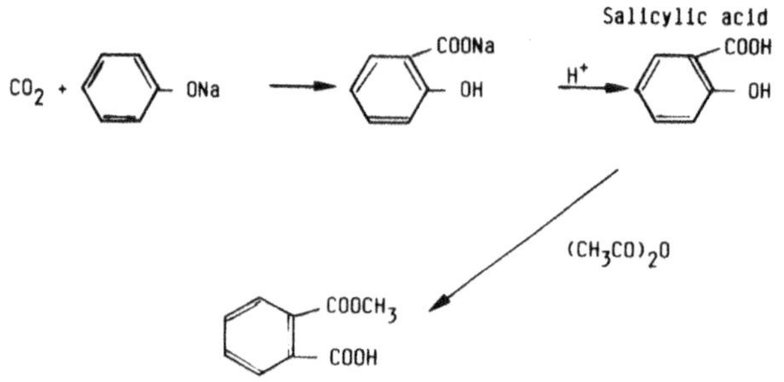

Fig. 6 Synthesis of Aspirin

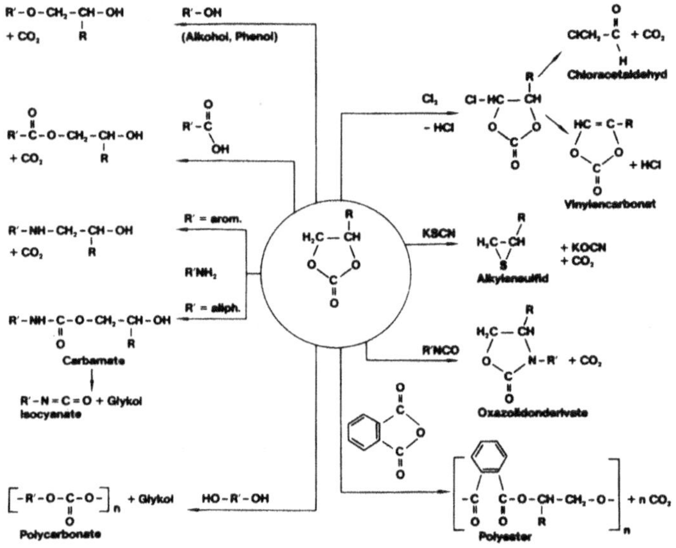

Fig. 7 Chemistry based on carbonates

Future chemical use of CO_2: It can be anticipated that in the future many more chemicals may be based on CO_2. Various reasons can be seen in support of such a development: a) environmental concern may lead to an increase in research focused on CO_2 utilization; b) increased availability of CO_2; c) lower cost of CO_2.
Often CO_2 is considered to be chemically innert. This is a myth as is evident from the urea synthesis. Table 2 lists a comparison of enthalpy values for synthesizing 4 important chemicals from CO/H_2 and CO_2/H_2.

Based on the $\Delta H°$ values, there is no great difference in useing CO/H_2 or CO_2/H_2. However, effects of entropy and kinetics have not been considered.

Table 2 $\Delta H°$ values comparing CO/H_2 with CO_2/H_2

Product	Reaction	$\Delta H°$ kJ/mol
CH_3OH	$CO(g)+2H_2(g) \longrightarrow CH_3OH(l)$	-128,5
	$CO_2(g)+3H_2(g) \longrightarrow CH_3OH(l)+H_2O(l)$	-131,0
HCOOH	$CO(g)+H_2O(L) \longrightarrow HCOOH(l)$	- 34,7
	$CO_2(g)+H_2(g) \longrightarrow HCOOH(l)$	- 37,7
CH_3COOH	$CO(g)+CH_3OH(l) \longrightarrow CH_3COOH(l)$	-135,2
	$CO_2(g)+H_2(g)+CH_3OH(l) \longrightarrow CH_3COOH(l)+H_2O(l)$	-138,1
CH_2=CH-COOH	$CO(g)+C_2H_2(g)+H_2O(l) \longrightarrow CH_2=CH\text{-}COOH(l)$	-214,7
	$CO_2(g)+C_2H_2(g)+H_2(g) \longrightarrow CH_2=CH\text{-}COOH(l)$	-217,6

g = gaseous; l = liquid

Finally, an attempt is made to single out various routes through which CO_2 in the future could enter into chemicals of interest. Figure 8 elucidates the synthesis of polymers using CO_2.

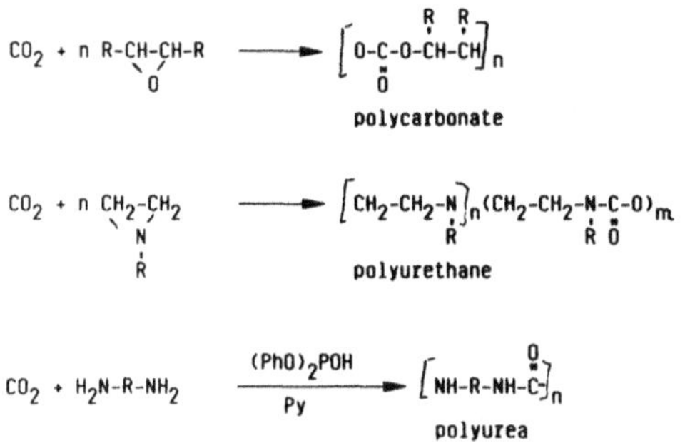

Fig. 8 Polymers from CO_2

Polycarbonates of 50.000 - 150.000 molecular weight can be made from epoxides and CO_2 with ZnR_2 catalysts. CO_2 reacts with aziridines yielding polyurethane, providing a phosgene free route to this polymer. Many patents are known in which CO_2 is hydrogenated to methane (Sabatier reaction) and higher olefins and paraffins (Haldor Topsøe DOS 3 103 207 (1981). Also the hydrogenation of CO_2 to CO has been reported. Compounds retaining at least one of the oxygen atoms (oxygenates) may even be of greater value. The synthesis of methanol from CO_2/H_2 has been discussed already.
Derivatives of formic acid can be derived according to the following equation

$$CO_2 + H_2 + HX \longrightarrow H\text{-}\overset{\overset{\displaystyle O}{\|}}{C}\text{-}X + H_2O$$

$$HX = H_2O,\ KOH,\ ROH,\ HNMe$$

An ester synthesis free of phosgene could proceed according to

$$CO_2 + 2\ ROH \xrightarrow{\text{cat.}} RO\underset{\underset{O}{\|}}{C}\text{-}OR + H_2O$$

The carboxylation of CH-acidic compounds could yield higher acids. The potential of such an approach is a synthesis of citric acid as shown in Figure 9 (Chem. Commun. (1966) 618).

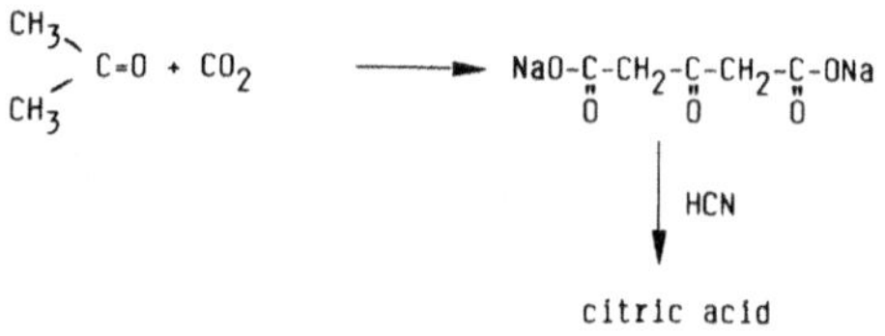

Fig. 9 Citric acid synthesis

Mention must also be given to the catalytic approach built on homogeneous transition metal chemistry. Our understanding of CO_2 coordination is increasing rapidly and it can be anticipated that catalytic spin-offs may result (see chapter A. Behr "Catalytic reactions of carbon dioxide with hydrocarbons").

Literature

Manusfacture and Industrial Applications of CO_2:

a) E. Hillenbrand in Ullmanns Enzyklopädie der technischen Chemie, Verlag Chemie, 4. Aufl. 14, 569 (1977)
b) A. Behr, Chem.-Ing.-Tech. 57, 893 (1985)
c) A. Behr, Bull. Soc. Chim. Belg. 94, 671 (1985)
d) R.E. Meissner, Hydrocarb. Proc. 113 (1980)
e) L. Hannon, Jr. Encycl. Chem. Proc. Des. 6, 280 (1978)
f) S. Inoue, Chem. Econ, Eng. Rev. 7, 31 (1975)
g) Various company brochures provided by: a) KWD (Kohlensäurewerk Deutschland, 5462 Bad Hönningen; b) AGEFKO (Kohlensäureindustrie GmbH, Postfach 1240, 4000 Düsseldorf; c) Alkylancarbonates Hüls AG, Postfach 1320, 4370 Marl).
h) Dr. A. Behr, Habilitationsschrift 1987

Extraction with CO_2:

a) U. Sievers, Chemie-Technik, 14, 54 (1985)
b) D.S. Gardner, Chem. Ind. 402 (1982)
c) F.M. Orr Jr., J.P. Heller, J.H. Taber, R.J. Card, Chem. Tech. 482 (1983)

ELECTROCHEMICAL APPROACHES TO THE REDUCTION OF CARBON DIOXIDE

Catherine O'Connell, Sven I. Hommeltoft, and Richard Eisenberg*
Department of Chemistry
University of Rochester
Rochester, New York 14627
USA

Abstract. Carbon dioxide reduction requires the input of energy or a high energy reductant to drive the reaction. Two possible energy sources which can be used in this capacity are electrochemical potential and light. In this paper approaches to CO_2 reduction by direct and indirect electrochemical methods are reviewed and analyzed. The indirect reductions involve the use of metal complexes in solution or on the electrode surface as electron transfer mediators and catalysts for reduction. The product distributions are sensitive to reaction media and conditions, but in many cases inadequate analyses are reported. A brief discussion of photochemical and photoelectrochemical reductions of CO_2 is also given.

The reduction of carbon dioxide and its conversion into fuels and metabolic feedstocks represents an important challenge to chemists of all types. Many factors, both thermodynamic and kinetic, must be considered in analyzing this problem.[1-5] Of particular note are the energetics of CO_2 reduction, the binding and activation of CO_2, and the disposition of oxygen released through C-O bond cleavage when reduction beyond the C 3+ oxidation state occurs.[2,3] In this paper, we examine the problem of CO_2 reduction using both electrochemical and photochemical methods with an emphasis on the former.

Thermodynamic Considerations for Chemical Reduction

Carbon dioxide is an extremely stable molecule, as evidenced by its ΔG_f° of -94.26 kcal/mole. Because of its stability, CO_2 serves along with H_2O as the major end product of metabolism and fossil fuel combustion. These two species thus function as the principal repositories or "sinks" of fixed oxygen in the biosphere. An analysis of CO_2 reduction must therefore begin with thermodynamics.

Reduction by Hydrogen. For favorable reduction of CO_2, either a high energy reductant or an external source of energy must be employed to drive the reaction. Of the possible chemical reductants, molecular hydrogen is clearly the most desirable since it is easily produced by

M. Aresta and G. Forti (eds.), Carbon Dioxide as a Source of Carbon, 33–54.

electrolysis of H_2O, and it serves as an effective O atom acceptor via the formation of H_2O. However, the simple reductions of CO_2 by H_2 as shown in eqns (1)-(4) reveal themselves to be thermodynamically unfavorable with small, positive ΔG°'s, although the reactions can be driven when H_2O is removed from the reaction system as it is formed. It is only when the reduction of CO_2 reaches methanol as the carbon product that H_2 reduction, eqn (4), becomes favorable. This analysis, however, does not take into account the intermediate oxidation states which the carbon atom of CO_2 must pass through before reaching the CH_3OH product.

	ΔG°, kcal/mole	
$CO_2 + H_2 = CO + H_2O\ (l)$	4.79	(1)
$CO_2 + H_2 = HCOOH\ (l)$	7.88	(2)
$CO_2 + 2\ H_2 = HCHO\ (g) + H_2O\ (l)$	10.57	(3)
$CO_2 + 3\ H_2 = CH_3OH\ (aq) + H_2O\ (l)$	-4.35	(4)

Also neglected in our consideration of eqns (1)-(4) is the possible stabilization which can be achieved by complexation of the C product to the metal center of the catalyst. For example, eqn (1), which is the reverse of the well-studied water gas shift reaction, is favorable only at temperatures above ~1000 °C, but it can be driven to the right at far milder temperatures through coordination of CO. Specifically, when Wilkinson's catalyst, $RhCl(PPh_3)_3$ in HMPA is treated with $CO_2 + H_2$ (1:1, 40 atm), the carbonyl complex $RhCl(CO)(PPh_3)_2$ forms in up to 70 % yield.[6] Similarly, the bicarbonate complex $RhH_2(O_2COH)(PPr^i_3)_2$ reacts at 298 °K in pyridine with CO_2 (1 atm) to give $Rh(CO)(O_2COH)(PPr^i_3)_2$ with CO coordination providing the driving force for the reaction.[7]

While the hydrogenation of CO_2 to formic acid, (2), is thermodynamically unfavorable with ΔG° of +7.88 kcal/mole for HCOOH in the liquid state, an equivalent reaction, the hydrogenation of aqueous bicarbonate to formate, eqn (5), is essentially thermodynamically neutral at ambient conditions. This reaction has been studied by

$$HCO_3^- + H_2 = HCOO^- + H_2O \qquad \Delta G^\circ = -0.33\ \text{kcal/mole} \qquad (5)$$

Wrighton and co-workers who find that under 1 atm H_2 at room temperature nearly equal concentrations of HCO_3^- and $HCOO^-$ exist at equilibrium.[8,9] The equilibrium was examined using as catalysts Pd on carbon and Pd with polymeric methylviologen on silica. This equilibrium has also been studied using the enzyme formate dehydrogenase.[10] Through the use of D and ^{13}C isotope labelling, Wrighton and co-workers found that the exchange rate of (5) using Pd on C was nearly comparable to the rate of H/D exchange between H_2O and D_2.

The direct hydrogenation of CO_2 to HCOOH can be drawn to the right by esterification. Thus CO_2 can be hydrogenated to alkyl formates in the presence of alcohols (MeOH, EtOH), eqn (6), the reaction being catalyzed by various transition metal carbonyl anions and hydrides such as $HRu_3(CO)_{11}^-$ and $HM(CO)_5^-$ (M = Cr, W)[11-13]. The reaction has been studied

$$CO_2 + H_2 + ROH \xrightarrow[MH(CO)_x]{} HCOOR + H_2O \qquad (6)$$

mechanistically by Darensbourg and Ovalles using the group 6 carbonyl anions, and they propose that the catalysis produces formic acid which subsequently undergoes esterification.[13] A separate study published earlier shows that CO_2 reduction can also be driven by formamide formation from CO_2 + H_2 + R_2NH at 800 psi using various group 8 complexes including $IrCl(CO)(PPh_3)_2$ and $CoH(dppe)_2$.[14]

Since catalyzed reductions of CO_2 by H_2 will invariably proceed via CO_2 interaction with M-H, the thermodynamics of this interaction are of importance. To date, only a few studies have dealt with this problem explicitly. Rathke and coworkers have examined the CO_2/Bu_3SnH system and have found that at 200 °C under 2 atm of CO_2 the equilibrium mixture contains roughly equal amounts of tin hydride and tin formate, eqn (7).[15]

$$Bu_3SnH \quad + \quad CO_2 \quad = \quad Bu_3Sn(OOCH) \qquad (7)$$

In the absence of a catalyst eqn (7) occurs fast enough at 115 °C - 175 °C to allow determination of ΔH = -18.3 kcal/mole and ΔS = -20.2 cal/mole-deg for this equilibrium. The negative entropy change is consistent with the reaction consuming one mole of gas. For the platinum hydride system trans-$PtH_2(PEt_3)_2$, Trogler finds that eqn (8) in toluene

$$trans\text{-}PtH_2(PEt_3)_2 + CO_2 \quad = \quad trans\text{-}PtH(OOCH)(PEt_3)_2 \qquad (8)$$

has an equilibrium constant of 2 atm^{-1} at 25 °C whereas in polar media, free formate and a cationic dimer, $Pt_2H_3(PEt_3)_4^+$, are observed to form.[16]

Other High Energy Reductants. In considering the use of other high energy reductants for reaction with CO_2, we note that these reductants must also be capable of serving as O atom acceptors. Possible reductants include oxophilic metals and their low oxidation-state compounds, phosphines, and silanes, but these systems can only be adapted for large-scale, catalytic reductions of CO_2 if they can be produced easily from a primary energy source such as by electrolysis or photosynthesis. Floriani's work using Cp complexes of the oxophilic metals Zr and Ti[17] is noteworthy regarding the employment of other high energy reductants, and is discussed elsewhere in this volume.

Because of the strength of Si-O bonds and the reaction chemistry of R_3SiH with metal complexes which in some respects parallels that of H_2, silanes are interesting candidates for reducing CO_2. The hydrosilation of CO_2 to silyl formates, eqn (8), has been reported using various Ru complexes including $HRu_3(CO)_{11}^-$ and $RuCl_2(PPh_3)$,[18,19] and the reduction of CO_2 to CO has been found in the reaction of $RhCl(PPh_3)_3$ with R_3SiH and CO_2.[20] In the latter reaction, the siloxane $R_3SiOSiR_3$ presumably forms as the oxygen sink. At Rochester, we have found that hydrosilation using primary and secondary silanes proceeds readily at 298 °K using Ir(CN)(CO)(dppe).[21] In addition to the simple silyl formate produced, other products are observed which may correspond to **A** and **B**. Species **B** is attractive since it represents reduction to the C(0) oxidation state.

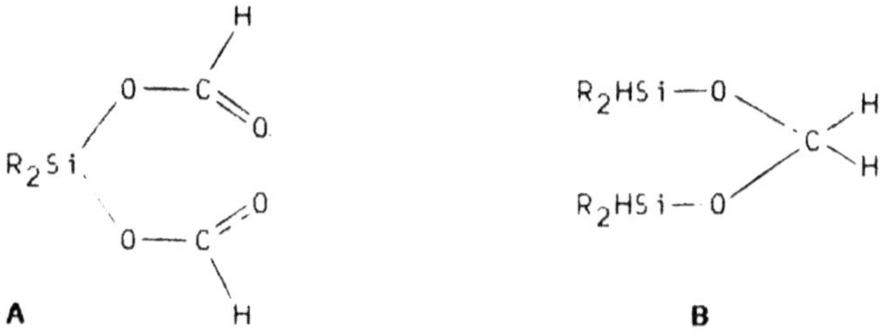

A B

Direct Electrochemical Reduction of CO_2

External energy sources which can be used to promote CO_2 reduction include both light and electrochemical potential. Both of these will be discussed in this paper with major emphasis given to electrochemical reductions of CO_2. The simplest half-cell reactions for CO_2 reduction are shown as eqns (9)-(14) along with half-cell potentials adjusted to pH 7. In analyzing these half-cell reactions, it should be remembered that $2\ e^- + 2\ H^+$ is stoichiometrically equivalent to H_2. The reduction to methanol shows itself to be least unfavorable but as with eqn (4), eqn (14) does not take into account intermediate oxidation states of C which must be traversed in the reduction process. In the $2\ e^-$ reductions, the formation of oxalate is perhaps the most interesting process since it involves CO_2 coupling and C-C bond formation, but it is also the most unfavorable of the $2\ e^-$ reductions thermodynamically. Which of the electrochemical reduction pathways is actually followed for a particular system depends more on numerous kinetic factors than on the reduction potentials of eqns (9)-(14).

			$E^{\circ\prime}$ (pH 7)	
$CO_2 + 2\ e^- + 2\ H^+$	=	$CO + H_2O$	-0.52	(9)
$CO_2 + 2\ e^- + H^+$	=	$HCOO^-$	-0.41	(10)
$2\ CO_2 + 2\ e^-$	=	$CO + CO_3^{2-}$	-0.55	(11)
$2\ CO_2 + 2\ e^-$	=	$C_2O_4^{2-}$	-0.90	(12)
$CO_2 + 4\ e^- + 4\ H^+$	=	$HCHO + H_2O$	-0.48	(13)
$CO_2 + 6\ e^- + 6\ H^+$	=	$CH_3OH + H_2O$	-0.38	(14)

Kinetic Considerations. CO_2 reduction electrochemically may be carried out directly at an electrode surface or indirectly using a complex in solution as an electron transfer mediator and catalyst. In direct electrochemical reduction, large differences exist between the thermodynamic potentials of eqns (9)-(14) and the actual potentials which must be applied to the electrodes to achieve CO_2 reduction. These differences correspond to overpotential and represent a measure of the kinetic barrier for the particular reduction reaction. The large overpotential for CO_2 reduction at an electrode surface arises because of the relative instability of the initial reduction product, CO_2^-, and the energy needed to generate it.

In the direct electrochemical reduction of CO_2, the particular pathway followed depends on the conditions of the electrolysis including the reaction medium. Saveant suggests that the initial reduction product, CO_2^-, partitions into different reaction channels which lead to formate, oxalate, and CO + CO_3^{2-} as shown in Scheme I.[23,24] Whereas neutral CO_2 is Lewis acidic at C and weakly basic at O, the species CO_2^- is radical-like at carbon and strongly basic at oxygen. Based on Scheme I, formate formation is favored in aqueous or wet media, reductive disproportionation to CO + CO_3^{2-} is facilitated when $[CO_2]$ is large, and oxalate formation can be made predominant under high current densities in strictly non-aqueous solvents. Scheme I thus serves as a guide to the formulation of conditions in the development of CO_2 reduction systems.

Scheme I

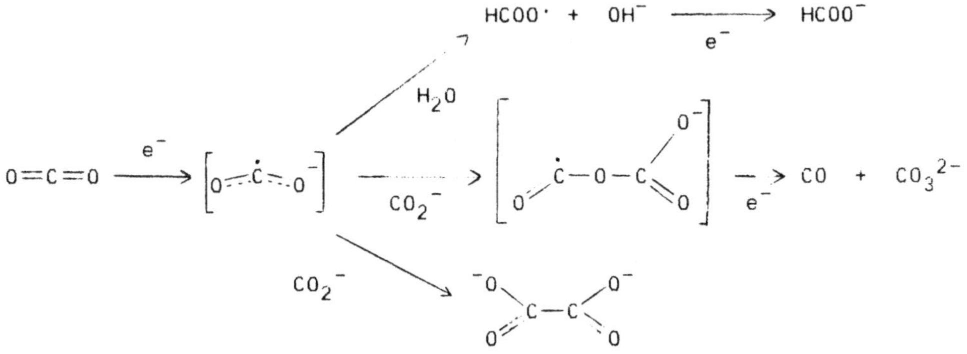

A Cautionary Note. In analyzing the many different reports on the electrochemical reduction of CO_2, we note that CO_2 reduction products are often not unambiguously identified qualitatively, and quantitation of these products has until recently been routinely neglected. It is disconcerting that qualitative tests for particular reduction products are not used in conjunction with confirmatory tests since structurally related compounds including different reduction products may give positive results to a specific qualitative assay. The lack of quantitation means that the effectiveness of any catalyst cannot be adequately assessed since current efficiency or faradaic yield is not available from the report. The reader must therefore exercise caution in evaluating the numerous reports of CO_2 electroreduction and electrocatalysis.

Analysis of gaseous products (CO, H_2) can be done routinely by gas chromatography, although a carrier gas other than He will be needed for accurate H_2 determination if a tc detector is used. Qualitatively, the presence of CO can also be indicated by FTIR spectroscopy[25] and by ^{13}C NMR spectroscopy.[26] The identification and quantification of formate and oxalate are best done by esterification followed by gc analysis.[25,27-30] However, because of differences in the reaction conditions, and in particular the solvent systems used, the esterification procedures vary widely and for the most part have not been optimized. In this regard the work by Saveant on electrochemical reduction in DMF deserves mention since the gas chromatographic analysis of oxalate was compared with a precipitation method based on CaC_2O_4, and conditions for the esterification were examined rigorously.[23] The success of the gravimetric method for oxalate hinges on fixing the pH of the reaction solution so that the corresponding carbonate does not precipitate as well.

Qualitatively, formate, formaldehyde, oxalate, and glycolic acids can be determined by colorimetric methods[31-33] using reagents such as chromotropic acid,[34,35] resorcinol,[36] and thiobarbituric acid.[34] Both oxalate and formate have been determined by isotachophoresis[37] and ion exchange chromatography. Enzyme assays have also been applied to this problem.[38] What appears needed at this time is a thorough evaluation of the relative accuracies of the different techniques, a determination of the interferences which the reaction conditions pose for any method, and the resolve of workers in the field to quantify all products arising from CO_2 reductions.

Recent Progress in Direct Electrochemical Reduction. Numerous studies of the direct electrochemical reduction of CO_2 have been reported, and the results through 1977 are summarized in a review.[39] In this section, we examine a few of the more current and important results.

The series of studies by Saveant provides the most detailed information on the kinetics of CO_2 reduction through analysis of product distribution as a function of current density, CO_2 pressure, and H_2O concentration as well as other operational parameters.[22-24] The basic Saveant scheme is shown above with three major reaction channels. However, the second electron transfer step in each channel can be provided by either the electrode or CO_2^-. The standard potential for the CO_2/CO_2^- couple was determined in dry DMF at a dropping mercury electrode to be -2.21v vs SCE, but this value may represent at least partly the potential for a couple involving the O-C coupled species $OCOCO_2^-$ as the reduced form since it is thought that amounts of this species and CO_2^- are comparable in aprotic media. The reduction potential corresponding to production of free CO_2^- would therefore be more negative since formation of the adduct $OCOCO_2^-$ represents net stabilization of the CO_2^- radical anion.[22]

Product distribution studies by Saveant and by Kaiser and Heitz in an earlier study[40] show that formate and glycolate occur when the reduction system contains H_2O, and that oxalate and CO predominate under anhydrous conditions. Glycolate results from reduction of oxalate which has been partially protonated in the presence of H_2O.[40] The combined yield of oxalate and glycolate represents the dimerization pathway as opposed to the pathways leading to formate and CO. Production of H_2 was negligible in all cases, the reduction of H_2O being minimal at water concentrations of less than 3%. As suggested by Scheme I, high CO_2 concentrations favor reductive disproportionation to CO and CO_3^{2-}, whereas high current densities and lower concentrations of CO_2 in strictly anhydrous media favor oxalate.[23] A quantitative analysis of the product distributions by Saveant led to the following rate constants for reaction of CO_2^-: $10^7\ M^{-1}\ s^{-1}$ for C-C self coupling, $3.2 \times 10^3\ M^{-1}\ s^{-1}$

for O-C coupling with neutral CO_2, and $7.7 \times 10^2\ M^{-1}\ s^{-1}$ for reaction with H_2O leading to formate.[24] Oxalate generation was greater in DMSO than in DMF at the expense of formate production, but less reduction of oxalate to glycolate was seen in aqueous DMSO than in aqueous DMF. Relatively little difference was seen between Hg and Pb as cathode materials in these studies.

A survey of voltammetric reduction waves for CO_2 in water, DMSO, MeCN, and propylene carbonate using mercury, platinum, gold, lead, and glassy carbon electrodes has been carried out by Eggins.[41] In general the potentials are quite negative (<-2.2 v vs SCE), although in H_2O peak potentials as positive as -1.12 v are seen for indium. The possible use of indium may therefore represent a significant lowering of the overpotential for CO_2 electroreduction. A study by Kapusta and Hackerman[42] shows that tin and indium electrodes are active to CO_2 reduction at applied potentials more negative than ~-1.3 v, giving formate as the major product. The high reactivity of these electrodes relative to Hg is ascribed to the possibility of adsorption of radical intermediates to the electrode surface, but the overall power efficiency of the reduction process is low because high overpotentials are needed to obtain reasonable current levels. Subsequent studies by others indicate that indium does not display the catalytic properties suggested by the earlier c.v. data.[43]

The reduction of CO_2 to MeOH on semiconductor electrodes has been reported by several different groups.[44-48] In the first of these communications, Monnier, et al., used TiO_2 and TiO_2-Ru electrodes in aqueous solution at -0.9 v vs SCE but neither the method of product analysis nor quantitation of the product is given in the report.[47] Frese and co-workers in a series of studies have described reduction of CO_2 to MeOH over GaAs at -1.2 to -1.4 v vs SCE in aqueous solutions with current efficiencies of up to 100 %.[44-46] Increasing the current density, however, resulted in a lowering of the faradaic yield. In general, the current efficiencies were highly variable, and with one set of conditions including high electrolyte purity, no MeOH was obtained.[45] In another study, Frese reports electrochemical reduction of CO_2 at Ru electrodes. The products included MeOH, CH_4, and CO obtained in variable but generally low yields at potentials in the range -0.55 to -0.65 v vs SCE. No details of analysis were given.[46]

While these recent results using indium, tin, ruthenium and semiconductor electrodes at low applied potentials are tantalizing, they must be substantiated by more quantitative studies involving detailed and unambiguous product analyses. In general, most of the direct electrochemical reductions occur at very negative potentials owing to the instability of the CO_2^- radical anion. Numerous mechanistic proposals have been made including adsorbed species, surface hydrides, and competitive pathways for H^+ and CO_2 reduction. However, there is little direct evidence at this point to support these proposals, and many of the questions regarding the intimate mechanism of CO_2 reduction remain unanswered.

Indirect Electrochemical Reduction of CO_2

Efforts to circumvent the principal difficulties in direct electro-reduction of CO_2, namely the high overpotentials required and the low current efficiencies observed, have centered on the use of homogeneous transition metal complexes as catalysts to mediate the reduction. These complexes function by activating CO_2 and/or stabilizing its 1 e^- reduction product through coordination. The electrocatalysts may also store the necessary reducing equivalents to convert CO_2 to specific and desired products without forming higher energy intermediates.

The complexes which can function as electrocatalysts are electro-chemically active in the range between the $E^{\circ\prime}$ values for CO_2 reduction and that needed for reduction directly at an electrode surface. Ideally, these complexes should show clean, reversible electrochemical behavior, and should possess long-term stability. The extent of catalysis can be readily determined from the observed decrease in required overpotential, the level of catalytic current generated, and the current efficiency for the formation of each product. While indications of interaction between CO_2 and the reduced metal complex can be easily ascertained by current enhancements in cyclic voltammetry and in electrolysis, the detailed nature of these interactions remains for the most part unclear.

Indirect Reductions Using Ni and Co Complexes. To date, work in this field has concentrated on complexes of tetraaza macrocycles including phthalocyanines and porphyrins. Most of the complexes studied have been of Ni and Co. The earliest report involved phthalocyanine complexes of Ni and Co adsorbed onto a graphite electrode for CO_2 electroreduction in aqueous media at ~-1.5 v (vs SCE).[34] Qualitative analytical tests showed positive for oxalic and glycolic acids but confirmatory tests were not applied and the gas phase above the electrolyte was not analyzed. When tetrasulfonated phthalocyanine complexes of these same metals were used in a second study,[49] current enhancement at ~-1.2 v vs SCE was observed at an amalgamated Pt electrode in aqueous solution but no product analyses were performed. Similar activity was found using solubilized Co tetraphenylporphyrin derivatives at -1.1 to -1.5 v vs SCE over Hg in aqueous solution.[36] Only qualitative tests were done, showing positive for formic acid, but small amounts of HCOOH were noted in the absence of catalyst as well.

A more recent study using Ni and Co phthalocyanines showed that the electrode material was important in the electrocatalysis of CO_2 reduction, indicating that the complexes bind in some way to the electrode surface.[50] In particular, graphite and glassy carbon seem specific in their ability to utilize the phthalocyanines as catalysts with an overpotential ~200 mv lower than the best metals. Some electrocatalytic behavior was observed on high overpotential metals (Hg and Pb) but none was detected on Pt or Au. With graphite or glassy carbon electrodes, Co and Ni phthalocyanines catalyzed the electro-reduction of CO_2 to formate in neutral aqueous solutions with a current efficiency of ~60 % between -1.2 and -1.4 v vs SCE. Deactivation of the catalyst occurs, however, on cycling to more negative potentials, and no analysis of the gas phase above the solution was given. Interestingly, small amounts of MeOH (<5 % current efficiency) were reported on electrolysis at pH less than 4.

In contrast to these results, Lewis and Lieber have recently studied the cobalt phthalocyanine (CoPc) system on carbon electrodes and have found that the only C-containing product is CO.[33] Controlled potential electrolysis at -0.95 to -1.2 v vs SCE of CoPc deposited on pyrolytic graphite or carbon cloth in aqueous solution at pH 5 under 1 atm of CO_2 yielded CO and H_2 (2:1) with an overall current efficiency for more than 10^5 turnovers of 90-95 %. Only trace amounts of oxalate and formate were detected by spot tests. The pH dependence of the CoPc reduction waves and the lack of current enhancement under CO_2 relative to Ar led the authors to propose that reduction occurs by protonation of $CoPc^-$ followed by a second reduction and subsequent reaction with CO_2. Use of CoPc dissolved in aqueous THF solution rather than adsorbed onto the C electrode resulted in poor stability and low catalytic efficiency for CO_2 reduction.

The first report of synthetic macrocycle complexes of Ni and Co for CO_2 electroreduction was made in 1980 from our laboratory.[27] This work which employed complexes such as **I-IV** was also the first to provide quantitative analysis of the gas phase products and their respective current efficiencies. CPC of **I-IV** in aqueous acetonitrile at a Hg pool electrode (-1.3 to -1.6 v vs SCE) under CO_2 yielded CO and H_2 with high overall current efficiencies (>90 %). A protic source appeared to be necessary for CO_2 reduction as no CO was detected upon electrolysis in dry DMSO. It was suggested that both CO and H_2 may arise from a common intermediate, possibly a metal-hydride formed by protonation of the reduced macrocycle.

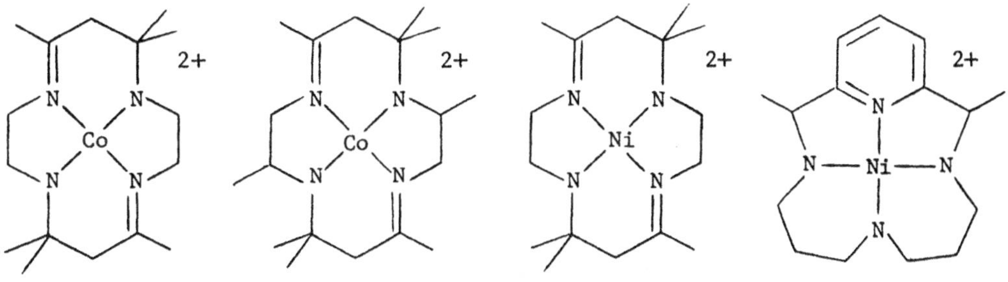

I II III IV

Two subsequent studies have used the same and closely related macrocycle complexes for promoting CO_2 reduction, and presumably they involve the same catalytic chemistry as that of Fisher and Eisenberg. The first study, by Bradley and coworkers, employed p-Si as the cathode instead of Hg which allowed the applied potential to be up to 0.6 v less negative than that required at a metal electrode.[51] The additional energy needed to promote reduction was provided by irradiation at 752 nm. The second study was by Tinnemans, *et al.*, who studied CO_2 reduction promoted by these macrocycles under conditions of electrochemical reduction at Hg in aqueous DMF or MeCN, and photochemical reduction using $Ru(bpy)_3^{2+}$ as sensitizer, ascorbic acid as sacrificial electron donor, and the macrocycle complex as electron transfer mediator and catalyst in aqueous solution.[52] Both CO and H_2 were obtained in each case with CO/H_2 selectivity generally low although in one instance a ratio of 26 was obtained. Positive qualitative tests for formate and oxalate after electrolyses were also observed. Thus while these macrocycle complexes of Ni and Co serve as electrocatalyts for CO_2 reduction, their selectivity is not great because of competitive H_2O reduction.

The most promising results to date using tetraaza macrocycle complexes for catalytic electroreduction of CO_2 are those of Sauvage and coworkers with $Ni(cyclam)^{2+}$, **V**.[31,32] The most remarkable aspect of this work is the extremely high selectivity ($>10^6$) to CO formation from CO_2 relative to H_2 production from water. Catalysis by **V** over a Hg pool electrode was fast (10^3 moles of CO produced per mole of complex per hour) and selective (the faradaic yield was >95%) at potentials of -1.25 v vs SCE or greater. Almost no H_2 was detected even in pure aqueous media. Through variation of pH, HCO_3^- and CO_3^{2-} were ruled out as precursors for CO formation; the reduction was thus shown proceed from CO_2.

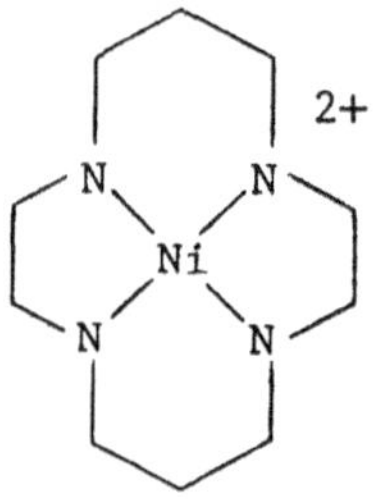

V

The fact that catalyst concentration had only a small effect on peak currents in c.v.'s of **V** under CO_2 and on the rate of CO production led Sauvage to suggest that adsorbed species are relatively more important than those in the bulk, i.e., reduction of CO_2 takes place at the electrode surface. It was proposed that adsorption of $Ni(cyclam)^+$ on Hg is followed by attack of CO_2 to give a Ni(III) species which is protonated and reduced to yield a Ni(II)-CO complex. Dissociation of CO then regenerates the starting $Ni(cyclam)^{2+}$ catalyst. Once the concentration of CO increases, several of the species in the proposed cycle may form carbonyl adducts which could be reduced more easily than the original species, helping to explain the relatively high activity of **V** as an electrocatalyst. The species $Ni(CO)(cyclam)^+$ was detected in the electrolysis cell by UV/vis, infra-red and ESR spectroscopies, thus lending support to the mechanistic proposal.[32]

In contrast to $Ni(cyclam)^{2+}$ which exhibited high CO/H_2 selectivity, excellent current efficiencies, and long-term stability, other macrocycle complexes proved disappointing. Nickel complexes of saturated 14-membered rings related to **I-IV** did show catalytic activity but were not selective to reduction for CO_2 relative to H_2O. In these systems, the methyl groups at the periphery of the macrocycles may render the Ni center less accessible to attack by CO_2 relative to attack by H^+. In

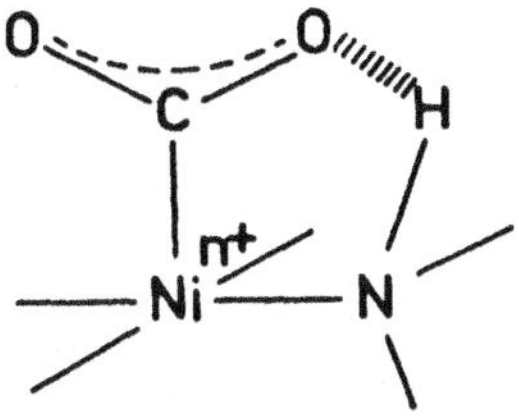

Figure 1. Proposed binding of CO_2 to Ni(cyclam)$^+$.[32]

addition, the Ni(cyclam)$^+$ system appears better suited to provide a stabilizing interaction for binding CO_2, shown as Figure 1, than the other macrocycle complexes. Complexes of unsaturated macrocycles and open chain ligands were found to be less active and showed poorer long-term stability.[32]

Indirect Reductions Using Complexes of Other Metals. In earlier reports dealing with phthalocyanine and porphyrin complexes, the inactivity of Cu and Fe systems as electrocatalysts was mentioned.[34,36,49] Recently, Becker and co-workers[25] have examined the electroreduction of CO_2 in anhydrous CH_2Cl_2 using various porphyrin complexes, **VI** and **VII**. Electrocatalysis was seen with Ag and Pd

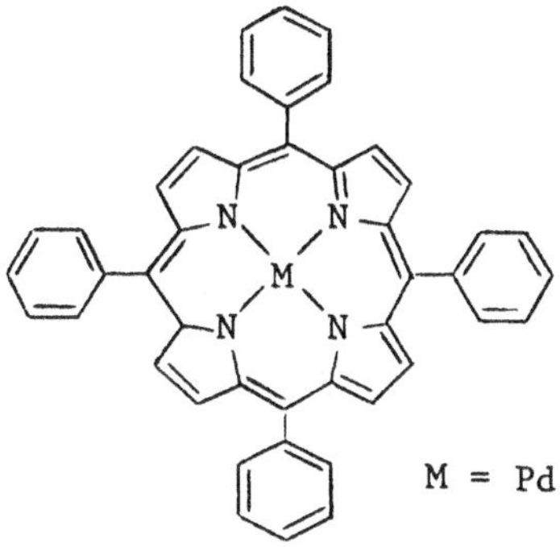

VI

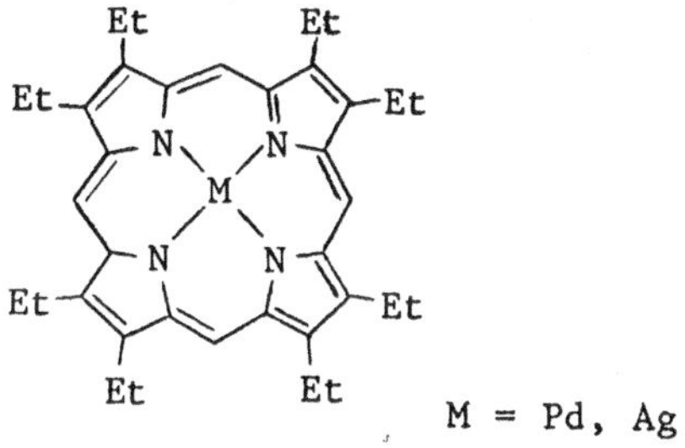

VII

systems at platinum and glassy carbon electrodes and applied potentials of -1.3 to -1.4 v vs SCE. A very significant aspect of this work was the observation of oxalic acid as the <u>major</u> CO_2 reduction product. Oxalate formation requires CO_2 coupling with C-C bond formation, and is probably the most interesting reduction channel for CO_2 utilization in the long run. In Becker's study, however, no current efficiencies were reported, H_2 (but no CO) was detected in the gas phase, and the metalloporphyrin catalysts exhibited decomposition. The particular combination of metal and ligand appears important since Ag(OEP), Pd(OEP) and Pd(TPP) showed catalytic activity whereas Ag(TPP), Cu(TPP), and Ni(TPP) did not (OEP = octaethylporphyrin; TPP = meso-tetraphenylporphyrin).[25]

A number of bpy (2,2'-bipyridine) complexes have come under scrutiny as electrocatalysts for CO_2 reduction. Lehn, et al., have reported that $ReCl(CO)_3(bpy)$ functions as an electrocatalyst at -1.5 v vs SCE in aqueous DMF with CO produced in >90 % faradaic yield.[26] The reaction was much slower in the absence of H_2O indicating the need for a protic source. Meyer and co-workers have examined the same complex in MeCN at -1.5 v vs SCE, and find only CO and CO_3^{2-} with current efficiencies of ~98 %.[29] Electrolyses at more negative potentials showed small amounts of NBu_3 indicative of Hofmann degradation of the supporting NBu_4PF_6 electrolyte, and no evidence for CO_3^{2-}. The Re complex undergoes a reversible bpy-based reduction at -1.35 v and an irreversible metal-based reduction at -1.7 v. While the mechanism of reduction is uncertain and may follow potential-dependent 1 e^- and 2 e^- paths, it appears that the reduced complex $Re(CO)_3(bpy)$ or its MeCN solvate reacts readily with CO_2.

Meyer and co-workers have examined other polypyridyl complexes as electrocatalysts including $Rh(bpy)_2(OTf)^{2+}$ (OTf = triflate) and $RuCl(terpy)(Ph_2PCH{=}CHPPh_2)^+$ (terpy = 2,2',2"-terpyridine).[30] The Rh system in dry MeCN gives formate production upon cpc with 64 % current efficiency. Other products include H_2 (12 % c.e.) and NBu_3 from the NBu_4PF_6 electrolyte by a route which appears to involve generation of a strong base and Hofmann degradation, eqn (15). It is interesting that

$$NBu_4^+ + B \longrightarrow BH^+ + NBu_3 + \quad (15)$$

H_2 only forms in the presence of CO_2. In contrast, the RuCl(terpy)-$(Ph_2PCH{=}CHPPh_2)^+$ yields CO as the only CO_2-derived product with no evidence for H_2 and a surprisingly positive test for NBu_3.

Another Rh complex examined as an electrocatalyst for CO_2 reduction is $Rh(dppe)_2^+$, the electrochemistry of which has been studied independently by several groups.[53a,b] Wagenknecht and Slater find that $Rh(dppe)_2^+$ promotes formate formation in dry MeCN at -1.2 to -1.3 v vs SCE over Hg.[28] Cyanoacetate was also detected, indicating generation of CH_2CN^-, which had been proposed earlier, followed by CO_2 trapping. No analysis of the gas phase is described. In a separate study of the reaction chemistry of $Rh(dppe)_2^o$ and $Rh(dppe)_2^-$ produced by Na naphthalenide reduction in benzene, we have found that both species react with CO_2 to produce CO as the only CO_2-derived product by gc and NMR spectroscopy.[53c]

Indirect electroreduction of CO_2 using the iron-sulfur cubane-like clusters $Fe_4S_4(SR)_4^{2-}$, **VIII**, (R = Ph, CH_2Ph) has been described recently by Tezuka.[37] The electrolysis is done in dry DMF at -1.7 to -2.0 v vs SCE over Hg and yields predominantly formate, some CO and small and variable amounts of oxalate. In the absence of the Fe-S cluster, cpc at -2.4 v vs SCE gave primarily oxalate although addition of H_2O or surfactant led to formate and CO at the expense of oxalate. With the Fe-S clusters, current efficiencies of up to 93 % for formate formation were observed, even in strictly anhydrous media. It was therefore proposed that CO_2^- is generated through electron transfer in the bulk

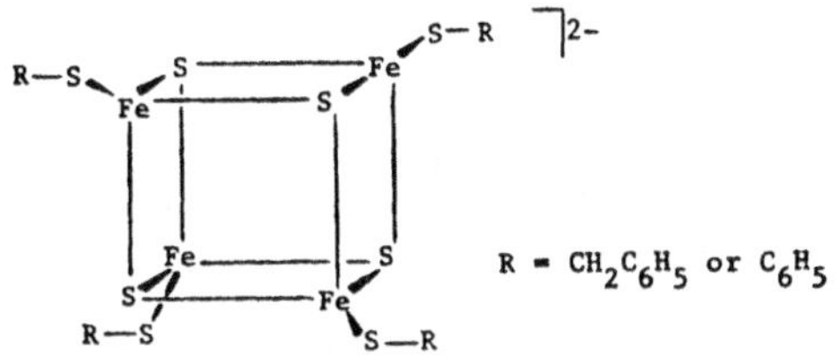

VIII

solution rather than at the electrode surface, leading to protonation and $HCOO^-$ formation rather than self-coupling and oxalate. The formate H atom may arise from R_4N^+ present as the cation of the cluster and the electrolyte.

In assessing the different studies of indirect electrochemical reduction of CO_2, we note that the success of any system (decrease in overpotential, product selectivity and current efficiency, level of catalytic current, and electrocatalyst stability) depend on many factors including the reaction medium, the supporting electrolyte, the electrode, and the applied potenial in addition to the nature of the electrocatalyst itself. Most of the studies have not been able to examine all of these variables in any systematic way. Unfortunately, the care with which the electroreductions have been done and the method(s) od product analysis also factor in any analysis of the results discussed above.

In most cases the formation of product is consistent with Saveant's original scheme for direct electrochemical reduction of CO_2 with CO and $HCOO^-$ as the major products seen to date.[23] In terms of catalytic efficiency, the most encouraging result remains that of Sauvage using the macrocycle $Ni(cyclam)^{2+}$ to promote reduction to CO.[31,32] The formation of oxalate as the major CO_2-derived product is observed in but one instance, in dry CH_2Cl_2 using Ag and Pd porphyrins.[25] It is surprising that in several studies using anhydrous media, the R_4N^+ electrolyte appears to enter as a reactant, giving up a proton and undergoing Hofmann degradation.[29,30,37]

Modified Electrodes. The lack of long-term stability of certain complexes as electrocatalysts, and the apparent differences when the complexes are physically adsorbed on the electrode surface relative to being in solution suggests that immobilization of the electrocatalysts using modified electrodes will be a fruitful avenue of inquiry for further indirect reductions of CO_2. To date, immobilization has been achieved by adsorption, electropolymerization, and incorporation into a polymer film by chemical coating methods. Through modified electrodes, local concentration of catalyst might channel CO_2 reduction selectively such as to oxalate formation via the dimerization of CO_2^-. The use of polymer modified electrodes for CO_2 reduction will require the synthesis of large modified electrodes for coulometry as well as more easily prepared smaller ones for voltammetry.

To date, Meyer and coworkers have studied a polymeric analog of the known CO_2 reduction catalyst $ReCl(CO)_3(bpy)$. Thus $ReCl(CO)_3(vbpy)$ (vbpy = 4-methyl-4'-vinyl-2,2'-bipyridyl), **IX**, was electropolymerized onto a platinum gauze electrode.[54] Experiments using modified rotating disc electrodes showed that currents were independent of rotation rate, suggesting that CO_2 penetrates the film rapidly and that its mass-transport is not rate limiting in catalysis. In dry MeCN at -1.55 v vs SCE, this electrode exhibited current efficiency for CO_2 reduction of 92 % which corresponded to 387 turnovers h^{-1} as compared to 15-23 turnovers h^{-1} by the monomer complex in solution under analogous conditions.[29] Whereas equal quantities of CO and CO_3^{2-} were generated in

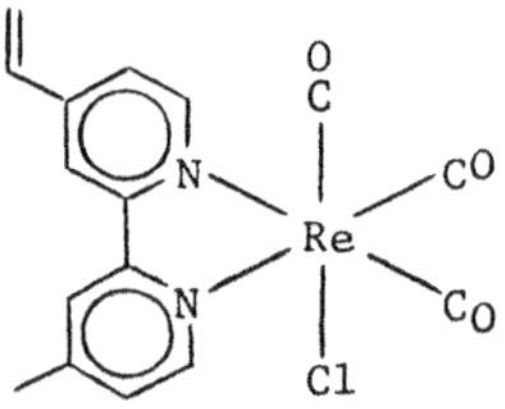

IX

the solution experiment, no CO_3^{2-} was detected using electropolymerized $ReCl(CO)_3(vbpy)$. It is possible, however, that carbonate remains trapped in the polymer film thus avoiding detection. Unfortunately, long-term stability remains a problem with this system. After 80 min of electrolysis, the electroactive material had been lost from the electrode surface.

Wrighton has followed up his studies of the $HCO_3^- + H_2 = HCOO^-$ system discussed above with polymer modified electrodes which can carry out the reduction of bicarbonate to $HCOO^-$ at potentials close to the thermodynamic limit for this couple.[38] The chemically derivatized electrodes consist of organosilane viologens, **X**, which are electropolymerized onto tungsten or platinum and then impregnated with Pd metal, as shown in Figure 2. These electrodes effect the reduction of

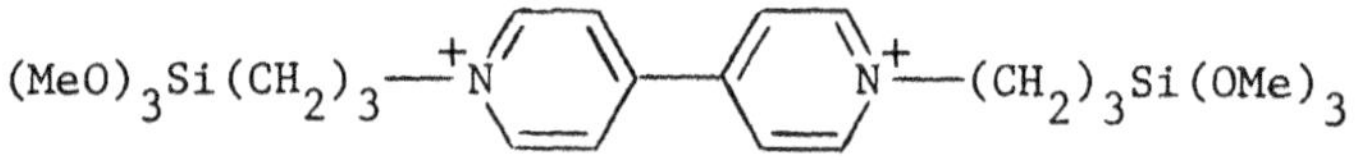

X

HCO_3^- to $HCOO^-$ in aqueous solution at -0.7 to -0.8 v vs SCE. Formate was the only reduced C-containing product detected by ^{13}C NMR spectroscopy after using $H^{13}CO_3^-$, and current efficiencies of up to 85 % were reported. However, significantly lower current efficiencies were also observed, and these were attributed to Pd hydride formation, reduction of the viologen polymer, and competing H_2 evolution. Degradation of activity occured with time due to reduction of the viologen polymer.

Photo- and Photoelectrochemical Approaches

The development of artificial photosynthesis in which CO_2 is selectively

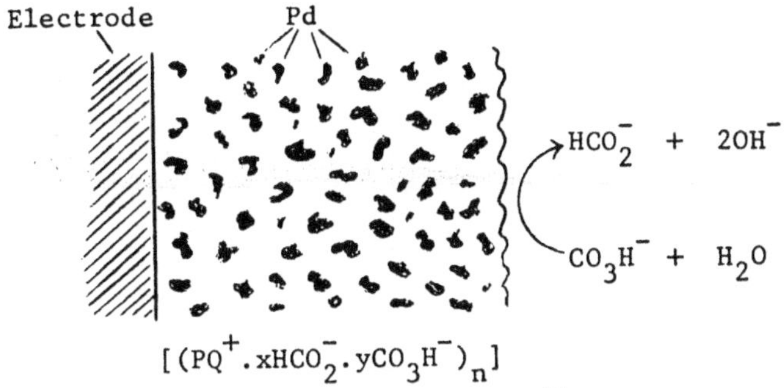

Figure 2. Modified electrode based on MV^{2+}/Pd for HCO_3^- reduction.[38]

reduced to a fuel or chemical feedstock material is an essential goal of present day chemistry. The use of light to drive CO_2 reduction involves the creation of an electron-hole pair, its separation into isolated charges, and the channeling of the energetic electron to a catalyst site at which the reduction can occur. Two different but related approaches will be discussed in this section.

The first involves systems based solely on photochemistry with solution electron transfer reactions serving to provide the means for electron-hole separation. The second involves systems which are photochemical and electrochemical hybrids and use irradiated semiconductor electrodes. The electron-hole separation is achieved in these systems in the semiconductor near the solution interface in the so-called space charge region. With these systems, the semiconductor electrode may be subject to an applied potential in addition to irradiation. A third type of photochemical system for CO_2 reduction, dicussed elsewhere in this volume, employs irradiated suspensions of semiconductor particles in aqueous solution. These suspensions lead to a variety of CO_2 reduction products including CO, HCOOH CH_2O, and CH_3OH with very low quantum yields and extremely poor energy efficiencies.

During the past decade, intense interest in solar energy storage and water splitting led to the development of a number of redox relay systems for H_2 production in which the well studied complex $Ru(bpy)_3^{2+}$ served as the photoactive agent.[55] In these systems the excited state of $Ru(bpy)_3^{2+}$ is quenched by methylviologen or paraquat (MV^{2+}) via electron transfer, with the resultant Ru(III) complex reacting rapidly with a so-called sacrificial donor while the highly reducing MV^+ transfers its electron to a catalyst center for H^+ reduction. A similar approach has been adopted by Lehn and Ziessel for CO_2 reduction.[56] They describe a relay system shown in Scheme II composed of $Ru(bpy)_3^{2+}/Co^{2+}$ in $MeCN/H_2O/Et_3N$ under CO_2 which generates CO and H_2 in almost equal amounts upon irradiation. The reaction proceeds with electron transfer quenching of $Ru(bpy)_3^{2+*}$ by Co^{2+} giving a reactive Co(I) species capable of reducing both H^+ and CO_2. Addition of bpy affects the relative product distribution, presumably by complexation to the Co catalyst.

Another $Ru(bpy)_3^{2+}$-based system has been reported by Tazuke,[57] but

Scheme II

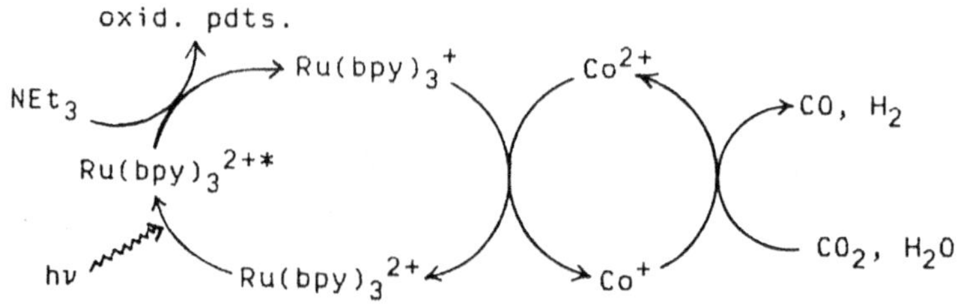

the quantum yields for CO_2 reduction are very low (0.01), and the report raises an interesting mechanistic question. The system is composed of $Ru(bpy)_3^{2+}$, MV^{2+} and triethanolamine. The major CO_2 reduction product is formic acid as determined by isotachophoresis. What is puzzling is that the catalyst usually provided in H_2O reduction systems[55] is not present; the species MV^+ is proposed to reduce CO_2 directly. This reaction is however prohibitively uphill ($E°_{MV^{2+/+}}$ is -0.44 v vs NHE while the reduction of CO_2 to CO_2^- requires at least -2.0 v), suggesting that an unidentified component of the reaction system adopts the catalyst role of charge accumulation and substrate activation.

Recent work by Lehn and co-workers may shed light on this question.[58] They find that $^{13}CO_2$ is photchemically reduced to $H^{13}COO^-$ in DMF or MeCN solutions containing triethanolamine (TEOA) and catalytic amounts of $Ru(bpy)_3^{2+}$ with H_2 and ^{13}CO formed as by-products. They proposed formation of a bis(bpy) complex of Ru as the active catalyst based on the observations that: a) mixtures of electrolytically generated $Ru(bpy)_3^+$and $Ru(bpy)_2H(CO)^+$ were found to reduce CO_2 to $HCOO^-$ in the dark whereas $Ru(bpy)_3^+$ alone failed to produce formate; b) $Ru(bpy)_2H(CO)^+$ did not photochemically reduce CO_2 to $HCOO^-$ in the absence of sensitizer but upon addition of $Ru(phen)_3^{2+}$ and triethanolamine, the reaction took place; and c) $Ru(phen)_3^{2+}$ and triethanolamine alone did not catalyze CO_2 reduction.

Lehn has also found that reduction of CO_2 can be promoted by $ReCl(CO)_3(bpy)$ which functions as both the photosensitizer and the reduction catalyst. The photolysis is done in triethanolamine-DMF with the system showing very high selectivity to CO formation. The mechanism of the reduction using $ReX(CO)_3(bpy)$ has been studied by Lehn, et al.,[59] and by Kutal, et al.[60] who report quantum yields for CO formation using the X=Br complex of 0.15 at 436 nm. There is agreement that the initial reaction of the excited state of $ReX(CO)_3(bpy)$ involves reductive quenching, eqn (16), and that the reduced species then rapidly reacts with CO_2 while the cation $TEOA^+$ undergoes oxidative decomposition.

$$ReCl(CO)_3(bpy)^* + TEOA \longrightarrow ReCl(CO)_3(bpy)^- + TEOA^+ \qquad (16)$$

Subsequent electrocatalytic studies by Meyer using $ReCl(CO)_3(bpy)$ support the intermediacy of the 19 e^- species $ReCl(CO)_3(bpy)^-$ and its rapid reaction with CO_2.[61] A related hydride complex, $ReH(CO)_3(bpy)$, has been examined by Sullivan and Meyer[62] who find that the complex reacts rapidly with CO_2 on photolysis leading to the corresponding formate species, $Re(HCOO)(CO)_3(bpy)$. In contrast, the same CO_2 insertion reaction occurs only slowly upon heating.

One of the first examples of photoelectrochemical reduction of CO_2 was made by Halmann using a p-GaP photocathode in aqueous buffer (pH 6.8) under CO_2.[63] A cathodic bias of -1.0 v vs SCE under illumination gave reasonable current levels, and products including formic acid, formaldehyde, and methanol. It is perhaps significant that reduction beyond the C(2+) state of HCOOH occurred in this system. Analysis of energy conversion vs applied potential showed that maximal energy conversion efficiencies were achieved at ~-0.8 to -0.9 v. Use of a photoanode, n-Si, in to addition the p-GaP photocathode removed the necessity for the external bias to achieve CO_2 reduction. A mechanism similar to that proposed for CO_2 reduction at metal electrodes involving surface adsorbed CO_2^- was invoked without supporting evidence.

The p-GaP photocathode for CO_2 reduction was also studied by Yoneyama and co-workers with Li_2CO_3 electrolyte.[64] The applied potential for significant current under CO_2 must be more negative than -0.6 v vs SCE with HCOOH, CH_2O, and MeOH obtained as the products. Current efficiencies for these products, which were rather low, improved greatly upon the addition of a crown ether to the system and led to the suggestion that the reduction proceeds by initial deposit of Li at the cathode which then reacts with CO_2 generating CO_2^-. However with the introduction of 0.1 M concentrations of crown ether into the system, confirmation of CO_2 reduction products using $^{13}CO_2$ as substrate seems advisable.

Another photoelectrochemical system based on p-CdTe has been reported which yields CO upon irradiation.[65] The solvent system is wet DMF with TBAP as the supporting electrolyte. The external bias required for CO_2 reduction was at least 0.7 v less negative than at metal electrodes, and the current efficiency for CO formation was as high as 70 %.

In a more recent study, Halmann and co-workers have examined a cell based on p-GaAs as the photocathode, and have found that the addition of a solution redox couple, in this case $V^{3+/2+}$, causes a marked shift in the current/voltage curve with much higher cathodic currents at an external bias of only -0.5 v vs SCE.[66] The CO_2 reduction products are as above HCOOH, CH_2O, and MeOH. The optical to chemical energy conversion increased by more than an order of magnitude with the $V^{3+/2+}$ couple present relative to that in its absence. The presumed formation of H_2 as a by-product of the photoelectroreduction was not confirmed or quantified, and the role of the $V^{3+/2+}$ couple in the system remains unclear.

One of the more interesting reports of the photoelectrochemical reduction of CO_2 involves a system that employs components with clearly identifiable roles. The system contains p-InP as the photocathode, MV^{2+} as an electrontransfer mediator, and the enzyme formate dehydrogenase as a catalyst to promote CO_2 reduction.[67] The system is illustrated in Figure 3. With an external bias of +0.05 v vs NHE and a solution pH of 6.8, formate was produced with a current efficiency in the range of 80-93 %. Again a labelling study using $^{13}CO_2$ seems advisable to confirm unambiguously that the formate comes from CO_2 and not from the enzyme preparation added to the system. A leveling off of the formate concentration upon electrolysis using a carbon electrode in place of the photocathode was attributed to enzyme decomposition but Wrighton's previous work[8] shows that an equilibrium is being established.

Final Comments and Conclusions

It is evident that significant progress has been made in direct and indirect electroreductions of CO_2 and in photo- and photoelectrochemical reductions during the past five years. A number of compounds have been found to act as electrocatalysts, and new systems have been devised

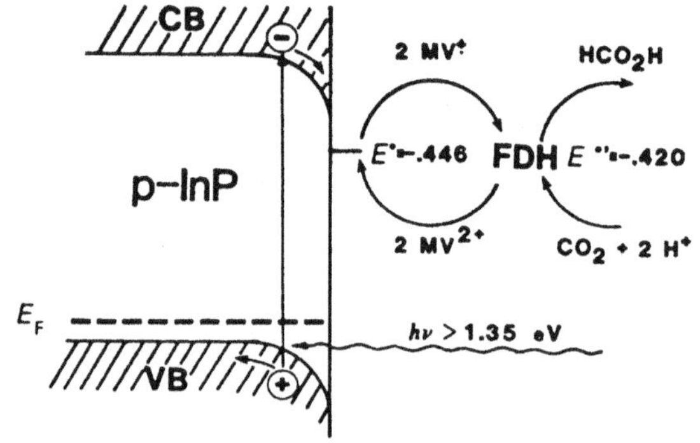

Figure 3. The semiconductor/enzyme system for photoelectrochemical reduction of CO_2 to formate.[67]

which in a few instances show high selectivity to a particular product. The study by Sauvage[31,32] using $Ni(cyclam)^{2+}$ is probably the most complete and informative one done to date on the indirect electroreduction of CO_2.

While Scheme I, a generalized scheme for direct electrochemical reduction of CO_2, suggests what conditions are needed for channeling to specific products, the detailed mechanism of direct electroreduction including the importance of surface adsorbed species is for the most part unknown. The situation is even more uncertain for the indirect electroreduction of CO_2 since the stabilization which makes indirect reduction possible by lowering the overpotential, requires that free CO_2^- radical anion not be involved in the mechanism. An understanding of the types of bonding and/or stabilizing interactions which can occur between CO_2, CO_2^-, and the electrocatalyst in oxidized and reduced forms is essential to the rational development of new electrocatalysts.

Of the possible products of CO_2 reduction, formate and CO are by far the most common. While formate formation is thought to occur in aqueous systems, the reduction using $Ni(cyclam)^{2+}$ in aqueous media yields primarily CO with only very small amounts of $HCOO^-$. The possible intermediacy of metal hydrides has been invoked in a few cases which lead selectively to CO. However, M-H intermediates should channel the system to formate rather than CO based on the well documented insertion of CO_2 into M-H bonds.

We have stated that oxalate formation is the most interesting and potentially most useful of the CO_2 reduction channels because it would provide a path for converting an abundant C_1 material into a C_2 precursor. From a catalysis standpoint, the controlled formation of a C-C bond by CO_2 coupling represents a real challenge. The selective channeling to oxalate must be kinetically controlled since head-to-tail coupling of CO_2 leading to CO is thermodynamically more favorable and occurs more readily. The key to oxalate formation is the development of radical character at the carbon atom of CO_2. However, all stabilization of CO_2 by coordination to a transistion metal metal complex is envisioned to occur between the CO_2 carbon atom and the metal by either an η^1 or η^2 interaction. It appears that the degree of stabilization needed for oxalate formation must be great enough to reduce the overpotential for CO_2^- formation significantly but weak enough to allow any M-C bond that is formed to be homolyzed easily. In this context, it is interesting to note that Co-C bonds found in alkyl cobalamins, cobaloximes, and Schiff base complexes are relatively weak and react primarily by homolytic bond scission.[68] Closely related to this is the observation by Floriani and co-workers[69] of coupling of CO_2-like molecules such as methyl pyruvate promoted by [NaCo(salen)].

If one key to oxalate formation is to have an electrocatalyst capable of forming a weak M-C bond, a second key is to have metal centers in close proximity. This can be accomplished in solution using binuclear and polynuclear complexes, but a more attractive and flexible approach would be to use modified electrodes containing high concentrations of the electrocatalyst. Moreover, through such assemblies, the necessary charge could be stored for reduction beyond the C(2+) oxidation state. The development of modified electrodes containing electrocatalysts thus represents an important direction to be followed in CO_2 reduction chemistry.

With regard to photo- and photoelectrochemical reduction of CO_2, the field is in its infancy. The attention given to the problem of light-to-chemical energy storage has focussed mainly on H^+ or H_2O reduction with only a few studies examining CO_2 as a substrate. Yet the progress has been significant with quantum yields of 0.15 at 436 nm for reduction to CO using $ReBr(CO)_3(bpy)$.[60] In future studies, examination of all components and steps is necessary, although those that are specific to CO_2 deal mainly with active site charge accumulation, substrate binding and activation, and selective product channeling. Questions regarding the photo-sensitizer, the electron transfer mediator, the quenching mechanism, and electron-hole separation will in most cases be the same as those posed in the development of water splitting systems. With proper catalyst development, use of photocathodes for photoelectroreduction of CO_2 appears promising, and significant progress along this line should be anticipated.

Acknowledgements

The support of the Office of Naval Research is gratefully acknowledged.

References and Footnotes

1. Volpin, M. E.; Kolomnikov, I. S. Pure Appl. Chem. **1973**, 33, 567.
2. Eisenberg, R.; Hendriksen, D. E. Adv. Catal. **1979**, 28, 79.
3. Denise, B.; Sneeden, R. P. A. Chemtech **1982**, 15, 108.
4. Darensbourg, D. J.; Kudaroski, R. A. Adv. Organomet. Chem. **1983**, 22, 129.
5. Behr, A. "Catalysis in C_1 Chemistry", W. Keim, Ed., D. Reidel Publishing, Boston, MA, 1983, pp. 169-217.
6. Koinuma, H.; Yoshida, Y.; Hirai, H. Chem. Lett. **1975**, 1223.
7. Yoshida, T.; Thorn, D. L.; Okano, T.; Ibers, J. A.; Otsuka, S. J. Am. Chem. Soc. **1979**, 101, 4212.
8. Stalder, C. J.; Chao, S.; Summers, D. P.; Wrighton, M. S. J. Am. Chem. Soc. **1983**, 105, 6318.
9. Chao, S.; Stalder, C. J.; Summers, D. P.; Wrighton, M. S. J. Am. Chem. Soc. **1984**, 106, 2723.
10. (a) Klibanov, A, M,; Albert, B. N.; Zale, S. E. Biotechnol. Bioeng. **1982**, 24, 25. (b) Ruschig, U.; Muller, U.; Willnow, P.; Hopner,T. Biochem. **1976**, 70, 325.
11. Evans, G. O.; Newell, C. J. Inorg. Chim. Acta **1978**, 31, L387.
12. Darensbourg, D. J.; Ovalles, C.; Pala, M. J. Am. Chem. Soc. **1983**, 105, 5937.
13. Darensbourg, D. J.; Ovalles, C. J. Am. Chem. Soc. **1984**, 106, 3750.
14. Haynes, P.; Slaugh, L. H.; Kohnle, J. F. Tetrahedron Lett. **1970**, 5, 365.
15. Klingler, R. J.; Bloom, I; Rathke, J. W. Organometallics **1985**, 4, 1893.
16. Paonessa, R. S.; Trogler, W. C. J. Am. Chem. Soc. **1982**, 104, 3529.
17. (a) Gambarotta, S.; Strologo, S.; Floriani, C.; Chiesi-Villa, A.; Guastini, C. J. Am. Chem. Soc. **1985**, 107, 6278. (b) Fachinetti, G.; Floriani, C.; Chiesi-Villa, A.; Guastini, C. J. Am. Chem. Soc. **1979**, 101, 1767.
18. Suss-Fink, G.; Reiner, J. J. Organomet. Chem. **1981**, 221, C36.
19. Koinuma, H.; Kawakami, F.; Kato, H.; Hirai, H. J. Chem. Soc., Chem. Commun. **1981**, 213.
20. Svoboda, P.; Hetflejs, J. Coll. Czech. Chem. Commun. **1975**, 40, 1746.
21. Eisenschmid, T.; Eisenberg, R. Unpublished results.
22. Lamy, E.; Nadjo, L.; Saveant, J.-M. J. Electroanal. Chem. **1977**, 78, 403.
23. Gressin, J.C.; Michelet, D.; Nadjo, L.; Saveant, J.-M. Nouv. J. Chim. **1979**, 3, 545.
24. Amatore, C.; Saveant, J.-M. J. Am. Chem. Soc. **1981**, 103, 5021.
25. Becker, J. Y.; Vainas, B.; Eger, R.; Kaufman, L. J. Chem. Soc., Chem. Commun. **1985**, 1471.
26. Hawecker, J.; Lehn, J.-M.; Ziessel, R. J. Chem. Soc., Chem. Commun. **1984**, 328.
27. Fisher, B. J.; Eisenberg, R. J. Am. Chem. Soc. **1980**, 102, 7361.
28. Slater, S.; Wagenknecht, J. H. J. Am. Chem. Soc. **1984**, 106, 5367.
29. Sullivan, B.P.; Bollinger, C.M.; Conrad, D.; Vining, W.J.; Meyer, T.J. J. Chem. Soc., Chem. Commun. **1985**, 1414.
30. Bollinger, C.M.; Sullivan, B.P.; Conrad, D.; Gilbert, J.A.; Story, N;

Meyer, T.J. J. Chem. Soc., Chem. Commun. **1985**, 796.
31. Beley, M.; Collin, J.-P.; Ruppert, R.; Sauvage, J.-P. J. Chem. Soc., Chem. Commun. **1984**, 1315.
32. Beley, M.; Collin, J.-P.; Ruppert, R.; Sauvage, J.-P. J. Am. Chem. Soc., submitted for publication.
33. Lieber, C. M.; Lewis, N. S. J. Am. Chem. Soc. **1984**, 106, 5033.
34. Meshitsuka, S.; Ichikawa, M.; Tamarai, K. J. Chem. Soc., Chem. Commun. 1974, **158.**
35. Ogura, K.; Yoshida, I. J. Mol. Catal. **1986**, 34, 67.
36. Takahashi, K.; Hiratsuka, K.; Sasaki, H.; Toshima, S. Chem. Lett. **1979**, 305.
37. Tezuka, Y.; Yajima, T.; Tsuchiya, A.; Matsumoto, Y.; Uchida, Y.; Hidai, M. J. Am. Chem. Soc. **1982**, 104, 6834.
38. Stalder, C.J.; Chao, S.; Wrighton, M.S. J. Am. Chem. Soc. **1984**, 106, 3673.
39. Russell, P.G.; Kovac, N.; Srinivasan, S.; Steinberg, M. J. Electrochem. Soc., **1977**, 124, 1329.
40. Kaiser, U.; Heitz, E.; Bunsen. Gesell. **1973**, 77,818.
41. Eggins, B.R.; McNeill, J. J. Electroanal. Chem. **1983**, 148, 17.
42. Kapusta, S.; Hackerman, N. J. Electrochem. Soc. **1983**, 130, 607.
43. Taniguchi, I.; Aurian-Blajeni, B.; Bockris, J.O'M. J. Electrochem. Soc. **1983**, 157, 179.
44. Canfield, D.; Frese Jr., K.W. J. Electrochem. Soc. **1983**, 130, 1772.
45. Frese, K. W.; Canfield, D. J. Electrochem. Soc. **1984**, 131, 2518.
46. Frese Jr., K.W.; Leach, S. J. Electrochem. Soc. **1985**, 132, 259.
47. Monnier, A.; Augustynski, J.; Stalder, C. J. Electroanal. Chem. **1980**, 112, 383.
48. Koudelka, M.; Monnier, A.; Augustynski, J. J. Electrochem. Soc. **1984**, 131, 745.
49. Hiratsuka, K.; Takahashi, K.; Sasaki, H.; Toshima, S. Chem. Lett. **1977**, 1137.
50. Kapusta, S.; Hackerman, N. J. Electrochem. Soc. **1984**, 131, 1512.
51. Bradley, M. G.; Tysak, T.; Graves, D. J.; Vlachopoulos, N. A. J. Chem. Soc., Chem. Commun. **1983**, 349.
52. Tinnemans, A.H.A.; Koster, T.P.M.; Thewissen, D.H.M.W.; Mackor, A. Recl. Trav. Chim. Pays-Bas **1984**, 103, 288.
53. (a) Pilloni, G.; Vecchi, E.; Martelli, M. J. Electroanal. Chem. **1973**, 45, 483. (b) Sofranko, J.A.; Eisenberg, R.; Kampmeier, J.A. J. Am. Chem. Soc. **1979**, 101, 1042. (c) Kunin, A. J.; Eisenberg, R. Unpublished results.
54. O'Toole, T.R.; Margerum, L.D.; Westmoreland, T.D.; Vining, W.J.; Murray, R.W.; Meyer, T.J. J. Chem. Soc., Chem. Commun. **1985**, 1416.
55. (a) Kalyanasundaram, K. Coord. Chem. Revs. **1982**, 46, 159., (b) Gratzel, M. Acc. Chem. Res. **1981**, 14, 376.
56. Lehn, J.-M.; Ziessel, R. Proc. Natl. Acad. Sci. USA **1982**, 79, 701.
57. Kitamura, N.; Tazuke, S. Chem. Lett. **1983**, 1109.
58. Hawecker, J.; Lehn, J.-M.; Ziessel, R. J. Chem. Soc., Chem. Commun. **1985**, 56.
59. Hawecker, J.; Lehn, J.-M.; Ziessel, R. J. Chem. Soc., Chem. Commun. **1983**, 536.
60. Kutal, C.; Weber, M. A.; Ferraudi, G.; Geiger, D. Organometallics

1985, 4, 2161.
61. Sullivan, B. P.;Bolinger, C. M.; Conrad, D.; Vining, W. J.; Meyer, T. J. J. Chem. Soc., Chem. Commun. **1985**, 1414.
62. Sullivan, B. P.; Meyer, T. J. J. Chem. Soc., Chem. Commun. **1984**, 1244.
63. Halmann, M. Nature **1978**, 275, 115.
64. Taniguchi, Y.; Yoneyama, H.; Tamura, H. Bull. Chem. Soc. Jpn. **1982**, 55, 2034.
65. Taniguchi, Y.; Audrian-Blajeni, B.; Bockris, J. O. J. Electroanal. Chem. **1983**, 157, 179.
66. Zafrir, M.; Ulman, M.; Zuckerman, Y.; Halmann, M. J. Electroanal. Chem. **1983**, 159, 373.
67. Parkinson, B. A.; Weaver, P. F. Nature **1984**, 309, 148.
68. Halpern, J. Acc. Chem. Res. **1982**, 15, 238.
69. Gambarotta, S.; Arena, F.; Floriani, C.; Gaetani-Manfredotti, A. J. Chem. Soc., Chem. Commun. **1982**, 835.

INTERACTIONS OF CO_2 WITH TRANSITION-METAL SYSTEMS, INCLUDING METAL SURFACES

James A. Ibers
Department of Chemistry
Northwestern University
Evanston, IL 60201, U.S.A.

ABSTRACT. The preparation and structures of known CO_2-transition metal compounds are reviewed. Reactions of transition-metal systems with liquid CO_2 serve to illustrate some of the factors that are responsible for the general instability of such complexes. The interactions of CO_2 with metal surfaces are reviewed briefly.

1. INTRODUCTION

That CO_2 is potentially the most abundant source of C_1 chemistry is well known, but the means to activate the CO_2 molecule is not. It is assumed, at least by transition-metal chemists, that such an activation chemistry of CO_2, if it exists, will be metal-mediated. There is no direct support for that assumption. But considerable research on the interactions of CO_2 with transition metals has transpired; the results of that research form the basis for this contribution.

2. KNOWN CO_2-TRANSITION METAL COMPLEXES

Scheme I shows possible modes of interaction of CO_2 with metals.
We first examine which of these modes has been found in isolated complexes. Because of the difficulties of spectroscopic characterization, the unpredictable effects of adventitious water, and the tendency of coordinated CO_2 to react further, it is prudent to accept structural characterization by diffraction methods as the criterion by which to judge the authenticity of a given class of CO_2-transition metal complexes. Table I lists the known complexes.

We next consider the method of preparation of these known complexes. Equation (1) illustrates a preparation for compound **I**.[1]

$$[Ni(PCy_3)_2]_2(\mu\text{-}N_2) + 2CO_2(g) \xrightarrow[\text{toluene}]{20^0C} \underset{\textbf{I}}{2Ni(CO_2)(PCy_3)_2} + N_2 \qquad (1)$$

M. Aresta and G. Forti (eds.), Carbon Dioxide as a Source of Carbon, 55–69.

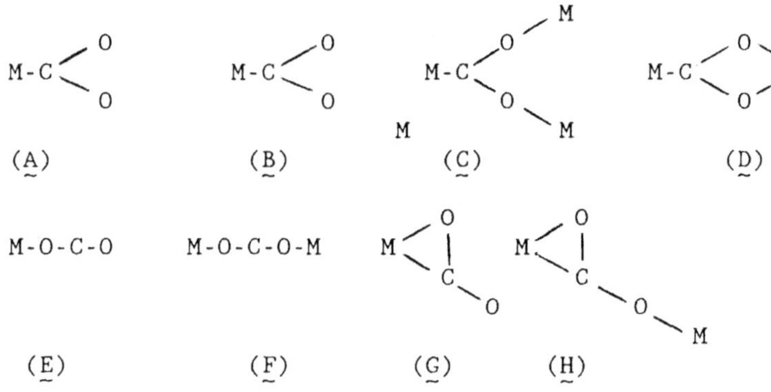

Scheme I*

*Only connectivity is shown. There is no attempt to define bond orders. M is a general symbol for any metal. In a sketch with more than one M the metals need not be the same.

Table I* Structurally Characterized CO_2-M Complexes

	Compound	Interaction Mode	Reference
I	$Ni(CO_2)(PCy_3)_2$	G	1
II	$Nb(CO_2)(\eta^5\text{-}C_5H_4Me)_2(CH_2SiMe_3)$	G	2
III	K[Co(CO_2)(n-Pr-salen)(THF)]	D	3
IV	$RhCl(CO_2)(diars)_2$	A	4
V	$Mo(CO_2)_2(CNPr^i)(PMe_3)_3$	G	5

*Abbreviations: Cy = cyclohexyl; Me = methyl; Pr = propyl; salen = N,N'-ethylenebis(salicylideneiminato); THF = tetrahydrofuran; diars = o-phenylenebis(dimethylarsine); $CNPr^i$ = isopropyl isocyanide.

Equation (1), while representing the overall reaction, probably does not provide a clear indication of how formation of the CO_2-Ni compound proceeds. In N_2-saturated toluene solution at room temperature the equilibrium

$$[Ni(PCy_3)]_2(\mu\text{-}N_2) \underset{-N_2}{\overset{+N_2}{\rightleftharpoons}} 2Ni(N_2)(PCy_3)_2 \qquad (2)$$

lies far to the right.[6] Thus the reactive species in the preparation of the CO_2-Ni complex is most probably $Ni(N_2)(PCy_3)_2$:

$$Ni(N_2)(PCy_3)_2 + CO_2 \longrightarrow Ni(CO_2)(PCy_3)_2 + N_2 \qquad (3)$$

Whether reaction (3) proceeds via initial loss of N_2 is not known. While N_2 in the language of transition-metal chemists is a "good leaving group", meaning that it is generally not strongly bound, the complex $Ni(N_2)(PCy_3)_2$ is already "coordinatively unsaturated", meaning that the Ni atom, formally in the zero oxidation state, is bound to less than the four ligands that would normally be expected for Ni(0) (cf. $Ni(CO)_4$). It is thus possible to form $Ni(N_2)(CO_2)(PCy_3)_2$, wherein the Ni center achieves coordinate saturation, and then the N_2 ligand could subsequently leave the Ni coordination sphere. Again in the parlance of transition-metal chemists PCy_3 is a very bulky ligand (or has a very large cone-angle[7]) and hence a species such as $Ni(N_2)(CO_2)$-$(PCy_3)_2$ might be too crowded to be stable. As we shall see the interplay of steric and electronic forces is extremely important in CO_2-transition metal chemistry as, indeed, it is in transition-metal chemistry in general.

The structure of I is illustrated in Figure 1. The interaction

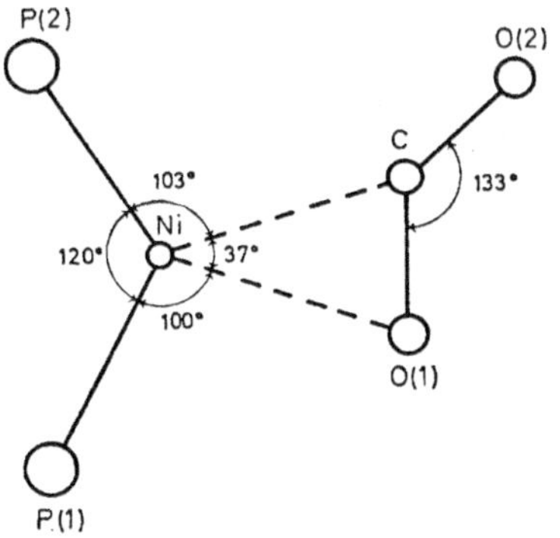

Figure 1. Sketch of the structure of $Ni(CO_2)(PCy_3)_2$ (from Reference 1).

mode G is apparent. For metal olefin complexes two limiting bonding descriptions of interaction G are possible: at one extreme one can describe the interaction as shown in (a)

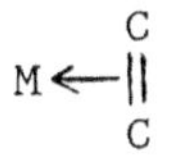

(a)

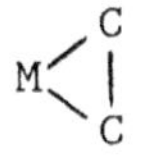

(b)

wherein the C-C π-bonds may interact with the requisite vacant metal orbital; in the alternative description (b) the complex is considered to be a metallocyclopropane. Probably the truth lies somewhere between these two extremes. Just as for the olefin complexes, the "side-on" CO_2 complexes have alternative descriptions. Thus the interaction shown in Figure 1 could be described in terms of the π-bonding of CO_2 to a Ni(0) center or as a metallo-oxirane involving a Ni(II) center.

Equation (4) illustrates the preparation of complex II.[2]

$$NbCl(\eta^5\text{-}C_5H_4Me)_2(CH_2SiMe_3) + CO_2(g) \xrightarrow[\text{THF, } 20^0C]{Na/Hg} Nb(CO_2)(\eta^5\text{-}C_5H_4Me)_2(CH_2SiMe_3) \quad (4)$$

How the reaction proceeds is not clear. The Nb-center in the starting complex is a "17-electron system" and hence has a tendency to add another electron. As shown in Figure 2 the CO_2-Nb interaction is again

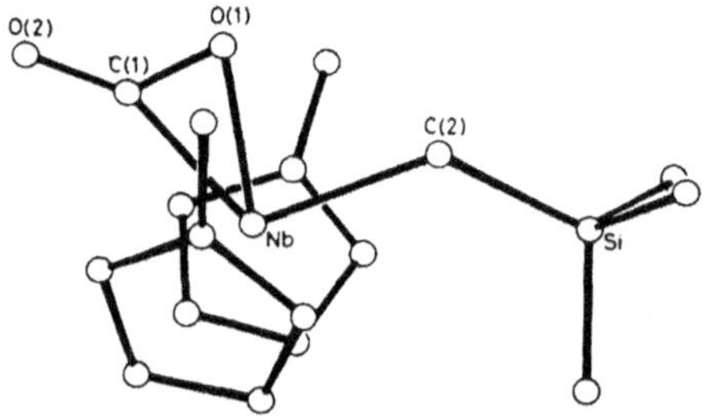

Figure 2. Sketch of the structure of $Nb(CO_2)(\eta^5\text{-}C_5H_4Me)_2(CH_2SiMe_3)$ (from Reference 2).

of type G. If one wishes to describe such an interaction as involving the donation of two electrons from CO_2 to Nb, then compound II is an 18-electron system; such systems are noted for their stability. Other reasons for the stability of complex II may involve the orientation of the CO_2 ligand (Figure 2). In that orientation, insertion of the CO_2 ligand into the metal-alkyl bond would produce not the usual and stable carboxylato complex $L_nNbOCOR$ but the rarer and presumably less stable metallocarboxylate ester L_nNbCO_2R. It is noteworthy that the compounds $Nb(\eta^5\text{-}C_5H_4Me)_2(CH_2SiMe_3)X$, X = CO, CO_3 are also produced in the above reaction (vide infra).

The preparation of compound III[3] is illustrated in equation (5) The structure of this rather complicated product is shown in Figure 3. Reduction of Co(n-Pr-salen) with K leads to a nucleophilic Co(I) center and an alkali cation that remain complexed within the same structure. This neutral intact ion-pair serves to anchor the CO_2 molecule, wherein the C atom is bound to the Co atom and the O atoms interact with the K atom (interaction type D). Thus both metals are involved in binding CO_2. The actual CO_2-Co interaction is of type A and probably involves a normal Co-C σ-bond.

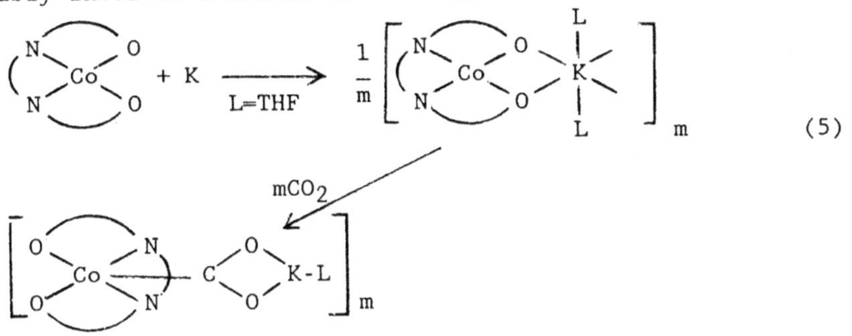

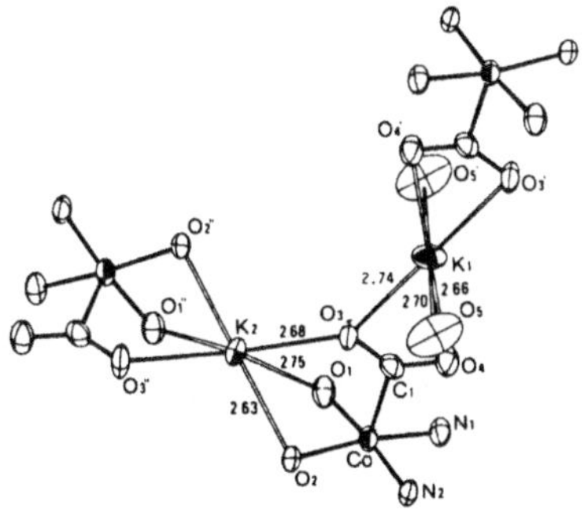

Figure 3. View of the repetitive unit in K[Co(CO_2)(n-Pr-salen)(THF)] (from Reference 3).

The preparation of compound IV is illustrated in equation (6).[4]

$$[Rh(diars)_2]Cl \xrightarrow[CH_3CN]{CO_2(g)} RhCl(CO_2)(diars)_2 \quad (6)$$

The structure of IV is illustrated in Figure 4; again the CO_2-metal interaction is of type A. Here we have gone from a Rh(I) center complexed to two diars ligands in a square-planar arrangement to a Rh(III) center in which Cl and CO_2 have been added trans to one another and cis to the diars ligands. Equation (6) is an example of an oxidative addition reaction.

The preparation of compound V is illustrated in equation (7).[5] The structure of compound V is illustrated in Figure 5. The CO_2-Mo interaction is of type G for both CO_2 molecules. Overall, the orientation of the C-O bonds involved with the Mo center is similar to that found in <u>trans</u>-$Mo(C_2H_4)_2(PMe_3)_4$. Complex <u>V</u> is very stable; it can be heated for several hours in vacuo or under $\tilde{N_2}$ at 80^0C without

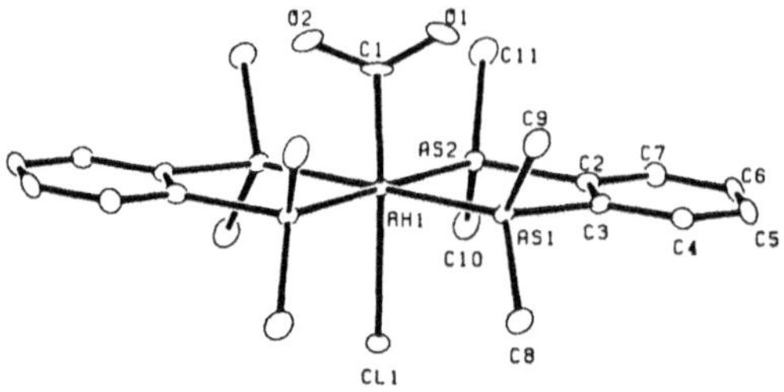

Figure 4. Sketch of the structure of $RhCl(CO_2)(diars)_2$ (from Reference 4).

$$\underline{cis}\text{-}Mo(N_2)_2(PMe_3)_4 + CO_2(g,\ 5\ atm) \xrightarrow[\text{pet ether}]{20^0}$$

$$\underline{trans}\text{-}Mo(CO_2)_2(PMe_3)_4$$

$$\downarrow CNPr^i$$

$$\underline{trans},\underline{mer}\text{-}Mo(CO_2)_2(CNPr^i)(PMe_3)_3 \quad + \quad PMe_3 \qquad (7)$$

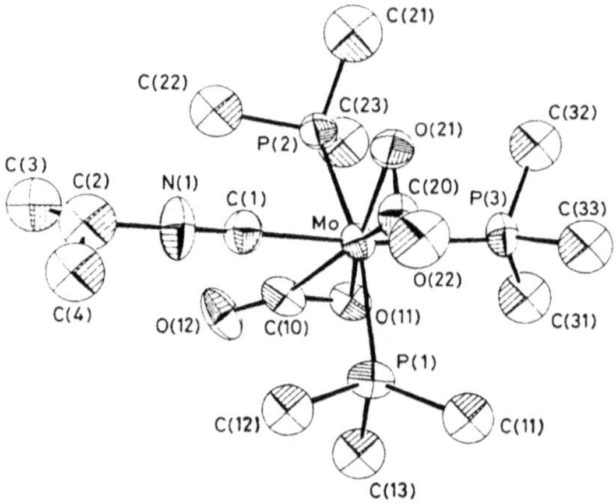

Figure 5. Sketch of the structure of $Mo(CO_2)_2(CNPr^i)(PMe_3)_3$ (from Reference 5).

decomposition. Thus the two CO_2 molecules are strongly bound to Mo, in part because of the oxophilic nature of molybdenum. It is again noteworthy that with the formation of trans-$Mo(CO_2)_2(PMe_3)_4$ there also are formed $Mo(CO_3)(CO)(PMe_3)_4$ and $[Mo(CO_3)(CO)(PMe_3)_3]_2$.[8]

3. "CO_2-COMPOUNDS" NOT MADE WITH CO_2

There are two cluster systems known in which CO_2 coordination in mode C occurs; however, the synthesis of these systems does not involve the use of CO_2. The preparation[9] of $[Ph_3PNPPh_3][Os_3H(CO)_{10}(\mu\text{-}CO_2)Os_6\text{-}(CO)_{12}]$ VI is illustrated in equation (8). The structure of the complex[10] is shown in Figure 6.

$$Os_3H(CO)_{11}^{-} + Os_6(CO)_{18} + Ph_3PNPPh_3 \xrightarrow[CH_2Cl_2]{}$$

$$[Ph_3PNPPh_3][Os_3H(CO)_{10})(\mu\text{-}CO_2)Os_6(CO)_{12}] \qquad (8)$$

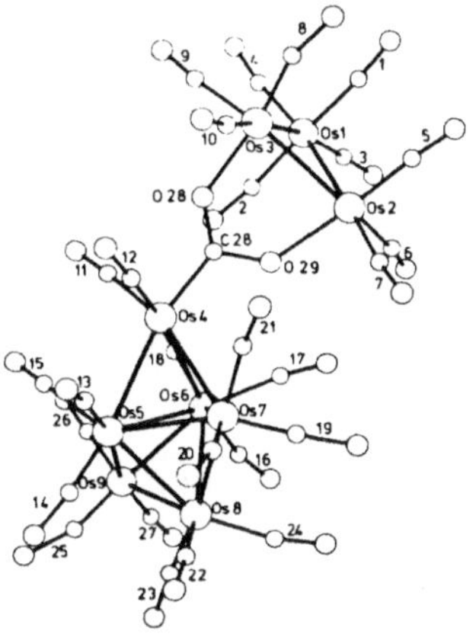

Figure 6. Sketch of the $[Os_3H(CO)_{10}(\mu\text{-}CO_2)Os_6(CO)_{12}]^-$ ion (from Reference 10).

The preparation of $\{[Re(CO)_4](\mu\text{-}CO_2)[Re(CO)_5]\}_2$ VII is shown in equation (9)[11]

$$4\text{"}Re(OH)(CO)_5\text{"} \xrightarrow[-2H_2O]{acetone} \{[Re(CO)_4](\mu\text{-}CO_2)[Re(CO)_5]\}_2 \qquad (9)$$

and the structure of the complex is shown in Figure 7. Here, as well for compound VI, identification of the bridge as CO_2 and not as a carboxylate or related species has depended as much on the absence of

O-H stretching modes as on X-ray diffraction methods, since the detection by such methods of an H atom attached to an O atom in the presence of an array of Os or Re atoms is very difficult, if not impossible.

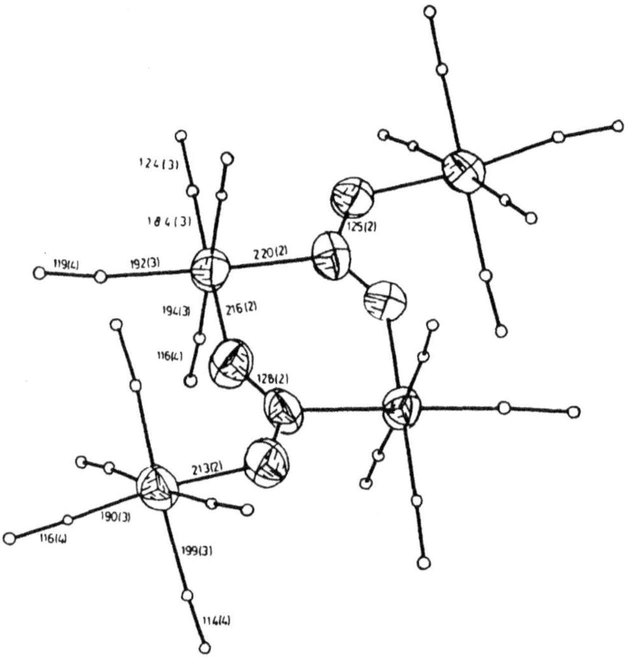

Figure 7. Sketch of the structure $\{[Re(CO)_4](\mu\text{-}CO_2)[Re(CO)_5]\}_2$ (from Reference 11).

4. SOME DETAILS OF CO_2-METAL STRUCTURES

Table II summarizes some pertinent metrical details on the CO_2-metal interactions in the various compounds discussed above. The CO_2 molecule itself (connectivity OCO) is linear, with a C-O bond length of 1.162 Å. It is thus apparent from the data of Table II that coordination of CO_2 to a metal, independent of the mode of interaction, brings about a lengthening of the C-O bonds and a decrease in the OCO angle. As one would normally describe the nature of the bonding in interaction mode G differently from that in interaction mode A it is clear that little insight into the description of the bonding may be obtained from the structural details of the bound CO_2 ligand. Moreover, the errors on bond distances and angles in Table II are not especially small. In fact the standard deviations are sufficiently large so that details in a given structure (e.g., the difference between C-O_1 and C-O_2 bond lengths in a structure with interaction type G) cannot be discerned. It should be kept in mind that if bond length d_1 has an estimated standard deviation $\sigma(d_1)$ and bond length d_2 a corresponding value $\sigma(d_2)$ then $\Delta = d_1 - d_2$ has a standard deviation $\sigma(\Delta) = (\sigma^2(d_1) + \sigma^2(d_2))^{1/2}$ and one is loathe to find significance in the difference Δ unless $\Delta/\sigma(\Delta)$ is at least 5. This is especially true

Table II. Bond Distances (Å) and Bond Angles (deg) in CO_2-Metal Systems

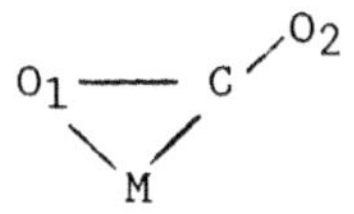

Cpd	M	Ref	M-O_1	M-C	O_1-C	C-O_2	∢O_1CO_2
I	(Ni)	1	1.99(2)	1.84(2)	1.22(4)	1.17(4)	133
II	(Nb)	2	2.173(4)	2.144(7)	1.283(8)	1.216(8)	132.4(7)
V	(Mo)	5	2.105(10)	2.147(7)	NA[a]	NA	133.5(10)

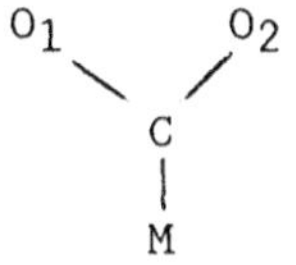

Cpd	M	Ref	M-C	C-O	∢O_1CO_2
III	(Co)	3	2.00(1)	1.20(2),1.24(2)	132
IV	(Rh)	4	2.05(2)	1.20(2),1.25(2)	126(2)

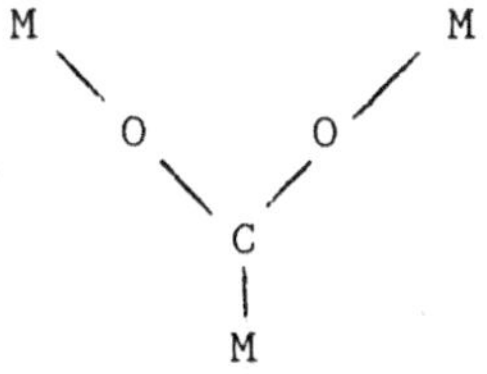

Cpd	M	Ref	M-C	C-O	M-O	∢ OCO
VI	(Os)	10	1.96(5)	1.28(6),	2.12(4),	NA
			1.32(6)	2.19(4)		
VII	(Re)	11	2.20(2)	1.25(2),	2.13(2),	NA
			1.28(2)	2.16(2)		

[a]NA = not available in original publication.

since it is well known that standard deviations from diffraction studies are typically underestimated by a factor of 2-3.

5. REACTIONS OF TRANSITION-METAL SYSTEMS WITH LIQUID CO_2

The CO_2-transition metal complexes described above involve a wide range of metals and an array of structural types. There is no pattern that is readily discerned. Why are these particular CO_2 complexes stable? Why are there so few isolable complexes? Have we looked at the wrong systems or have we missed the detection of complexes? How might we increase the probability of CO_2-metal interactions? One such way is to use liquid CO_2, rather than CO_2 gas. Liquid CO_2 can be used either in the absence of solvent[6] or in a mixed CO_2/solvent system.[12] The advantages are (i) a greatly increased CO_2 concentration and (ii) elimination or decrease of solvent competition for a vacant coordination site on the metal. The disadvantage, of course, is the increased difficulty of characterization. While one can monitor some of these high-pressure reactions through the use of spectroscopic methods,[12] isolation of compounds, unless stable under reduced CO_2 concentration, is not feasible. Nevertheless, the method is a convenient way to survey for possible complexes.

Table III illustrates the reactivity of some transition metal systems toward liquid CO_2.[6] Except as noted, no solvents were involved. Reactions were carried out at room temperature under a pressure of 60-70 atm. There was roughly a 1000-fold molar excess of CO_2 to transition-metal complex. In the wet-CO_2 experiments the M:H_2O ratio was 1:1.

5.1 No Reaction.

As Table III indicates, a number of the complexes studied show no reactivity toward liquid CO_2. These compounds were recovered unchanged from the pressure vessel, although the fact that they were generally dispersed over the quartz liner of the vessel argues for their having been dissolved in liquid CO_2. At least for the N_2-containing complexes of Table III, the persistence of N_2 as a ligand argues against CO_2 reactivity at high pressure. Thus, CO_2 did not displace N_2 from $IrCl(N_2)(PPh_3)_2$. There is no evidence that CO_2 will insert into the Ir-R bond of $IrR(CO)(PPh_3)_2$, R = Me, Ph. Moreover, in a separate experiment, the addition of CH_3I to $IrCl(CO)(Ph_3)_2$ in liquid CO_2 proceeded smoothly to afford $IrClI(CH_3)(CO)(PPh_3)_2$. Thus, there is no evidence of CO_2 insertion into the Ir-CH_3 bond.

Although $[Ni(PCy_3)_2]_2(\mu\text{-}N_2)$ reacts with CO_2(g) to afford $Ni(CO_2)(PCy_3)_2$ (equation (1)), and indeed this same complex is formed with liquid CO_2 (vide infra), the closely related complexes $Pd(PCy_3)_2$ and $Pt(PCy_3)_2$ do not react with dry CO_2, although the Pt compound does react with wet CO_2. Nor does the related Rh system, $[RhH(PCy_3)_2]_2(\mu\text{-}N_2)$, display any reactivity toward CO_2. Finally, the earlier transition-metal systems, $M(CO)_3(PCy_3)_2$, M = Mo, W, do not react with dry CO_2, although the W complex does react with wet CO_2.

Table III. Reactivity of Some Transition-Metal Systems Toward Liquid CO_2

Complex	reacts with dry CO_2	reacts with wet CO_2
$IrCl(N_2)(PPH_3)_2$	no	
$IrR(CO)(PPh_3)_2$, R = Me, Ph	no	
$IrR(O_2)(CO)(PPh_3)_2$ R = Me, Ph	yes	
$[RhH(PCy)_3]_2(\mu\text{-}N_2)$	no	
$[Ni(PCy_3)_2]_2(\mu\text{-}N_2)$	yes	
$Pd(PCy_3)_2$	no	
$Pt(PCy_3)_2$	no	yes
$Mo(CO)_3(PCy_3)_2$	no	
$W(CO)_3(PCy_3)_2$	no	yes
$[Co(PPh_3)_3]_2(\mu\text{-}N_2)$	no	
$[Co(PEt_2Ph_3]_2(\mu\text{-}N_2)$	yes	
$Na[Co(N_2)(PEt_2Ph)_3]$	yes	
$IrMe(Ph_2PCH_2CH_2PPh_2)_2$	yes*	

*CO_2(liq)-CH_2Cl_2 solvent

5.2 Reactions with Dry CO_2

The compound $IrMe(O_2)(CO)(PPh_3)_2$ reacts with dry liquid CO_2 to form what has been characterized spectroscopically to be a peroxocarbonate, $IrMe(OCO_3)(CO)(PPh_3)_2$, by analogy with $Pt(OCO_3)(PCy_3)_2$[13] and the structurally characterized complex[14] $Rh(OCO_3)(4\text{-}MeC_6H_4)(PhP(CH_2\text{-}CH_2CH_2PPh_2)_2)$. In the last instance, formation of the Rh peroxocarbonate probably resulted from interactions with minor amounts of O_2 fortuitously present in the reaction mixture.

The ready displacement of the N_2 ligand from $[Ni(PCy_3)_2]_2(\mu\text{-}N_2)$ led us to examine other first-row transition-metal dinitrogen complexes. The compound $[Co(PEt_2Ph)_3]_2(\mu\text{-}N_2)$ reacts with liquid CO_2 to yield a highly air-sensitive red oil. On the basis of spectroscopic results the presence of CO and CO_3^{2-}could be confirmed. Similar chemistry was seen for $Na[Co(N_2)(PEt_2Ph)_3]$ in liquid CO_2. The reductive disproportionation of CO_2 to CO and CO_3^{2-} has been observed previously in $[Mo(CO)_3)(CO)(PMe_2Ph)_3]_2$,[15] $[Ti(C_5H_5)_2]_4(CO_3)_2$,[16] and the reaction of $Na[Fe(C_5H_5)(CO)_2]$ with CO_2.[17] The intermediacy of a C_2O_4 ligand, as observed in $IrCl(C_2O_4)(PMe_3)_3$,[18] (Figure 8), is an attractive idea but has not been conclusively demonstrated for any system to date.

The compound $IrMe(Ph_2PCH_2CH_2PPh_2)_2$[12] actually reacts neither with pure liquid CO_2 nor with CO_2(g). However, it does react slowly with CO_2 in a liquid CO_2-CH_2Cl_2 mixed solvent. The product probably

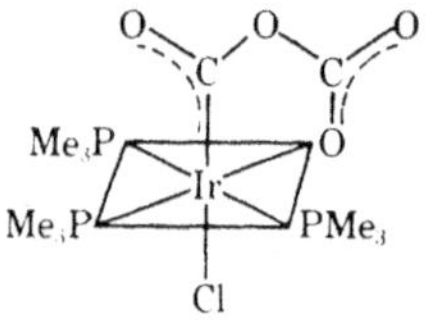

Figure 8. Sketch of $IrCl(C_2O_4)(PMe_3)_3$ (from Reference 18)

involves insertion of CO_2 into the Ir-Me bond, although the nature of the product could not be determined by spectroscopic methods. Six-coordinate Ir complexes involving the $Ir\text{-}OOCCH_3$ fragment as well as the $Ir\text{-}COOCH_3$ fragment are known and their carboxylate frequencies appear in the same region (1600-1700 cm^{-1}).[19,20]

5.3 Reactions with Wet CO_2

The presence of adventitious water has resulted in misformulations of several "CO_2" compounds.[21-25] In part this is because CO_2, whether as a liquid or as a gas, is very hard to dry and if the concentration of CO_2 to metal complex is high very small amounts of water can cause problems. As Table III indicates both $Pt(PCy_3)_2$ and $W(CO)_3(PCy_3)_2$ react with wet CO_2 to form what are believed to be the bicarbonato complexes _trans_-$PtH(O_2COH)(PCy_3)_2$ and $WH(O_2COH)(CO)_3(PCy_3)_2$. The question of how bicarbonato complexes are formed has been addressed in some detail for rhodium-hydrido systems.[25] Here one need only point out that three pathways seem possible: (i) formation of a CO_2-M complex that subsequently undergoes nucleophilic attack by H_2O at the coordinated CO_2 group; (ii) oxidative addition to the metal of H_2CO_3, formed by the reaction of H_2O with CO_2; (iii) oxidative addition to the metal of HOH with subsequent insertion of CO_2 into the M-OH bond. As we note below, different pathways probably apply to different systems.

5.4 Discussion of the Results with Liquid CO_2

Even under the forcing conditions of liquid CO_2 with no solvent present and a CO_2:M ratio of perhaps 1000:1, most representative metal complexes remain unreactive toward CO_2 (Table III). But in the course of this survey of the reactivity of some transition-metal systems toward liquid CO_2, several interesting reactions were found, as deduced from the nature of the products.

The formation of $IrR(OCO_3)(CO)(PPh_3)_2$ from $IrR(O_2)(CO)(PPh_3)_2$ is probably an example of external attack by CO_2 on a coordinated species. The compounds $IrR(O_2)(CO)(PPh_3)_2$ are coordinatively saturated, and even if an open coordination site were provided by PPh_3 dissociation, CO_2 coordination to the Ir atom would seem unlikely because of the demonstrated inertness of $IrCl(N_2)(PPh_3)_2$ and $IrR(CO)(PPh_3)_2$ to CO_2.

Similarly, the formation of $WH(O_2COH)(CO)_3(PCy_3)_2$ from $WH(OH)(CO)_3(PCy_3)_2$ and liquid CO_2 probably results from external attack of the CO_2 molecule on the W-OH linkage, as its seems unlikely that prior to insertion CO_2 could coordinate to a seven-coordinate W atom. But the formation of trans-$PtH(O_2COH)(PCy_3)_2$ from the reaction of $Pt(PCy_3)_2$ with wet liquid CO_2 could proceed through CO_2 attack on a trans-$PtH(OH)(PCy_3)_2$ intermediate, formed by oxidative addition of H_2O to $Pt(PCy_3)_2$.

Spectroscopic evidence for CO and CO_3^{2-} in the products of the reactions of $[Co(PEt_2Ph)_3]_2(\mu\text{-}N_2)$ and $Na[Co(N_2)(PEt_2Ph)_3]$ with liquid CO_2 suggests that in these instances intermediate CO_2 complexes were formed. Such complexes, not stabilized by steric bulk, are open to attack by additional CO_2, ultimately leading to reductive disproportionation to CO and CO_3^{2-}. The use of steric bulk to stabilize highly reactive CO_2 complexes is clearly important. Even when stable CO_2 complexes are formed (e.g. equation (4) for compound **II** and equation (7) for compound **V**) CO and CO_3 complexes frequently are formed at the same time.

But the results for the Ni triad provide the most dramatic illustration of the importance of electronic factors in CO_2 coordination. The compound $[Ni(PCy_3)_2]_2(\mu\text{-}N_2)$, which in solution is in equilibrium with $Ni(N_2)(PCy_3)_2$ (equation (2)), readily forms the modestly stable compound $Ni(CO_2)(PCy_3)_2$ with CO_2(gas) or liquid CO_2. It does not matter whether the CO_2 is dry or wet. The analogous Pd complex, $Pd(PCy_3)_2$, is unreactive toward either dry or wet liquid CO_2. And the complex $Pt(PCy_3)_2$, while unreactive toward dry liquid CO_2, reacts with wet liquid CO_2 to afford trans-$PtH(O_2COH)(PCy_3)_2$.

It seems clear then that the formation of a stable CO_2-transition metal complex results from a delicate balance of steric and electronic forces. Until these can be defined more precisely, the discovery of such stable complexes will remain in large measure a matter of luck.

6. INTERACTION OF CO_2 WITH METAL SURFACES

The nature of the interactions of CO_2 with metal surfaces is not without controversy, probably because of the intrinsic difficulties of defining the nature of a given surface. CO_2 apparently is physically adsorbed on Au[26,27] as well as on Ni, Cu, and Ag.[27] On the other hand, CO_2 apparently dissociates on Mo according to Scheme II.[26]

$$CO_2(g) \longrightarrow CO_{ads} + O_{ads} \quad 77\ K$$

$$\downarrow 298\ K$$

$$C_{ads} + O_{ads}$$

Scheme II

The interaction of CO_2 with Rh surfaces has been most studied. A recent paper[28] summarizes the literature and the controversies. Apparently at about 215 K CO_2 dissociates into CO_{ads} and O, with O diffusing into the bulk Rh structure. It is speculated that CO_2 adsorption on Group VIII metal surfaces first occurs by formation of a M-C bond (mode A). However, what occurs next depends upon the particular surface, most especially on the presence of steps or terraces. It is speculated that under certain conditions an interaction mode akin to H occurs with subsequent dissociation of CO_2 (Figure 9).

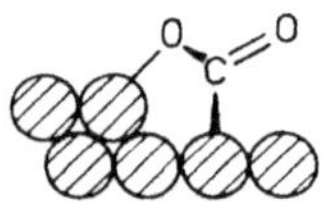

Figure 9. Possible structure of CO_2 near a step on a metal surface (from Reference 28).

References

1. Aresta, M.; Nobile, C.F.; Albano, V. G.; Forni, E.; Manassero, M. J. Chem. Soc. Chem. Commun. 1975, 636-637.

2. Bristow, G.S.; Hitchcock, P.B.; Lappert, M.F. J. Chem. Soc. Chem. Commun. 1981, 1145-1146.

3. Fachinetti, G.; Floriani, C.; Zanazzi, P.F. J. Am. Chem. Soc. 1978, **100**, 7405-7407; Gambarotta, S.; Arena, F.; Floriani, C.; Zanazzi, P.F. J. Am. Chem. Soc. 1982, **104**, 5082-5092.

4. Calabrese, J.C.; Herskovitz, T.; Kinney, J.B. J. Am. Chem. Soc. 1983, **105**, 5914-5915.

5. Alvarez, R.; Carmona, E.; Gutierrez-Puebla, E.; Marín, J.M.; Monge, A.; Poveda, M.L. J. Chem. Soc. Chem. Commun. 1984, 1326-1327.

6. Mason, M.G.; Ibers, J.A. J. Am. Chem. Soc. 1982, **104**, 5153-5157.

7. Tolman, C.A. Chem. Rev. 1977, **77**, 313-348.

8. Alvarez, R.; Carmona, E.; Poveda, M.L.; Sánchez-Delgado, R. J. Am. Chem. Soc. 1984, **106**, 2731-2732.

9. Eady, C.R.; Guy, J.J; Johnson, B.F.G.; Lewis, J.; Malatesta, M.C.; Sheldrick, G.M. J. Chem. Soc. Chem. Commun. 1976, 602-604.

10. Guy, J.J.; Sheldrick, G.M. Acta Crystallogr. 1978, **B34**, 1718-1722.

11. Beck, W.; Raab, K.; Nagel, U.; Steimann, M. Angew. Chem. Int. Ed. Engl. 1982, **21**, 526-527.

12. Lilga, M.A.; Sohn, Y.S.; Ibers, J.A. Organometallics 1986, **5**, 766-771.

13. Tatsuno, Y.; Otsuka, S. J. Am. Chem. Soc. 1981, **103**, 5832-5839.

14. Dahlenburg, L.; Prengel, C. Organometallics 1984, **3**, 934-936.

15. Chatt, J.; Kubota, M.; Leigh, G.J.; March, F.C.; Mason, R.; Yarrow, D.J. J. Chem. Soc., Chem. Commun. 1974, 1033-1034.

16. Fachinetti, G.; Floriani, C.; Chiesi-Villa, A.; Guastini, C. J. Am. Chem. Soc. 1979, **101**, 1767-1775.

17. Evans, G.O.; Walter, W.F.; Mills, D.R.; Streit, C.A. J. Organomet. Chem. 1978, **144**, C34-C38.

18. Herskovitz, T.; Guggenberger, L.J. J. Am. Chem. Soc. 1976, **98**, 1615-1616.

19. Smith, S.A.; Blake, D.M.; Kubota, M. Inorg. Chem. 1972, **11**, 660-662.

20. Deeming, A.J.; Shaw, B.L. J. Chem. Soc. A 1969, 443-446.

21. Komiya, S.; Yamamoto, A. J. Organomet. Chem. 1972, **46**, C58-C60.

22. Krogsrud, S.; Komiya, S.; Ito, T.; Ibers, J.A.; Yamamoto, A. Inorg. Chem. 1976, **15**, 2798-2805.

23. Ito, Y.; Tsuchiya, H.; Yamamoto, A. Chem. Lett. 1976, 851-854.

24. Crutchley, R.J.; Powell, J.; Faggiani, R.; Lock, C.J.L. Inorg. Chim. Acta 1977, **24**, L15-L16.

25. Yoshida, T.; Thorn, D.L.; Okano, T.; Ibers, J.A.; Otsuka, S. J. Am. Chem. Soc. 1979, **101**, 4212-4221.

26. Atkinson, S.J.; Brundle, C.R.; Roberts, M.W. Faraday Discuss. Chem. Soc. 1974, **58**, 62-79.

27. Kamath, P.V.; Rao, C.N.R. J. Phys. Chem. 1984, **88**, 464-469.

28. Hendrickx, H.A.C.M.; Jongenelis, A.P.J.M.; Nieuwenhuys, B.E. Surface Science 1985, **154**, 503-523.

PHOTOSYNTHETIC ELECTRON TRANSPORT: THE SOURCE OF ELECTRONS FOR CO_2 REDUCTION IN PHOTOSYNTHESIS

G. Forti
Centro di Studio del C.N.R. sulla Biologia Cellulare e Molecolare delle Piante, Dipartimento di Biologia dell'Università, Via Celoria 26, 20133 Milano, Italy

ABSTRACT. An up-to-date outline of photosynthetic electron transport and photophosphorylation is given.

1. INTRODUCTION

Photosynthetic electron transport of green plants and Cyanobacteria utilizes water as the electron donor (at the E_m of 810 mV at pH 7) to reduce the pyridinenucleotide coenzyme NADP (E_m = -320 mV), against the apparent electrochemical gradient of 1.13 electron volt. The free energy for such electrochemical work is provided by light quanta absorbed by the chlorophylls and other photosynthetic pigments. The synthesis of ATP from ADP and inorganic phosphate (P_i) is coupled to electron transport so that the overall stoichiometry of the process of oxygenic photosynthesis can be described by the equation:

$$(1) \quad 2H_2O + 2NADP + nADP + nP_i + 8h\nu \longrightarrow O_2 + 2NADP + nATP$$

(the stoichiometric ratio ATP/NADPH may be variable, as will be discussed later).

ATP and NADPH are then utilized in the enzymatic process of CO_2 fixation and reduction to the carbohydrate level.

The photochemical apparatus of photosynthesis, including the pigments bound in complexes to proteins which constitute the light--harvesting "antennas", the centres of photochemical reactions, and the electron carriers involved are embedded into the photosynthetic membranes (the "thylakoids"). The thylakoids occupy a relevant part of the internal volume of the chloroplasts, which are the organelles where photosynthetis occurs (see Fig. 1).

They constitute a continuum of membranes including an internal aqueous phase, separated from the external water phase by the thickness of the membrane (ca. 75 Å), which is very scarcely permeable to most ionized species, particularly protons. This feature of the thylakoids is very important in the mechanism of ATP synthesis, as will be discussed later.

M. Aresta and G. Forti (eds.), Carbon Dioxide as a Source of Carbon, 71–81.

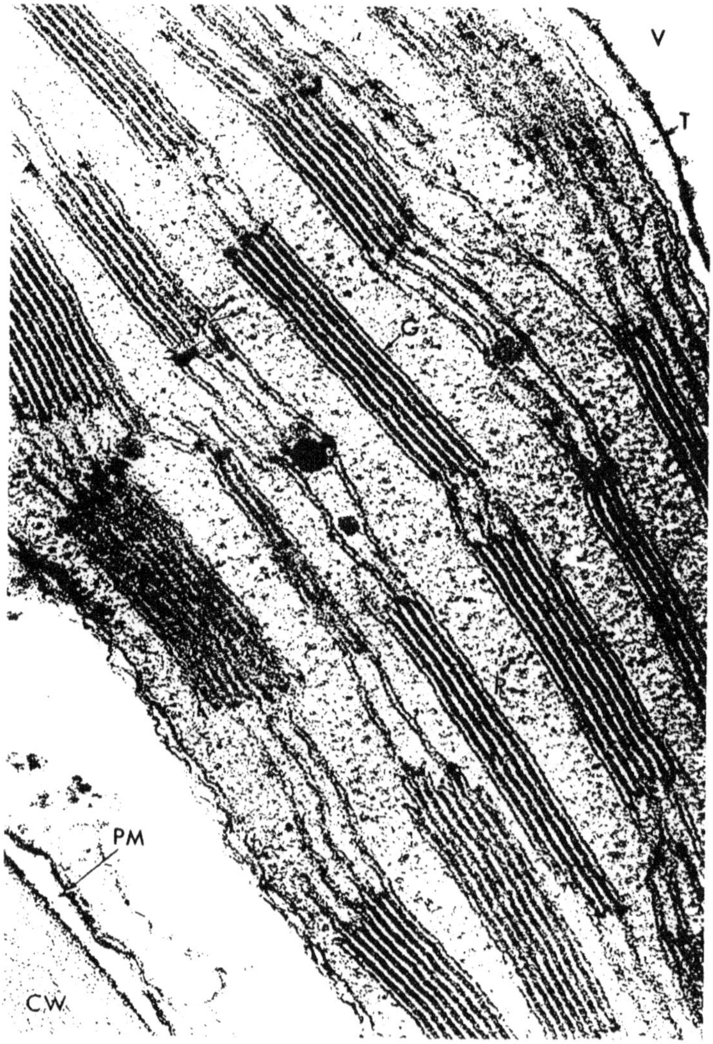

Figure 1. A chloroplast (partial view, cross section) at the electron microscope. The thylakoid membranes are stacked to form "grana" (G). The internal space ("lumen") of the membranes is clearly a continuum from one granum to the other, through the connecting membranes ("stroma membranes").
The chloroplast is separated from the cytoplasm by a double membrane enveloppe.
The tonoplast (T) membrane separates the vacuole (V) from the cytoplasm. The cell is limited towards the outside by a plasma membrane (PM) and a cell wall (CW).

The light energy utilized is absorbed by the photosynthetic pigments, i.e. chlorophyll a,chl.b. and the carotenoids in green plants, the phycocyanins and phycoerytrin in cyanobacteria and red algae respectively, and the bacteriochlorophylls in photosynthetic bacteria. The whole visible spectrum and near infrared region is covered by the photosynthetic pigments of the different photosynthetic organisms, and is utilized for photosynthesis.

2. PHOTOSYNTETIC ELECTRON TRANSPORT

The pathway of electron transport from water to NADP is represented in fig. 2, an up-to-date version of the Hill and Bendall (1) scheme of photosynthetic electron transport.

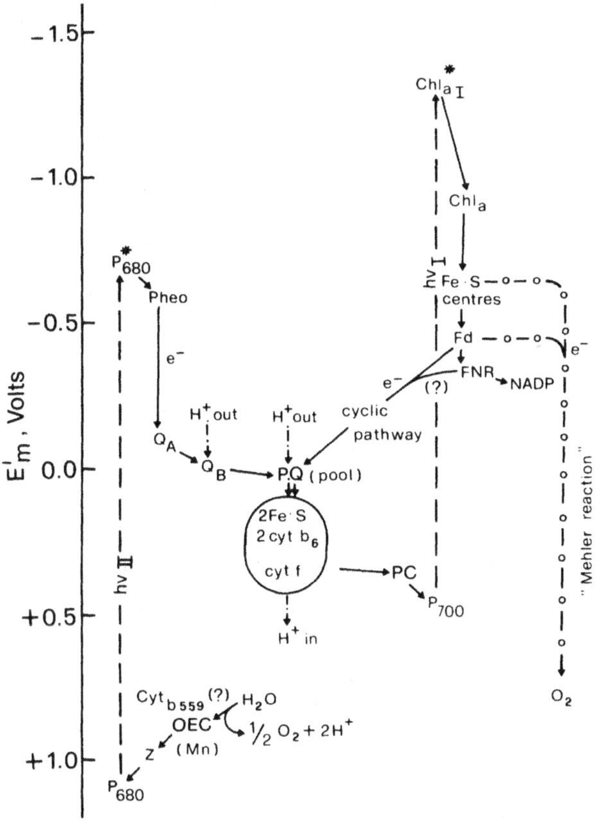

Figure 2. Scheme of the photosynthetic electron transport in chloroplasts. Dashed lines: photochemical reactions. Continous lines: electron transport reactions (the arrows indicate the direction of electron flow). --o--o--: electron transport to O_2 at the reducing side of PS I. The electrochemical potential scale can be used to locate the midpoint potential at pH 7 of the individual electron carriers.

Two photochemical reactions, connected in series by a chain of electron carriers, drive electrons against the apparent thermodynamic gradient. At the oxidizing end of the system, an enzyme complex (OEC, oxygen evolving complex) oxidizes water and evolves O_2, while at the reducing end a flavoprotein reduces NADP. The two photochemical reactions are catalyzed by two distinct chlorophyll a molecules, called P_{680} (or chl a_{II}) and P_{700} (or chl a_I) respectively, after the wavelenghts at which oxidation causes bleaching. P_{680} and P_{700} are the "reaction centres" of photosytem II (PS II) and photosystem I (PS I), respectively. Each is "served" by a bed of light-harvesting pigment molecules, ca. 250 to 300, which constitutes its "antenna". When a photon (a light quantum) of appropriate wavelenght hits a chlorophyll molecule, this is lifted to an excited state acquiring the quantum energy as electronic excitation energy. The excitation energy (the "exciton") then migrates from molecule to molecule within the photosystem, or also among neighboring photosystems, till it reaches a reaction centre, where it is "trapped" and photochemistry is initiated by the transfer of the excited electron to a "primary" acceptor. The reaction centre chlorophyll is rapidly reduced by a donor closely bound to it. The basic mechanism of both photosystems is represented by the following scheme:

$$\mathrm{D.Chl.A.} \xrightarrow{h\nu} \mathrm{D.Chl^{*}.A} \longrightarrow \mathrm{D.Chl^{+}.A^{-}} \longrightarrow \mathrm{D^{+}.Chl.A^{-}}$$

(where Chl* represents the excited state of the reaction centre chl).

The probability for an exciton to be used for photochemistry rather than for the competing processes (thermal decay, fluorescence) has been described in terms of the ratio of the kinetic constant of the primary photochemical reaction to the sum of the kinetic constants of all the processes which compete for the same excitons (2).

In the case of PS II, the primary acceptor is a pheophytin molecule, which is reduced in less than a picosecond by the excited P_{680}. The next acceptor is a bound molecule of plastoquinone, Q_A, and the primary electron donor, Z, is also thought to be a quinone. The sequence of events at PS II is represented as follows:

$$\mathrm{Z \cdot P_{680} \cdot Pheo\ Q_A} \xrightarrow{h\nu} \mathrm{Z \cdot P^{*}_{680} \cdot Pheo \cdot Q_A} \xrightarrow{\sim 1\,ps} \mathrm{Z \cdot P^{+}_{680} \cdot Pheo^{-} \cdot Q_A} \xrightarrow{20\text{-}300\,ns} \mathrm{Z^{+} \cdot P_{680} \cdot Pheo^{-} \cdot Q_A}$$

$$\xrightarrow{\sim 400\,ps} \mathrm{Z^{+} \cdot P_{680} \cdot Pheo \cdot Q_A^{-}}$$

Z^+ is then reduced by OEC, which oxidizes water, and evolves O_2 (3). Oxygen evolution has been shown to occur every 4th flash, when short flashes allowing only one turnover of the reaction centre are fired (flash duration of ca. 1 μsec.), separated by dark periods long enough to allow the reoxidation of the acceptors to be completed (4). The periodicity of O_2 evolution has been interpreted by the following scheme:

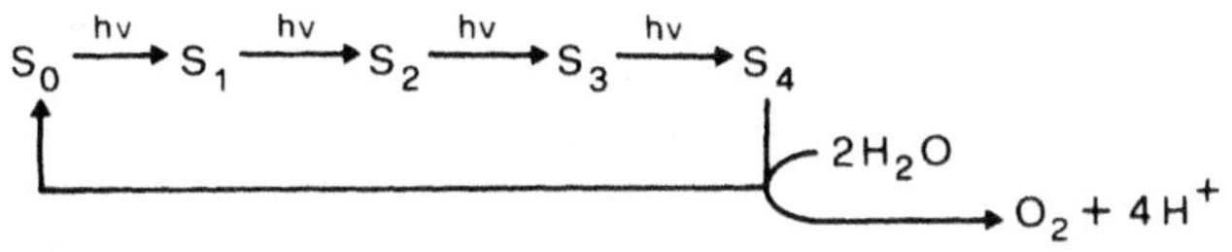

After a number of flashes, the system is randomized because of "double hits" and "misses", and continous production of O_2 is observed (fig. 3).

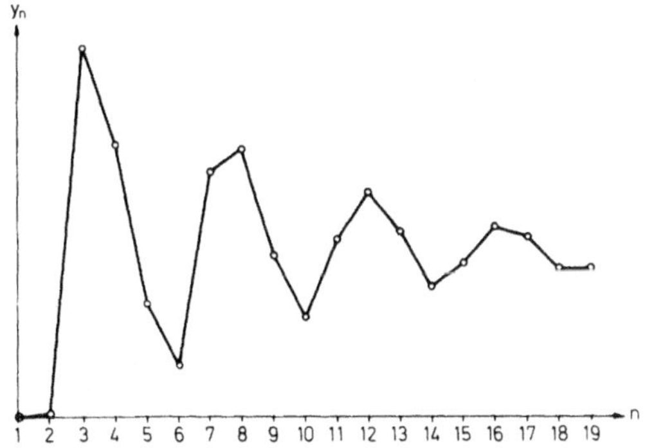

Figure 3. O_2 evolution as function of the number of single turnover flashes of saturating intensity.
Ordinate: Y_n, O_2 per flash; abscissae: flash number (the flash duration is a few μs; the dark time between two flashes is long enough to let the electron transport system restore the initial redox state of the primary electron acceptors and donors).

The oxygen evolving complex consits of several proteins, all of them bound to the membranes. Manganese is bound to some of these proteins and is responsible for the accumulation of the 4 positive charges i the $S_0 \rightarrow S_4$ transition (the "S states"). When 4 positive charges have been accumulated, the oxidation of 2 water molecules gives rise to an O_2 molecule. Four protons are liberated inside the thylakoid lumen during the S states transitions, with a different periodicity (5,6).

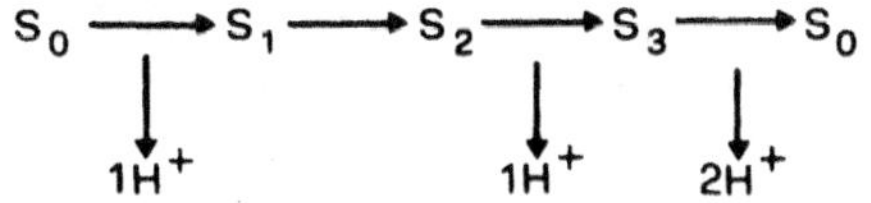

The detailed mechanism of water oxidation is still unknown, and is one of the most important problems in photosynthesis research. It has been demonstrated that $Mn^{3+} \longrightarrow Mn^{4+}$ transitions are involved (7,5), whilst the $Mn^{2+} \longrightarrow Mn^{3+}$ transition is under debate (3,8). A scheme of H_2O oxidation according to Witt et al. (8) is given in fig. 4.

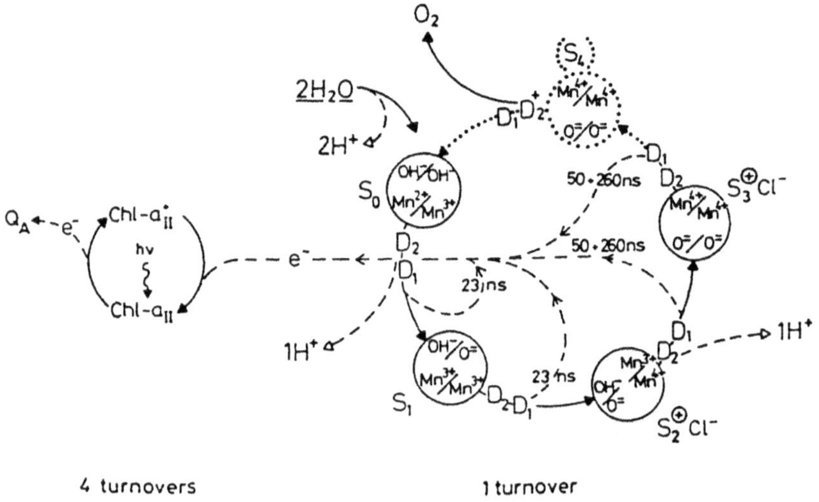

Figure 4. A scheme of water oxidation in photosynthesis (Witt et al.,8) D_1, D_2: primary electron donors to $chl^+ a_{II}$ (also called P^+_{680}); S_o to S_4: the "states" of the O_2 evolving enzyme complex (see text). The redox state of manganese is indicated. The figures indicate the half times of any particular reaction in nanoseconds.

On the reducing side of PS II, the protein-bound quinone, Q_A, accepts one electron from pheophytin. It has been identified as a plastoquinone molecule, and is reduced in 0.6-1 nanosecond. Q reduction is temperature insensitive (it occurs at high rates at 77° K) (see reviews, 3,6,9,10).

Q_A^- is reoxidized by another molecule of plastoquinone, PQ_B, bound to a 32 KD protein on the outer surface of the thylakoid membrane. PQ_B is reduced in two steps while bound to the protein, and is protonated by protons from the external water phase (6,10). PQ_BH_2 then dissociates from the protein into the soluble pool of plastoquinone, and is substituted on the 32 KD protein by a molecule of PQ (the dissociation constant is much lower for the oxidized PQ, as compared to PQH_2). The reoxidation of PQH_2 has been shown to be strongly dependent on the presence of HCO_3^- (or CO_2) which has been proposed to accelerate the PQ-PQH_2 exchange on the 32 KD protein (3). This step is the one--electron-two electrons "gate" of electron transport on the reducing side of PS II. The regulatory role of CO_2 is an important example of "feed forward" regulation by the final electron acceptor of the photosynthetic electron transport system.

The electron acceptor for PQH_2 reoxidation is the cytochrome f-b_6 complex, a five polypeptides complex containing the Fe-S centre, one cyt.f and two molecules of cyt.b_6 . Cyt.f is reduced by the Fe.S centre, at the internal side of the membrane (6,10): as a consequence, two H^+ are released into the inner aqueous phase, contributing to the generation of a H^+ gradient across the membrane. Cyt.f (E_m = 360-365 mV) is reoxidized, with an half-time of ca. 70 μs, by the copper protein plastocyanin (PC, E_m = 370 mV) (6).

PC is in solution in the inner water phase (the thylakoid lumen), and binds to the membrane at the PS I internal site, where it is oxidized by P^+_{700} according to a biphasic kinetics. A fast component ($t_{½}$ of ca. 20 μs) is predominant, and the slow component has an half time of ca. 20 ms (6).

P_{700} (E_m = 450 mV) is oxidized in 30-50 picoseconds in the photochemical reaction of PS I. The acceptor seems to be a chlorophyll a molecule, which is reduced to the anionic radical (6). This reduces a series of three bound Fe.S centres, then the electron is transferred to ferredoxin (Fd, E_m = -440 mV), a 10 KD iron sulfur protein in solution in the external water phase (the "stroma" of chloroplasts) (11). Ferredoxin forms a 1:1 complex with the flavoprotein ferredoxin-NADP reductase (FNR) bound on the surface of the membranes.

NADP also binds to the flavoprotein and is reduced and released into the external water phase, where it is utilized by the CO_2 reducing enzyme systems (the Calvin cycle in "C_3" plants or the NADPH-specific malate dehydrogenase in C_4" plants).

3. PHOTOSYNTHETIC PHOSPHORYLATION

The synthesis of ATP from ADP and inorganic orthophosphate (P_i) is coupled to photosynthetic electron transport (see eq. 1), and is now understood on the basis of the chemiosmotic theory of Mitchell (12).

Mitchell's theory holds that an elecrochemical proton gradient across the membrane (itself only slightly permeable to many ionized species and particularly to H^+) is formed by the vectorial transport of H^+ into the thylakoid lumen coupled to electron transport, as a consequence of the alternate disposition across the membrane of electron carriers which can bind protons and others which cannot be protonated (see fig. 2).

In this way, the loss of redox free energy ocurring during electron transport is partially conserved as electrochemical potential energy of the proton gradient.

The synthesis of ATP occurs when the protons accumulated inside the thylakoid lumen are transported out into the external water phase by an anisotropic, proton translocating ATP synthase-ATPase (the complex

CF_0-CF_1), which catalyzes the reaction

$$(2) \quad ADP + P_i + nH^+_{in} \longrightarrow ATP + H_2O + nH^+_{out}$$

The free energy change of ATP synthesis is given by

$$(3) \quad \Delta G_p = \Delta G^\circ_p + RT \ln \frac{[ATP]}{[ADP]\,[P_i]}$$

and the free energy change of H^+ efflux is

$$(4) \quad \Delta G_{H^+} = RT \ln \frac{[H^+_{out}]}{[H^+_{in}]} + F\,(\Psi_{out} - \Psi_{in})$$

(where Ψ is the electric potential and F is Faraday's constant). ΔG°_p is dependent upon pH, Mg^{2+} concentration, H_2O concentration, ionic strenght and temperature. At pH 8, Mg^{2+} 1 mM, ionic strenght 0.1 M and 25° C, ΔG°_p = 32.2 kJ/mole (see 13).

Synthesis of ATP can only occur when $\Delta G_p + \Delta G_H < 0$, while ATP hydrolysis occurs when the reverse is true, provided that the ATPase is activated.

As shown by equation (4), ΔG_{H^+} (which is also called Δu_{H^+}) can be separated into two components: the chemical potential of protons and the electric field across the membrane, to which all charged species present are contributing. The source of free energy for ATP synthesis is therefore the "proton motive force",

$$\Delta p = \frac{\Delta \mu H^+}{F}$$

or $\quad \Delta p = \Delta I - 2.3 \frac{RT}{F} \Delta pH$

where $\Delta pH = pH_{out} - pH_{in}$, $\Delta\Psi = \Psi_{out} - \Psi_{in}$).

At 30° C, $\Delta p = \Delta\Psi - 60\,\Delta pH$ (if $\Delta\Psi$ is expressed in mV).

A large $\Delta\Psi$ is generated by each of the two primary photochemical reactions (negative on the outside), due to the fact that the primary electron acceptors are on the outside of the membrane, and the primary donors on the inside (see review 9). The rise time of the electric field generation is therefore very fast (ca. one nanosecond), and is conveniently measured by the red shift of the absorption band of endogenous pigments (chl b, carotenoids) when subjected to a large electric field (see review 9).

The formation of Δ pH across the membrane is a much slower process, linked to electron transport along the chain (9,13,5,14). On the other hand, $\Delta\Psi$ decays rather rapidly in thylakoids (see 9) owing to the diffusion of counter-ions (such as Cl^-), so that while in the pre-steady state period at the onset of illumination Δp is mainly made up by $\Delta\Psi$, in the steady state regime, $\Delta\Psi$ in vanishingly small and Δp is mostly due to ΔpH (9,13,14,5).

The synthesis of ATP starts 4-5 ms after the onset of illumination with saturating light intensity (13,14), which is the turnover time of the ATP synthase. This means that ΔI <u>or</u> ΔpH can fulfill the energy requirement for ATP synthesis.

ATP formation has been demonstrated in the absence of light, if ΔpH is imposed artificially across thylakoid membrane, or imposing a $\Delta\Psi$ large enough to supply the energy required (see review 13). In both cases, the activity of the ATP synthase complex is required and ATP synthesis is concomitant with the transfer of protons from the internal water space of the thylakoid lumen to the external bulk phase.

All available evidence indicates that the synthesis of ATP is not <u>directly</u> coupled to electron transport, but is dependent only on the proton motive force. If an uncoupler (a substance which equilibrates H^+ across the membrane) is added in continous light, ATP synthesis is decreased or abolished, while electron transport is accelerated, due to the release of the control exerted by $\Delta\mu_{H^+}$ on the rate of electron transport.

A large body of evidence indicates that the generation of the proton motive force utilized for ATP synthesis is the cooperative result of the activity of a large number of electron transport chains. When the ionophore gramicidin (which opens channels for all monovalent cations, including protons) is added in the ratio of one gramicidin channel (two molecules of gramicidin form a conducting channel) per 10^5 chlorophylls (ca. 200 electron transport chains) the synthesis of ATP in a single turnover flash is inhibited by a factor of more than two (13). Further evidence on this point has been provided recently by Hangarter and Ort (15): ATP synthesis was measured in a series of single turnover flashes of saturating intensity under conditions where $\Delta\Psi$ was abolished by the presence of the K^+-specific ionophore nonactin and was therefore only dependent on ΔpH. The uptake of ca. 60 nmol H^+/mole of chl was required before ATP synthesis could be observed at constant yield of ATP/flash, independently of which part of the electron transport chain is activated. These experiments are easily interpreted, according to Mitchell's theory, on the basis of a delocalized pool of protons available to the ATP synthase complex.

The ATP synthesising enzyme complex consists of two parts: a readily extractable complex of five polypeptides (CF_1), protruding out the external surface of the membranes and bound (non covalently) to an extremely lipophylic complex (CF_0), which spans across the membrane (see 13,16). CF_0 is responsible for proton translocation from the inner water phase (the thylakoid lumen) to CF_1. CF_1 "senses" the proton electrochemical gradient, changing its conformation to one favorable to the binding of ADP and P_i from the external water phase, and forms ATP which is liberated into the external phase together with 3 protons (13,14,5). The conformation of the enzyme goes through a series of changes occuring along the catalytic cycle, starting when CF_0 binds protons from the lumen. The entire catalytic cycle is reversible, and ATP hydrolysis occurs (under the appropriate thermodynamic conditions) coupled to the translocation of protons into the inner phase.

The ATP synthase-ATPase is in an inactive form unless a high electrochemical proton gradient is present: the value of the proton gradient needed for enzyme activation is higher than that needed for ATP synthesis (13). This is why the ATP synthase does not hydrolyse ATP in the dark, when no ΔG_{H^+} exists, nor in the prevailing conditions of ATP, ADP and Pi concentrations. It may, however, function as an ATPase when the activated form of the enzyme is stabilized by reduction with thiol compounds (13). Such a situation may well occur in intact leaves, and may play a role in the regulation of ATP/ADP concentration in the chloroplasts.

The stoichiometry of proton efflux coupled to ATP synthesis, i.e. the ratio H^+/ATP (see eq. 2), has been the object of debate for years (13,14,5). On the basis of the thermodynamic requirement, obviously, the ratio H^+/ATP may vary depending upon the value of ΔG_{H^+}. With the commonly observed values of ΔpH of ca. 3.2–3.5 and $\Delta\Psi$ of 10–30 mV (in the steady-state) not less than 3 protons per ATP must be translocated (see 13,14,5). This is indeed the most likely value, reported in a number of observations: the interpretation of the experiments is, however, complicated by the concomitant low rate of H^+ efflux unrelated to ATP synthesis (13).

If one assumes that the H^+/ATP ratio is 3, and the ratio of protons injected into the thylakoid lumen to electrons transported is 2 ($H^+/e^-=2$), then an ATP/2e^-, or ATP/NADPH, ratio of 1.33 is calculated. Most experimental results confirm ATP/2e^- ratios of 1 to 1.33 (11,14, 13), and the ratio may in fact be flexible in response to several factors regulating the membrane permeability to protons, the internal volume and its changes, the rate of ATP utilization.

As the assimilation of CO_2 requires 3 ATP and 2 NADPH per CO_2 in the plants utilizing the Calvin cycle (the ATP requirement is 5 ATP/CO_2 in "C_4" plants), additional light quanta are necessary if the ratio ATP/2e is lower than 1.5 to produce ATP in the amount needed. This can be provided by coupling ATP synthesis to cyclic electron transport around PS I ("cyclic photophosphorylation"), or to the reoxidation of ferredoxin by O_2 at the reducing side of PS I ("Mehler reaction", see fig. 1).

Both these processes have been demonstrated to occur during photosynthesis in isolated, intact chloroplasts as well as in intact leaves (see review 11). This is in agreement with observations on quantum requirement for CO_2 assimilation in intact isolated chloroplasts, which indicated that not less than 10–12 quanta are required per molecule of CO_2 assimilated (17). In intact chloroplasts electron flow at the reducing end of PS I is diverted to the cyclic pathway (or to the Mehler reaction) when ATP is deficient: lack of ATP (or low ATP/ADP ratio) causes the drop in concentration of 1.3-bis phosphoglycerate (1.3-bis PGA), the final electron acceptor in Calvin cycle photosynthesis, the accumulation of NADPH (which is anyway in large excess with respect to NADP in steady-state photosynthesis) and the almost complete disappearing of NADP. Lack of NADP will prevent the reoxidation of Fd^{red} by FNR; therefore reoxidation of Fd^{red} can only occur through the cycle around PS I or the Mehler reaction (Fd^{red} is readily oxidized by O_2), until ATP regenerated by one of these processes allows the synthesis of 1.3-bis PGA, then reoxidation of NADPH and restoration of linear electron flow.

In agreement with this regulatory mechanism, Heber (17) has observed that upon a sudden decrease of light intensity, steady state CO_2 assimilation and ATP concentration were sharply decreased in intact chloroplasts, whilst NADPH concentration was unchanged or increased.

The problem of the ATP/NADPH ratio is still unsolved, and the possibility exists that it might be higher than 1.5 in intact leaves. Recent research has indicated the recycling of electrons across the membrane at the level of PQH_2 reoxidation by the cyt.f-b_6 complex (the so-called "Q-cyt.b cycle") (10,14). This would give a H^+/e^- ratio higher than 2, and therefore increase the ATP/NADPH ratio (a H^+/e^- ratio of 3 would produce an ATP/NADPH ratio of 2). This would allow an overall quantum requirement for CO_2 reduction in photosynthesis of 8 hv/CO_2.

REFERENCES

(1). Hill, R. and Bendall, F. (1960) Nature (London) **186**, 136-137
(2). Butler, W.L. (1978) Ann. Rev. Plant Physiol. **29**, 345-378
(3). Govindjee, Kambara, T. and Coleman, W. (1985) Photochem. Photobiol. **42**, 187-210
(4). Joliot, P. and Kok, B. (1975) In: Bioenergetics of Photosynthesis, Govindjee Ed., pp. 387-412, Academic Press, N.Y.
(5). Junge, W. and Jackson, J.B. (1982) In: Photosynthesis, Govindjee Ed., Vol. 1, pp. 589-646, Academic Press, N.Y.
(6). Hahenel, W. (1984) Ann. Rev. Plant. Physiol. **85**, 659-693
(7). Dekker, J.P., van Gorkom, H.J., Wensink, J. and Ouwehand, L. (1984) Biochim. Biophys. Acta **767**, 1-9
(8). Witt, H.T., Saygin, O., Brettel, K. and Schlödder, E. (1986) Proc. VII Int. Congress on Photosynthesis, Providence, U.S.A., in press
(9). Witt, H.T. (1979) Biochim. Biophys. Acta **505**, 355-370
(10). Cramer, W.A. and Crofts, A.R. (1982) In: Photosynthesis, Govindjee Ed., Vol. I, pp. 387-467, Academic Press, N.Y.
(11). Forti, G. (1987) In: New Comprehensive Biochemistry, Elsevier, North Holland, Amsterdam, in press
(12). Mitchell, P. (1977) FEBS Letters **78**, 1-20
(13). Schlödder, E., Gräber, P. and Witt, H.T. (1982) In: Electron Transport and Photophosphorylation, J. Barber Ed. pp. 105-175, Elsevier Biomedical Press, Amsterdam
(14). Ort, D.R. and Melandri, B.A. (1982) In: Photosynthesis, Godvinjee Ed. Vol. I, pp. 537-587, Academic Press, N.Y.
(15). Hangarter, R. and Ort, D.R. (1985) Eur. J. Biochem. **149**, 503-510
(16). Strotmann, H. and Bickel-Sandkötter, S. (1984) Ann. Rev Plant Physiol. **35**, 97-120
(17). Heber, U. (1973) Biochim. Biophys. Acta, **305**, 140-152.

RIBULOSE-1,5-BISPHOSPHATE CARBOXYLASE/OXYGENASE (RuBisCO) (ISOLATION, STRUCTURE, AND REGULATION)

T. Akazawa, A. Incharoensakdi, and T. Takabe
Research Institute for Biochemical Regulation
School of Agriculture, Nagoya University
Chikusa, Nagoya 464
Japan

Ribulose-1,5-bisphosphate (RuBP) carboxylase/oxygenase (RuBisCO) catalyzes the photosynthetic CO_2 fixation in the Calvin-Benson cycle (reductive photosynthetic carbon cycle) as well as the initial oxygenation reaction in the photorespiratory glycolate pathway (oxidative photosynthetic carbon cycle).

Since it is well known that a close relationship exists between the amount of RuBisCO in green leaf tissue and the rate of photosynthesis, it is the prevailing idea that RuBisCO is one of the decisive elements of the biomass production and/or agricultural productivity. Thus in the rapidly advancing field of biotechnology, RuBisCO is an important target of intensive research by numerous investigators, as typically exemplified in the work dealing with site-directed mutagenesis of RuBisCO. The final aim of long range research along such lines is to improve the CO_2 fixation efficiency of crop plants through technological manipulation of RuBisCO molecules.

1. ISOLATION

Historically, RuBisCO was first isolated as the soluble leaf protein of various plant species and was designated as Fraction-1-protein. Its enzymic entity was identified in later years (1,2). In the early investigation on RuBisCO, its isolation, purification and structural characterization was hampered by numerous technical problems. However, with the advances of laboratory techniques of protein purification, such as ion exchange column chromatography, gel filtration, isoelectric focusing, HPLC, among many others. It is now relatively easy to obtain high yield of RuBisCO in pure form for structural and enzymatic studies (3). In green plants, RuBisCO constitutes the major portion of the soluble phase of the chloroplast stroma (frequently it is called the world's most abundant single protein) (4). On the other hand, in most photosynthetic microorganisms, the content of RuBisCO is not necessarily high; but the use of new techniques such as antibody-bound affinity column chromatography has greatly facilitated an effective purification of the enzyme molecule from such sources. However, it must be emphasized also that the recent trend of molecular biological research on RuBisCO relies on the effective use of nucleic acids (cDNA or mRNA) rather than on RuBisCO molecules. Nevertheless, it is prerequisite to characterize

M. Aresta and G. Forti (eds.), Carbon Dioxide as a Source of Carbon, 83–91.

RuBisCO molecules or its subunits immunochemically, and for that purpose purified preparation of RuBisCO from specific origins is required.

2. STRUCTURE AND FUNCTION

Regardless of the origin of RuBisCO from either photosynthetic prokaryotes or eukaryotic green plants, it is established that the protein is comprised of two different subunit types: large (A or L), M_r ca. 50,000 and small (B or S), M_r ca. 15,000 (2,4). The holoenzyme molecule has a symmetric oligomeric structure, designated as A_8B_8, based on several experimental grounds. Only one exception to this general rule is the enzyme from Rhodospirillum rubrum, which lacks the subunit B, and the quaternary structure is A_2 (2).

Based on several experimental grounds it has been known for many years that the catalytic site of the RuBisCO molecule resides in the subunit A (2). In contrast, the functional role of subunit B has remained unclear. Employing the reversible dissociation system of RuBisCO molecules from some photosynthetic bacteria, such as Aphanothece halophytica, Synechococcus and Chromatium vinosum, the essentiality of subunit B in the catalytic activities of both the carboxylation and oxygenation reactions has been demonstrated (5-10). The dissociation-association system of RuBisCO was proved to be useful for the study of the interactions between subunits A and B, as explicitly demonstrated by the formation of the catalytically active heterologous active hybrid molecules derived from different organisms. The formation of such heterologous hybrids can be represented in Eq. 1:

$$A^x{}_8 + 8B^y = A^x{}_8B^y{}_8 \qquad (1)$$

where x and y denote the species origin of the subunit. Figure 1 illustrates the formation of the enzymatically active hybrid ($A^a{}_8B^c{}_8$) from the catalytic core (A_8) derived from Aphanothece and subunit B derived from Chromatium, whereas the reverse combination does not make the active hybrid.

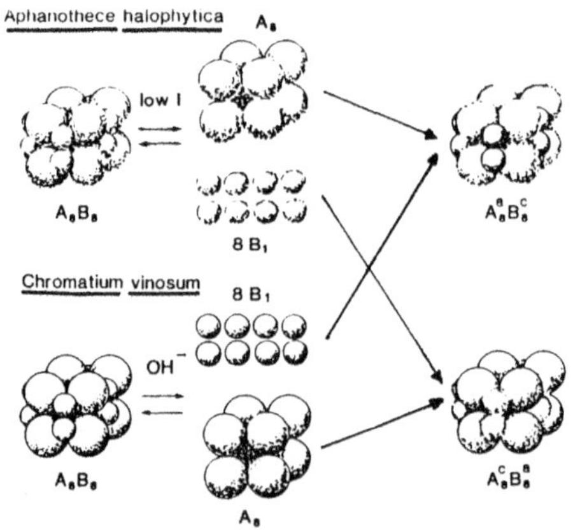

Fig. 1 Formation of heterologous hybrid from Aphanothece and Chromatium RuBisCOs ($A^a{}_8B^c{}_8$ can be formed, but not $A^c{}_8B^a{}_8$)

Recently, it has been shown that the catalytic core (an octamer of subunit A) derived from Aphanothece RuBisCO is able to reassociate with subunit B of spinach RuBisCO isolated by sucrose gradient centrifugation at pH 11.2, producing a catalytically competent enzyme molecule, whereas no active enzyme is formed when the catalytic core was derived from Chromatium (11) (Fig. 2). Furthermore, from our work using Aphanothece and Chromatium RuBisCOs, it has been implicated that subunit B serves as a stabilizer of [enzyme-activator CO_2-Mg] complex (12). The interaction of constituent subunits (A and B) in the RuBisCO molecule is schematically presented in Fig. 3.

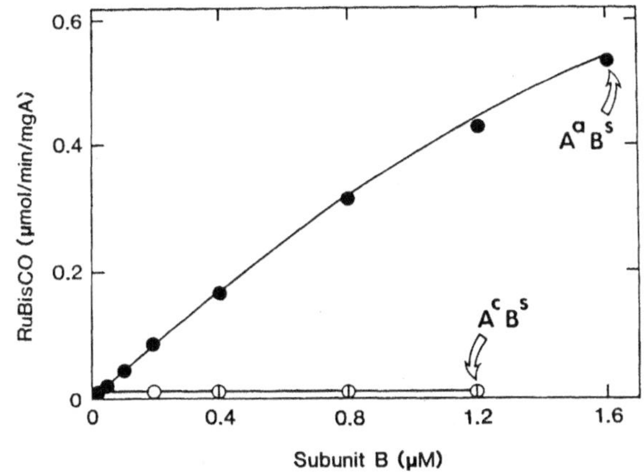

Fig. 2 Reassembly of Aphanothece and Chromatium catalytic cores with spinach subunit B

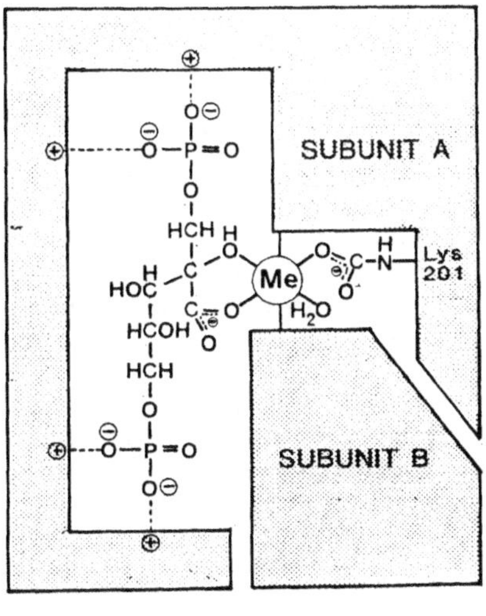

Fig. 3 Interaction between subunits A and B of RuBisCO molecule

3. BIOSYNTHESIS

Research progress on the biosynthesis of RuBisCO has one of the most fascinating historical backgrounds in modern plant molecular biology, and even now many active investigations are being made world-wide (13,14). In green plants, subunit A is encoded by chloroplastic DNA and is synthesized on the chloroplastic 70S ribosomes. On the other hand, subunit B is encoded by nuclear DNA and is synthesized on the cytoplasmic 80S ribosomes as a precursor, which is transported across the chloroplast envelope into the stroma. After being cleaved of the signal peptide, the resulting subunit B is assembled with subunit A to form the mature RuBisCO molecule (A_8B_8). The assembly mechanism (s) of RuBisCO has received an intensive research effort particularly in relation to the possible presence of the binding protein(s) in chloroplast which may be crucial in this step (15). It seems that the newly synthesized subunit A is bound to this binding protein before assembling with subunit A. However, it is not certain whether the combination of subunit A with the binding protein is an obligatory step in the assembly of the holoenzyme in the higher plant. In this context, RuBisCO from the prokaryotic origins appears to be of considerable interest. Using several microbial systems, it is now firmly established that genes for both A and B subunits make up a single operon (16,17,18). Moreover, the expression of genes for the constituent subunits of <u>Chromatium</u> RuBisCO in <u>E. coli</u> has been achieved using vectors for unfused proteins and subsequent production of the enzymically active protein has been demonstrated (20). It was also proved that the enzyme protein produced has a quaternary structure similar to that of the original plant-type (A_8B_8) native enzyme. Basically similar results have been obtained by two independent groups showing the expression of RuBisCO genes from <u>Anabaena</u> 7121 (20) and <u>Anacystis nidulans</u> (21) in <u>E. coli</u>.

4. REACTION MECHANISM

The detailed mechanism of RuBP carboxylase reaction has been facilitated by the availability of the total primary sequence of the subunit A moiety of the enzyme protein (22,23,24). RuBisCO is activated by CO_2 plus a divalent metal ion, usually Mg^{2+}. Using spinach RuBisCO, the activation reaction was shown to proceed via the carbamate formation by the binding of CO_2 to the $-NH_2$ group of Lys^{201}, followed by the subsequent stabilization with Mg^{2+} (c.f. Fig. 4). The catalytic site [substrate (CO_2 and/or O_2 and RuBP)-binding sites] of RuBisCO is believed to exist stereochemically in a close juxtaposition of the CO_2 activation site in the subunit A molecule. However, an important finding is that the CO_2 molecule involved in the activation ($^{A}CO_2$) is different from the substrate CO_2 ($^{S}CO_2$).

Little is known about the mechanism of the oxygenase reaction, in comparison to the carboxylation reaction of RuBisCO. This is chiefly due to the lack of sensitive method routinely available for laboratory assays. Since CO_2 and O_2 are mutually competitive in the carboxylation and oxygenation reactions catalyzed by RuBisCO, it is likely that each molecule attacks the same site of the enzyme molecule, but their binding sites have not been determined. Recently, it has been shown by NMR measurements and isotope-trapping experiments that carboxylation of RuBP proceeds via the ordered addition and enolization of RuBP followed by reaction with CO_2; that is, RuBP binds to the enzyme molecule prior to the binding of substrate CO_2, producing a six-carbon intermediate which is finally hydrolyzed at C-3 position to yield two molecules of 3-PGA (25). More recently, the partition of carboxylation and oxygenation is shown to occur at the step prior to the formation of a six-carbon reaction intermediate (26).

The establishment of the three dimensional structure of RuBisCO in conjunction with the primary amino acid sequences is obligatorily necessary for reaching the final answer to the enzyme reaction mechanisms concerned. Several attempts have been aimed at ways of obtaining crystals of purified RuBisCO having A_2-type structure from Rhodospirillum rubrum (27) as well as A_8B_8-plant type structure from tobacco leaves (28), Alcaligenes eutrophus (29) and Chromatium vinosum (30). Undoubtedly, data from the X-ray crystallographic studies will provide fruitful information not only for elucidating the enzyme reaction mechanism in more detail but also for the possible manipulation by recombinant DNA technology to alter the efficiency of photosynthesis (i.e., increasing the ratio of carboxylation to oxygenation).

5. MOLECULAR EVOLUTION

As can be readily envisaged from the catalytic role of subunit A in the RuBisCO molecule, its highly conserved structural (primary sequence) homology among diverse enzyme sources has been substantiated by many investigations. In the early structural studies of RuBisCO, the primary sequence determination was performed employing the classical stepwise degradation of RuBisCO molecule. However, recent advances in cloning techniques have greatly enlightened our knowledge of the molecular structure of RuBisCO and, from many collected data, further deepened our insights into the molecular evolution of RuBisCO (17). The evolution of the gene of RuBisCO is hypothesized that the cyanelle, which exists in the protista (Cyanophora paradoxa) and contains both genes for subunits A and B, bridges between the evolution from bacteria to higher plants. Figure 4 represents data of amino acid sequences around the activator site (Lys^{201}) in the subunit A of RuBisCO molecules from various sources. The sequences are well conserved in passing from the bacterium (Rhodospirillum rubrum), through the cyanobacterium (Synechococcus), and the green alga (Chlamydomonas), to the higher plants. Whereas there are indications

	↓						200		↓	↓						210			
Maize	Arg	Gly	Gly	Leu	Asp	Phe	Thr	Lys	Asp	Asp	Glu	Asn	Val	Asn	Ser	Gln	Pro	Phe	Met	Arg
Spinach	Arg	Gly	Gly	Leu	Asp	Phe	Thr	Lys	Asp	Asp	Glu	Asn	Val	Asn	Ser	Gln	Pro	Phe	Met	Arg
Chlamydomonas	Arg	Gly	Gly	Leu	Asp	Phe	Thr	Lys	Asp	Asp	Glu	Asn	Val	Asn	Ser	Gln	Pro	Phe	Met	Arg
Synechococcus	Arg	Gly	Gly	Leu	Asp	Phe	Thr	Lys	Asp	Asp	Glu	Asn	Ile	Asn	Ser	Gln	Pro	Phe	Gln	Arg
Chromatium	Arg	Gly	Gly	Leu	Asp	Phe	Thr	Lys	Asp	Asp	Glu	Asn	Ile	Asn	Ser	Gln	Pro	Phe	Met	Arg
Rhodospirillum	Trp	Leu	Gly	Gly	Asp	Phe	Ile	Lys	Asn	Asp	Glu	Pro	Gln	Gly	Asn	Gln	Arg	Phe	Ala	Pro

Fig. 4 Conserved amino acid sequences around the activator site (Lys^{201}) of RuBisCOs from various sources

that purple non-sulfur bacterium (Rhodospirillum rubrum) and higher plants have common genetic origin of subunit A, the origin and evolution of subunit B are not clear. For example, it is not certain whether R. rubrum had lost the gene of subunit B during evolution or whether the gene of subunit B did evolve later than that of subunit A.

6. REGULATION

The regulatory mechanism of activation of RuBisCO is mediated by light. This light-dependent activation of RuBisCO is attributed to the increase in pH and Mg^{2+}, which has been shown to occur in the chloroplast stroma upon illumination (31,32). By employing a mutant of Arabidopsis which is incapable of activating RuBisCO upon activation, it has been demonstrated that a protein factor, RuBisCO activase, may be responsible for activation in vivo (33,34).

Another aspect of light-modulated regulation of RuBisCO in Phaseolus vulgaris is recently reported to be due to the changes in the concentration of an inhibitor, phosphate ester, capable of tightly binding with the enzyme catalytic site (35). It is claimed that the levels of this inhibitor in vivo varying with light intensity could account for light-dependent changes in RuBisCO activities. More recently, this phosphorylated inhibitor was identified as 2-carboxyarabinitol-1-phosphate (36). A phosphorylated inhibitor was also independently reported by Servaites (37) in tobacco leaves and again the inhibitor compound was later identified to be the same as that in pea (38).

It should be pointed out, however, that the modulation of RuBisCO activity is not necessarily mediated by light but also by other factors such as environmental temperature and cellular levels of various metabolites, notably NADPH and phosphogluconate. This is especially the case where RuBisCO molecule is readily dissociated under specified conditions. It has been shown that an elevated temperature and a quaternary ammonium compound, betaine, reduce the magnitude of the dissociation of RuBisCO molecule of Aphanothece under a condition of low ionic strength (39,40). Recent experiments have shown that betaine together with KCl greatly stabilize RuBisCO molecule against heat and cold inactivation (Fig. 5).

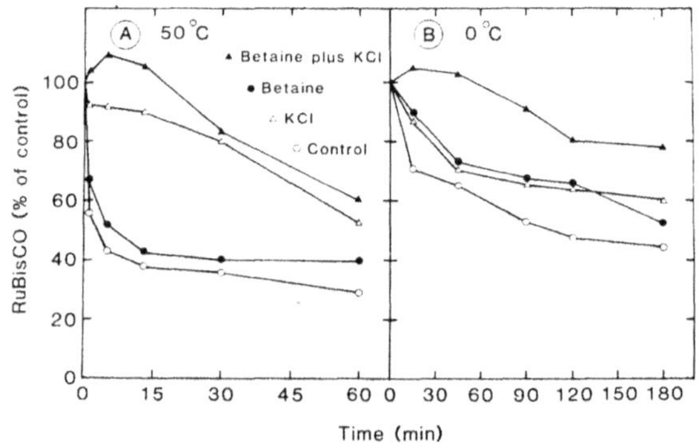

Fig. 5 Protective effect of betaine and KCl against heat and cold inactivation of RuBisCO from Aphanothece halophytica

It is conceivable that intracellular levels of certain anions such as orthophosphate and chloride can regulate the function of RuBisCO possibly by interacting with the RuBP-binding site of the enzyme molecule.

Apart from light-dependent activation of RuBisCO, there have been ample evidence that light regulates the biosynthesis of RuBisCO via the changes of the concentrations of its mRNA in a number of higher plants (41) as well as in a unicellular green alga, Chlorogonium elongatum (42,43). Now it appears that light affects the synthesis of subunit A at the post-transcriptional level since the increase in mRNA for subunit A as influenced by light is related largely to the increase in chloroplast genome copy number (44). The effect of light on the synthesis of subunit B, on the other hand, is under transcriptional control since the copy number of subunit B gene in the nucleus is not significantly affected by light (45).

It should be recognized that the recent tendency of research effort has been focused on the regulation of RuBisCO synthesis at the gene level. For example, Portis's and Ogren's groups at the University of Illinois are trying to isolate the nuclear gene coding for the RuBisCO activase mentioned above in order to study the developmental expression of the gene in relation to other photosynthetic genes. Another notable recent achievement made by Fluhr and Chua (46) has been the phytochrome control of two genes encoding the small (B) subunit of RuBisCO (rbc 3a and rbc 3c) in etiolated pea leaves and in the transgenic petunia plant. In mature green leaves, however, gene expression has been shown to be modulated by blue light.

judging from the recent interest to study RuBisCO by employing recombinant DNA technology, it is not surprising that RuBisCO, one of the most intensively studied proteins in the plant kingdom, will remain as the "ever-green" enzyme in years still to come.

REFERENCES

1. Siegelman, H.W. and Hind, G. (1978) Photosynthetic Carbon Assimilation. Plenum Press, New York.
2. Akazawa, T. (1979) in "Encyclopedia of Plant Physiology, Vol. 6, Photosynthesis II, Photosynthetic Carbon Metabolism and Related Process" (Ed. Gibbs, M. and Latzko, E.) pp. 208-229, Springer-Verlag, Berlin-Heidelberg-New York.
3. Nishimura, M. and Akazawa, T. (1982) in "Methods in Chloroplast Molecular Biology" (Ed. Hallick, R.B., Edelman, M. and Chua, N-H) pp. 751-765. Elsevier Biomedical Press, Amsterdam.
4. Ellis, R.J. (1979) Trends Biochem. Sci. 4, 241-244.
5. Asami, S., Takabe, T., Akazawa, T. and Codd, G.A. (1983) Arch. Biochem. Biophys. 225, 713-721.
6. Andrews, T.J. and Ballment, B. (1983) J. Biol. Chem. 258, 7514-7518.
7. Takabe, T., Rai, A.K. and Akazawa, T. (1984) Arch. Biochem. Biophys. 229, 202-211.
8. Andrews, T.J. and Ballment, B. (1984) Proc. Natl. Acad. Sci. (USA) 81, 3660-3664.
9. Takabe, T., Incharoensakdi, A. and Akazawa, T. (1984) Biochem. Biophys. Res. Commun. 122, 763-769.
10. Incharoensakdi, A., Takabe, T., Takabe, T. and Akazawa, T. (1985) Biochem.

Biophys. Res. Commun. 126, 698-704.
11.Incharoensakdi, A., Takabe, T., Takabe, T. and Akazawa, T. (1986) Biochem. Biophys. Res. Commun. 138, 118-124.
12.Incharoensakdi, A., Takabe, T. and Akazawa, T. (1986) Arch. Biochem. Biophys. 248, 62-70.
13.Dean, C., van den Elzen, P., Tamaki, S., Dunsmuir, P. and Bedbrook, J. (1985) Proc. Natl. Acad. Sci. (USA) 82, 4964-4968.
14.Dean, C., van den Elzen, P., Tamaki, S., Dunsmuir, P. and Bedbrook, J. (1985) EMBO J. 4, 3055-3061.
15.Hemmingsen, S.E. and Ellis, R.J. (1986) Plant Physiol. 80, 269-276.
16.Kobayashi, H. and Akazawa, T. (1982) Arch. Biochem. Biophys. 214, 531-539.
17.Akazawa, T., Takabe, T. and Kobayashi, H. (1984) Trends Biochem. Sci. 9, 1-4.
18.Shinozaki, K. and Sugiura, M. (1985) Mol. Gen. Genet. 200, 27-32.
19.Viale, A., Kobayashi, H., Takabe, T. and Akazawa, T. (1985) FEBS Lett. 192, 283-288.
20.Gurevitz, M., Somerville, C.R. and McIntosh, L. (1985) Proc. Natl. Acad. Sci. (USA) 82, 6546-6550.
21.Tabita, F.R. and Small, C.L. (1985) Proc. Natl. Acad. Sci. 82, 6100-6103.
22.Lorimer, G.H. and Andrews, T.J. (1981) in "The Biochemistry of Plants, Vol. 8, Photosynthesis" (Ed. Hatch, M.D. and Boardman, N.K.) pp. 329-374, Academic Press, New York.
23.Lorimer, G.H. (1981) Annu. Rev. Plant Physiol. 32, 349-383.
24.Miziorko, H.M. and Lorimer, G.H. (1983) Annu. Rev. Biochem. 52, 507-535.
25.Pierce, J., Lorimer, G.H. and Reddy, G.H. (1986) Biochemistry 25, 1636-1644.
26.Pierce, J., Andrews, T.J. and Lorimer, G.H. (1986) J. Biol. Chem. 261, 10248-10256.
27.Janson, C.S., Smith, W.W., Eisenberg, D. and Hartman, F.C. (1984) J. Biol. Chem. 259, 11594-11596.
28.Schneider, G., C-I, Branden. and Lorimer, G.H. (1981) J. Mol. Biol. 175, 99-102.
29.Bowien, B., Mayer, F., Spies, E., Pahler, A., English, U. and Saenger, W. (1980) Eur. J. Biochem. 106, 405-410.
30.Nakagawa, H., Sugimoto, M., Kai, Y., Harada, S., Miki, K. and Kasai, N. (1986) J. Mol. Biol. (submitted).
31.Portis, A.R., Chon, C.J., Mosbach, A. and Heldt, H.W. (1977) Biochim. Biophys. Acta. 461, 313-325.
32.Purczeld, P., Chon, C.J., Portis, A.R., Heldt, H.W. and Heber, U. (1978) Biochim. Biophys. Acta. 501, 488-498.
33.Salvucci, M.E., Portis, A.R. and Ogren, W.L. (1985) Photosynthesis Res. 7, 193-201.
34.Salvucci, M.E., Portis, A.R. and Ogren, W.L. (1986) Plant Physiol. 80, 655-659.
35.Seemann, J.R., Berry, J.A., Freas, S.M. and Krump, M.A. (1985) Proc. Natl. Acad. Sci. (USA) 82, 8024-8028.
36.Berry, J.A., Lorimer, G.H., Pierce, J., Seemann, J.R., Meek, J. and Freas, S. (1986) Proc Natl. Acad. Sci. (USA) (in press)
37.Servaites, J.C. (1985) Plant Physiol. 78, 839-843.
38.Gutteridge, S., Parry, M.A.J., Bueton, S., Keys, J.A., Mudd, A., Feeney, J., Servaites, J. and Pierce, J. (1986) Nature (in press)
39.Incharoensakdi, A., Takabe, T. and Akazawa, T. (1985) Arch. Biochem.

Biophys. 237, 445-453.
40.Incharoensakdi, A., Takabe, T. and Akazawa, T. (1986) Plant Physiol. 81, 1044-1049.
41.Tobin, E.M. and Silverthorn, J. (1985) Annu. Rev. Plant Physiol. 36, 569-593.
42.Westhoff, P. and Zetsche, K. (1981) Eur. J. Biochem. 116, 261-267.
43.Steinbiß, J. and Zetsche, K. (1986) Planta 167, 575-581.
44.Inamine, G., Barbara, N., Weissbach, A. and Brot, N. (1985) Proc. Natl. Acad. Sci. (USA) 82, 5690-5694.
45.Sasaki, Y., Sakihama, T., Kamikubo, T. and Shinozaki, K. (1983) Eur. J. Biochem. 133, 617-620.
46.Fluhr, R. and Chua, N-H. (1986) Proc. Natl. Acad. Sci. (USA) 83, 2358-2362.

CHLOROPHYLL FUNCTION IN PLANT PHOTOSYNTHESIS AND NONBIOLOGICAL APPLICATIONS. ISOTOPE EFFECTS IN PHOTOREDUCTION OF CARBON DIOXIDE BY WATER

Michael S. Showell, Jun-Ling You, Gerald D. Fong, Karen A. Butcher, and Francis K. Fong, Department of Chemistry, Purdue University, West Lafayette, Indiana 47907, U.S.A.

I. Introduction

Photosynthesis as represented by the reaction stoichiometry

$$CO_2 + H_2O \xrightarrow[\text{Chl } \underline{a}]{\text{visible light}} (1/n)(CH_2O)_n + O_2 \qquad (1)$$

is of great importance in carbon dioxide chemistry. In green plants the oxygen evolution from water oxidation and carbon dioxide reduction evidently involve the participation of two Chl a photosystems respectively known as PSII and PSI. In nonbiological applications the observed effects of reaction 1 are attributed to a single Chl a light reaction. An important problem in photosynthesis research is a description of the intrinsic properties of Chl a in the photocatalysis of reaction 1. The goal is to find the optimized parameters for the reduction of carbon dioxide by water outside of the living plant.[1] This chapter deals with the current progress in the use of Chl a model systems towards the attainment of that goal.

On account of its biological significance photosynthesis research has evolved around developments on subchloroplast particles enriched in PSI and PSII reaction centers. In vitro Chl a work has received comparatively less attention for two principal reasons: (1) The structural properties of in vitro Chl a aggregates that are photocatalytically active have turned out to be unexpectedly complex even in their simplest form. Only in recent years have questions been answered concerning the hydration stoichiometry[2-5] and stereospecificity[6] of bonding interactions with respect to the asymmetric carbon at C-10. (2) A description of the intrinsic properties of Chl a based on observations of in vivo photosynthesis is exceedingly difficult to formulate because the experimental behavior reflects the collective role of the many elaborate mechanisms crucial

M. Aresta and G. Forti (eds.), Carbon Dioxide as a Source of Carbon, 93–111.

to the physiological functions of plant photosynthesis, but may not be applicable to a nonbiological demonstration of reaction 1.

In spite of the many difficulties, studies on _in vitro_ Chl _a_ systems have in recent years made steady progress. Inquiries directed by established _in vivo_ behavior have provided clues for seeking out relevant information from the vast literature on Chl _a_. The work on modeling the PSI reaction center dimer[7], _P_700, has continued, based on the properties of a C_2-symmetric monohydrate dimer, (Chl _a_.$H_2O)_2$, obtained in rigorously dried nonpolar solutions.[8-12] Several recent studies on the oligomers of Chl _a_ dihydrate, (Chl _a_.$2H_2O)_n$,[9,13-17] have strengthened the consideration of the dihydrate dimer, (Chl _a_.$2H_2O)_2$, as a viable model for the PSII reaction center, _P_680, on account of the _in vitro_ demonstrations of water splitting,[18] carbon dioxide photoreduction[19] and superoxide formation[20] using (Chl _a_.$2H_2O)_n$ as photocatalyst. The _P_680 is a Chl _a_ dimer,[21] so-named because of its maximum absorption at 680 _nm_.

The field of nonbiological chlorophyll photosynthesis is relatively new, having only recently been established on experimental grounds. Little is known of the molecular mechanisms underlying the _in vitro_ photosynthetic reaction reported by Fruge _et al_.[19] The observation of carbon dioxide reduction products was made without the biological apparatus provided in plant photosynthesis. The product yields were comparatively low. Isotope effects, observed by Fruge _et al_[19] when deuterated water was used in a nonbiological demonstration of reaction 1, evidently played a significant role in the reaction mechanism. In current _in vitro_ work in this laboratory, pronounced isotope exchange effects involving the carbon dioxide oxygen atoms were observed when ^{18}O-enriched water was used in reaction 1. Comparable isotope effects have not been reported in the study of natural photosynthesis.

It seems desirable to provide a description of the path of carbon dioxide reduction _in vivo_ and related physiological effects, and to consider such a description as the point of departure for discussing the growing body of observables relevant to the role of the chlorophyll in photocatalyzing reaction 1 outside of green plants. The purpose of this comparison is to disseminate the details of the diverse behavior, so the investigation can be narrowed to a focus not only on the fundamental problems peculiar to Chl _a_ as the photocatalyst for reaction 1, but also on those for reaction paths that are deleterious to reaction 1.

II. Path of carbon reduction in plant photosynthesis.

In plant photosynthesis specialized catalytic mechanisms are provided for water oxidation by the Chl _a_ light reaction, which is followed by CO_2 reduction to yield carbohydrates. The elaborate mechanisms are of universal importance to life on earth. The path of carbon reduction, for example, is an intricate balance of light and thermochemical reactions. The one-electron photooxidation/reduction cycles of P680 are kinetically coupled to the four-electron oxygen evolution process in concerted mechanisms provided by a Mn-protein complex. The harmful effects of superoxide formation from molecular oxygen as the result of a Chl a light reaction are minimized by the presence of superoxide dismutase (SOD). Under steady-state conditions natural photosynthesis is thus an interconnection of concurrent processes. Kinetically coupled reaction cycles spanning thirteen decades of time domain, beginning with the 10^{-12}-s events of Chl a primary light reactions and ending with the 10-s events of the carbon reductive cycle, are sustained by the light absorbed by the chlorophyll. The reaction cycles in plant photosynthesis are the subject of extensive research, an example of which is provided in the following analysis of carbon dioxide reduction.

Carbon dioxide is assimilated in the dark by hydrolytic splitting of the 1,5-ribulose bisphosphate (RuBP) carboxylation adduct to yield two phosphoglyceric acid (PGA) molecules

$$\text{RuBP} + CO_2 \longrightarrow 2\text{PGA} \tag{2}$$

The PGA is reduced to C_3 sugars by reduced nicotinamide adenine dinucleotide phosphate (NADPH) assisted by adenosine triphosphate (ATP):

$$\text{PGA} + \text{NADPH} + \text{ATP} \longrightarrow C_3 + \text{NADP} + \text{ADP} \tag{3}$$

in a sequence of photochemical and thermochemical steps involving coupled kinetic relationships between RuBP and PGA.[1]

The kinetic data reported by Bassham and Calvin[22] for in vivo carbon reduction were reproduced from Figs. 10 and 12 of Ref. 22 by photographic methods. The enlarged figures were then digitized by means of a CDC2 6600 computer. Graphs of the digitized kinetic curve fits are shown in Fig. 1. The least-squares fits were obtained using standard nonlinear regression techniques. The dark decay of RuBP concentration, C_D, is given by the first-order kinetic expression

$$C_D = C_D^0 \exp(-t/\tau_D) \tag{4}$$

where $\tau_D = 11 \pm 3$s and $C_D^0 = 133 \pm 13$ are respectively the lifetime and relative initial reservoir size of RuBP. The broken curve in Fig. 1 represents the rate of RuBP decay in the light when the CO_2 pressure is rapidly raised from 0.003 to 1%. The decay of RuBP follows the CO_2 pressure increase on a time scale <1s. A comparison of the observed rates shows that the lifetime, $\tau_L \approx 1$s, of RuBP in the light is more than one order of magnitude smaller than that, $\tau_D = 11 \pm 3$ s, in the dark, which suggests a concerted mechanism, precluding reaction 2 as a precursor to reaction 3.

The stoichiometry of reaction 2 obtained under photosynthetic conditions has been discussed by earlier writers.[23-25] Bassham and

Kirk[23] suggested that RuBP carboxylation in Chlorella may result in one molecule of PGA and one molecule of some other 3-carbon compound,

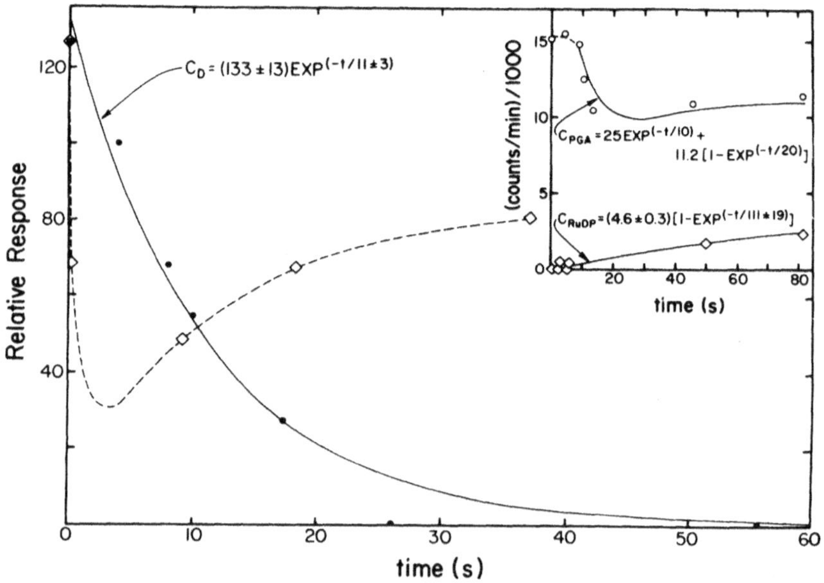

Fig. 1. Changes in PGA and RuBP concentrations in light and dark carboxylation: ----decay of RuBP concentration in the dark;- - -change in RuBP concentration in the light when CO_2 pressure is increased. Inset: Coupled kinetic relationship of PGA (0) and RuBP ( )concentrations in transition from dark to light.

possibly a triose phosphate. On the other hand, Galmiche[25] presented evidence that the stoichiometry of RuBP carboxylation obtained in the light is the same as that in the dark. From Fig. 1 we observe that the equilibrium concentration of RuBP in the dark is low. A consideration of the kinetic sequence for RuBP formation in the reductive pentosephosphate (RPP) cycle[22] for carbon reduction may be helpful in a discussion on the stoichiometry of RuBP carboxylation.

The inset in Fig. 1 shows the kinetic data of PGA (upper curve) and RuBP (lower curve) using ^{14}C tracer techniques in going from conditions of dark to light.[22] The increase in RuBP concentration,

$\underline{C}_{RuBP}$, is fitted to the expression

$$\underline{C}_{RuBP} = \underline{C}^{O}_{RuBP}[1 - \exp(-t/\underline{\tau})] \quad (5)$$

where $\underline{C}^{O}_{RuBP}$ = 4600 ± 300 and $\underline{\tau}$ = 111 ± 19 $\underline{s}$ are the steady-state concentrations of RuBP in counts/$\underline{min}$ and the rise time of $\underline{C}_{RuBP}$, respectively. The decrease in PGA concentration, $\underline{C}_{PGA}$, is preceded by ~5-$\underline{s}$ lag time, during which the relative reservoir size of PGA remains constant. This delay is attributed to the time required for the accumulation of NADPH from the Chl $\underline{a}$ light reactions to initiate the thermochemical reduction of PGA. The kinetic behavior of $\underline{C}_{PGA}$ in the light is given by

$$\underline{C}_{PGA} = \underline{A}e^{-t/\underline{\tau}_1} + \underline{B}(1 - e^{-t/\underline{\tau}_2}) \quad (6)$$

where $\underline{\tau}_1$ = 10$\underline{s}$ and $\underline{\tau}_2$ = 20$\underline{s}$ are respectively the decay and rise times of $\underline{C}_{PGA}$, whereas $\underline{A}$ = 25,000 and $\underline{B}$ = 11,200 are respectively the initial and steady-state reservoir sizes of PGA in counts/$\underline{min}$. Based on the above-described kinetic effects for the dark <---> light transitions, three stages of the RPP cycle (with corresponding lifetimes) are given as follows:

(i) Cycle initiation on transition from dark to light

$$5PGA \xrightarrow[5ATP]{10[H]} 5C_3 \qquad (\underline{\tau}_1 = 10\ \underline{s})$$

$$2C_3 \longrightarrow C_6$$

$$C_6 + 2C_3 \longrightarrow C_5 + C_7$$

$$C_7 + C_3 \longrightarrow 2C_5$$

$$2C_5 \xrightarrow{2ATP} 2RuBP$$

$$5PGA + 10[H] \xrightarrow{7ATP} C_5 + 2RuBP \quad (7)$$

(ii) Cycle sustenance

$$3CO_2 + 3RuBP \xrightarrow{6[H]} 3C_3 + 3PGA \qquad (\underline{\tau}_L \approx 1\ \underline{s})$$

$$3PGA \xrightarrow[3ATP]{} 3C_3$$

$$2C_3 \longrightarrow C_6$$

$$C_6 + 2C_3 \longrightarrow C_5 + C_7$$

$$C_7 + C_3 \text{ -----}\rangle\ 2C_5$$

$$3C_5 \xrightarrow{3ATP} 3RuBP$$

$$3CO_2 + 12[H] \xrightarrow{6ATP} C_3 + 3H_2O \qquad (8)$$

(iii) Cycle termination on transition from light to dark

$$2RuBP + 2CO_2 \text{ -----}\rangle\ 4PGA \qquad (\underline{\tau}_D = 11 \pm 3\ \underline{s}) \qquad (9)$$

In 7 and 8 the details of C_3 condensation to give C_5, C_6, C_7 sugars and RuBP follow the usual carbon path in the RPP cycle. The RuBP photoreductive carboxylation step in 8 follows Bassham and Calvin's conclusion based on stoichiometry considerations (see p. 53 of ref. 22). The thermochemical reduction of PGA, involving ATP and [H] from NADPH, as the cycle initiation step in 7 is manifested by the initial decay of PGA given by $\underline{\tau}_1 = 10\ \underline{s}$ on transition from dark to light (see inset of Fig. 1). The relatively slow buildup of RuBP in 7 and the rapid RuBP photoreductive carboxylation ($\underline{\tau}_L \approx 1\ \underline{s}$) in 8 accounts for the simultaneous growth of PGA and RuBP after cycle initiation, as evidenced by the coupled increases in PGA and RuBP following the initial PGA decay shown in the inset of Fig. 1. The carbon balance in 8 is that obtained under stable conditions of photosynthesis. Combining 7 and 9 the carbon balance from cycle initiation and termination on completion of the dark <---> light dark sequence is obtained.

$$PGA + 2CO_2 + 10[H] \xrightarrow{7ATP} C_5 + 2H_2O \qquad (10)$$

from which it is observed that PGA is consumed, along with CO_2, as a primary reactant upon photoinitiation and completion of the RPP cycle. We note that the cycle initiating and terminating mechanisms underlying reactions 7 and 9 also provide a means for regulating the photosynthetic rate in response to intensity variations in the incident light. The reservoir size of PGA increases as the light intensity is reduced to 9. This reservoir also makes available an instantaneous supply of PGA to increase the photosynthetic rate according to 7 when the light intensity is increased.

III. Nonbiological photosynthesis: Observations of isotope effects

That water is the source of O_2 evolution in plant photosynthesis was suggested by the early writers around 1930.[26-29] Shortly afterwards experimental support for that suggestion was provided by several research groups.[30-32] More recently the Chl $\underline{a}$ water splitting reaction in vitro was demonstrated[18]

$$2H_2O \xrightarrow[(\text{Chl } \underline{a}.2H_2O)_n]{\text{visible light}} 2H_2 + O_2 \tag{11}$$

using platinized polycrystals of $(\text{Chl } \underline{a}.2H_2O)_n$ deposited on Pt electrodes. The demonstration of reaction 11 led to the photoreduction of CO_2 by H_2O to formic acid using platinized $(\text{Chl } \underline{a}\cdot 2H_2O)_n$. The reduction was shown to be photocatalyzed by the chlorophyll and may additionally produce formaldehyde[19]:

$$2CO_2 + 2H_2O \xrightarrow[(\text{Chl } \underline{a}.2H_2O)_n]{\text{visible light}} 2H_2CO_2 + O_2 \tag{12}$$

The demonstrations of oxygen evolution in both reactions 11 and 12 had been accomplished by mass spectrometric observations of $^{16}O^{18}O^+$ (mass 34) and $^{18}O_2^+$ (mass 36) lines from ^{18}O-enriched solutions, providing evidence of molecular oxygen as the oxidation product. Of current interest are the isotope effects observed in connection with the carbon dioxide and its reduction products in reaction 12.

The experimental methods for observing reaction 12 were described by Fruge _et al._[19] A Pt foil was platinized by passing a 30-_mA_ current for 10 _min_ through a 7×10^{-2} _M_ chloroplatinic acid solution containing 6×10^{-4} _M_ lead acetate. Chl _a_ was prepared in the usual manner from fresh spinach leaves by chromatographic methods and suspended in _n_-pentane as polycrystalline chlorophyll _a_ dihydrate.[4] The Pt black electrode was plated with Chl _a_ by suspending the electrode in an aluminum trough filled with 10^{-4} _M_ Chl _a_ and subjecting it to a 1000 $\underline{Vcm^{-2}}$ potential under the illumination of a 200-_W_ tungsten lamp. The Chl _a_ plated electrode, still under illumination, was platinized again for 15 _s_ at 30 _mA_ in the chloroplatinic acid solution. The platinized $(\text{Chl } \underline{a}\cdot 2H_2O)_n$ sample was immersed in warm water (40-60°C) for 2 _h_ and then allowed to equilibrate in the water at room temperature overnight. The electrode was then placed in a cell, fitted with Kel-F valves, containing a mixture of $H_2{}^{16}O$, $D_2{}^{16}O$, and/or $H_2{}^{18}O$. The sample cell containing doubly distilled ^{18}O- and/or D-enriched water was deaerated by purging with Ar gas for 20 _min_. The water was then saturated with CO_2. Research grade CO_2, purchased from Matheson, was further purified by passing the gas over a red-hot platinum coil to eliminate organic impurities and then through a Pyrex U-tube containing BTS Cu-Mn pellets for oxygen removal. Following CO_2 saturation the photoelectrolytic cell was closed, and the sample was illuminated with a 1000-_W_ xenon arc lamp. Corning CS 3-69 and water filters were used to eliminate short wavelength (<520 _nm_) and IR photons, respectively. After 30 _min_ of illumination, the gaseous mixture above the aqueous solution was analyzed using a Consolidated Electrodynamics Model 21-110-B mass spectrometer. During the entire procedure the Pt/Chl _a_ electrode was not allowed to dry in extended exposure to air.

An important isotope effect in the _in vitro_ reduction of CO_2 by

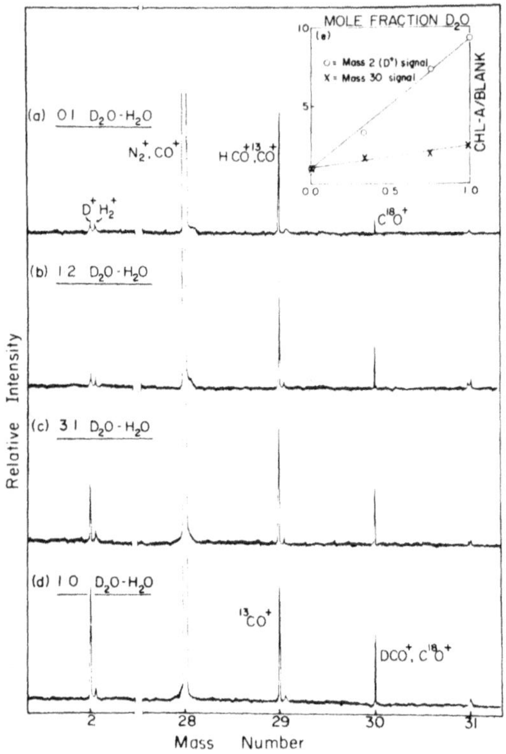

Fig. 2. Mass spectral determination of the dependence of product distribution on the mole fraction of D_2O in H_2O. The Pt/Chl a sample was subjected to 30 min of illumination with IR and short-wavelength filtered light with CO_2-saturated solutions containing (a) pure $H_2{}^{16}O$; (b) 1:2 $D_2{}^{16}O$-$H_2{}^{16}O$; (c) 3:1 $D_2{}^{16}O$-$H_2{}^{16}O$; and (d) pure $D_2{}^{16}O$. The spectra in the mass 28-31 and mass 2 regions were obtained with the ion source focused at mass 44 and 2, respectively. The mass 29($^{13}C^{16}O^+$) signal was observed to be ≈ 1% of the mass 28 ($^{12}C^{16}O^+$) signal consistent with the natural abundance ratio of $^{13}C/^{12}C$. Mass intensities in the 28-31 region were scaled to a constant mass 29 signal. In the mass 2 region the H_2^+ signal was arbitrarily chosen as the reference. The monotonic trend of increasing D^+ and mass 30 ($D^{12}C^{16}O^+$) line intensities with increasing x_D, the mole fraction of D_2O, is evident in a-d. In c the ratio, Chl a/blank, is that of the scaled signal intensity at masses 30 and 2 obtained from the illuminated sample and its corresponding unilluminated blank. The variations of Chl a/blank intensity ratios at masses 30 and 2 are linear to x_D.

water is manifested by the data shown in Fig. 2. According to the following analysis, the observed variation in the product distribution as the mole fraction of $D_2{}^{16}O$ in H_2O, $\underline{x}_D$, is varied indicates that only deuterated photosynthetic products are obtained.

Both D^+ and mass 30 intensities evidently increase linearly with $\underline{x}_D$ (see Figs. 2a-d). The observed mass 30/28 intensity ratio of 0.006 in pure D_2O (Fig. 2d) corresponds to a total mass 30 line intensity about three times the expected contribution from $^{12}C^{18}O^+$ owing to the natural abundance of ^{18}O (≈0.2%). The rise in the mass 30 line with $\underline{x}_D$ indicates that $D^{12}C^{16}O^+$ is the primary source of contribution to the observed intensity enhancement. Standardized mixtures of DCOOD in 3 ml of CO_2-saturated D_2O were examined by mass spectrometry. Under 1 atm of CO_2, a sample/blank intensity ratio of 5 ± 1 was obtained from the D^+ line from solutions 0.001 and 0.01 M in DCOOD. No significant intensity changes at mass 30 were observed. However, on reduction of the CO_2 pressure by a factor of 2 in the 0.001-M DCOOD sample and increasing the DCOOD concentration to 0.01 M, a 50-fold increase in the $D^+/H_2{}^+$ intensity ratio and a 2-fold increase in the mass 30/29 ratio relative to a blank prepared at 1 atm in CO_2 were observed. The simultaneous increases of the D^+ and mass 30 lines thus observed are comparable in magnitude to the corresponding increases observed in Fig. 2. This observation suggests that the CO_2 pressure is reduced during the Pt/Chl a light reaction, resulting in the $D^{12}C^{16}O^+$ fragment. This conclusion[19] of reduction in the CO_2 pressure is corroborated by the observation of isotope exchange effects described in the discussion to follow. Significantly, in contrast with the simultaneous increases with increasing $\underline{x}_D$ in the line intensities at masses 2 and 30 for the deuterated products, the line intensity at mass 29 for $H^{12}C^{16}O^+$ remains virtually constant as $\underline{x}_D$ is varied (see Figs. 2a-d). The conclusion is thus reached that only deuterated products of reaction 12 are observed in the series of experiments recorded in Fig. 2.

The mass spectrometric measurements of vapor samples extracted from a photoelectrolytic cell before and after 30 min of visible-light illumination of a Pt/(Chl a.$2H_2O)_n$ electrode immersed in an ^{18}O- and D-enriched water (H/D = 1:0.7; $^{16}O/^{18}O$ = 1:0.5) sample are shown in Fig. 3. In the photoinduced process ^{18}O atoms in the isotopically enriched water mixture could conceivably undergo equilibration with the ^{16}O atoms of unlabeled carbon dioxide molecules in the vapor phase, resulting in a statistical distribution of ^{16}O and ^{18}O in the gaseous CO_2 compatible with the isotope composition in the ^{18}O-enriched aqueous solution. In Fig. 3 the observed mass spectrometric intensity ratios of $^{13}C^{16}O^{18}O^{++}/^{13}C^{18}O_2{}^{++}$ = 4.0 and $^{12}C^{16}O^{18}O^{++}/^{12}C^{18}O_2{}^{++}$ = 4.4 are in reasonable agreement with a binomial distribution of ^{16}O and ^{18}O.[18] However, the observed ratios $^{13}C^{16}O_2{}^{++}/^{13}C^{18}O_2{}^{++}$ = 5.9 and $^{12}C^{16}O_2{}^{++}/^{12}C^{18}O_2{}^{++}$ = 6.2 are significantly greater than that, 4, expected for a random isotopic mixing.

The following study was undertaken in order to provide a quantitative analysis of the isotope exchange effects shown in Fig. 3.

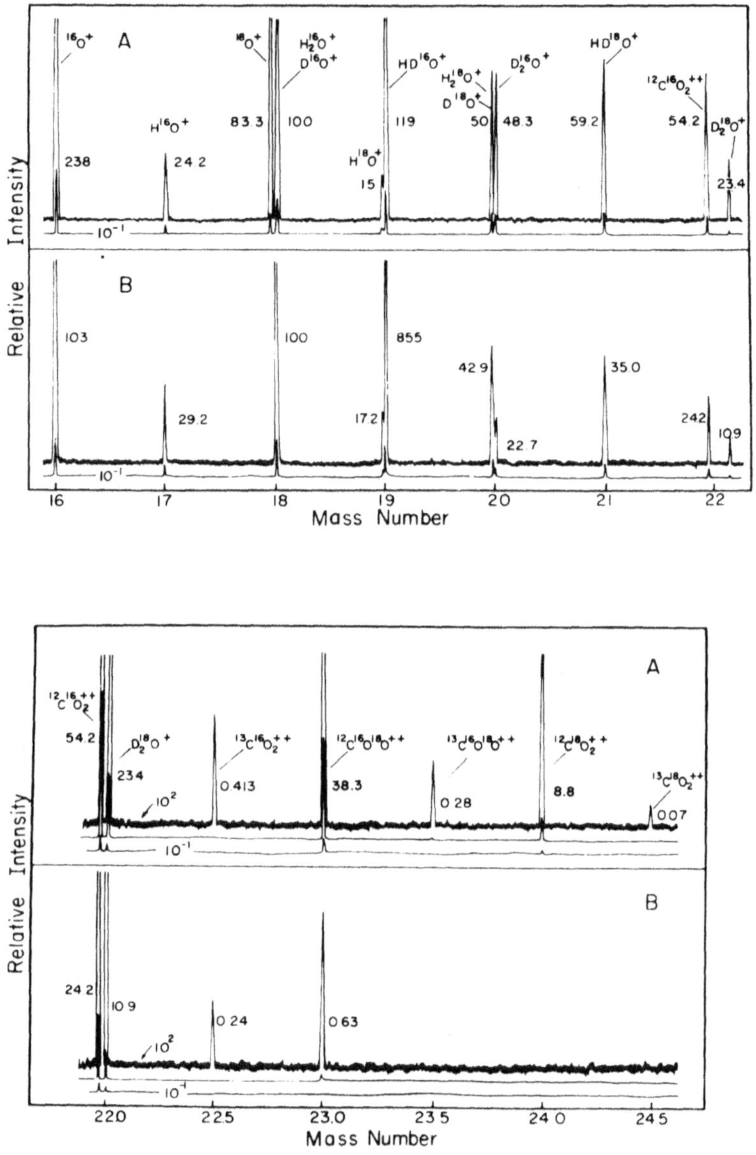

Fig. 3. Isotope exchange effects induced by illumination of Pt/(Chl $\underline{a}.2H_2O)_n$ in an ^{18}O- and D-enriched water mixture having the isotope ratios H/D = 1:0.7 and $^{16}O/^{18}O$ = 1:0.5; A and B respectively denote sample after and before illumination.

Samples of ^{18}O-enriched ($^{16}O/^{18}O = 1.82$) water saturated with unlabeled carbon dioxide were placed in photoelectrolytic cells of the design described by Fruge et al.[19] Table I lists the mass spectrometric intensity ratios $^{18}O^+/^{16}O^+$, $^{12}C^{18}O^+/^{12}C^{16}O^+$ and $^{12}C^{18}O_2{}^+$ measured from liquid (1) and vapor (v) aliquots extracted from the sample cells after 30-min illumination of the CO_2-saturated solutions with filtered visible light, with (Chl *a*) or without (blank) an immersed Pt/(Chl *a*.$2H_2O)_n$ electrode.

Table I. Mass spectrometric intensity ratios of carbon dioxide fragments containing oxygen isotopes.

Sample	$^{18}O^+/^{16}O^+$	$^{12}C^{18}O^+/^{12}C^{16}O^+$	$^{12}C^{18}O_2{}^+/^{12}C^{18}O^{16}O^+$
Chl *a* (v)	0.35	0.27	0.16
Chl *a* (1)	--	0.21	0.19
blank (v)	$\approx$0.0	0.02	0.04
blank (1)	--	0.30	0.17

The observations of isotope exchange effects in Fig. 3 and Table I may be expressed in terms of the equilibria

$$CO_2 + H_2{}^*O \overset{K_1}{\longleftrightarrow} C^*OO + H_2O \tag{13}$$

$$CO^*O + H_2{}^*O \overset{K_2}{\longleftrightarrow} C^*O_2 + H_2O \tag{14}$$

where $^*O$ and O respectively denote ^{18}O and ^{16}O. Given the fact that the equilibrium constants for 13 and 14 are approximately equal, i.e., $K_1 \approx K_2 = K$, we have

$$[C^*OO]/[CO_2] = ([H_2{}^*O]/[H_2O])K \approx (1/1.82)K \tag{15}$$

and

$$[C^*O_2]/[CO_2] \approx (1/1.82)^2K^2 \tag{16}$$

where we have made use of the isotope composition of the water samples. The isotope exchange equilibrium constant may be written

$$K = \exp(-\Delta\Delta G^0/RT) = \exp[-(\Delta G_p^0 - \Delta G_r^0)/RT] \tag{17}$$

where

$$\Delta\Delta G^0 \approx [\hbar\omega(C-{}^*O) - \hbar\omega(C-O)] + [\hbar\omega(H-O) - \hbar\omega(H-{}^*O)] \tag{18}$$

is approximately given as the difference in the vibrational zero-point energies of the bonds involved in the isotope exchange equilibria. Given $\omega(H-O) \approx 3490\ \underline{cm}^{-1}$ for the O-H stretch frequency in liquid water[33] and $\omega(C-O) \approx 2350\ \underline{cm}^{-1}$ for the asymmetric vibration of CO_2,[34] we relate the ratios for the vibrational frequencies to those of the corresponding reduced masses in the usual manner: $\omega(C-{}^*O)/\omega(C-O) \approx [\mu(C-O)/\mu(C-{}^*O)]^{-1/2} = 0.964$ and $\omega(H-{}^*O)/\omega(H-O) \approx [\mu(H-O)/\mu(H-{}^*O)]^{-1/2} = 0.943$, where we have assumed that the reduced mass, $\mu(H-O)$, for the stretching frequency in liquid water is equal to the atomic mass of the oxygen. We thus obtain $\Delta\Delta G^0 \approx 115\ \underline{cm}^{-1}$, yielding the equilibrium constant $\underline{K} = 0.58$ at 298 K.

The apparent equilibrium constant, $\underline{K}_a$, can be readily obtained from the experimental values listed in Table I. Let $\underline{x}_0$, $\underline{x}_1$ and $\underline{x}_2$ be the mole fractions of CO_2, C^*OO and C^*O_2, respectively, so $\underline{x}_1 + \underline{x}_2 + \underline{x}_3 = 1$. We now assume that any isotope effects in the cracking of CO_2 are negligbly small. The intensity ratios for the carbon dioxide fragments are thus obtained:

$$[C^*O^+]/[CO^+] = (\underline{x}_1 + 2\underline{x}_2)/(\underline{x}_1 + 2\underline{x}_0) \approx 0.27(v) \text{ or } 0.21(1) \quad (19)$$

and

$$[C^*O_2{}^+]/[C^*OO^+] = \underline{x}_1/\underline{x}_2 \approx 0.16(v) \text{ or } 0.19(1) \quad (20)$$

The equilibrium composition for a random isotopic distribution and that obtained from the zero-point energy model are compared in Table II with the experimental values obtained from eqs. 19 and 20. This comparison underscores the failure of the random mixing interpretation discussed in connection with the results shown in Fig. 3. The good agreement

Table II. Comparison of equilibrium compositions for random distribution and the model using zero-point energies (z-p-e model) with experimental values (vapor and liquid) calculated from eqs. 19 and 20.

Compositions	Vapor	Liquid	Z-p-e model	Random Mixing
$\underline{x}_0(CO_2)$	0.65	0.69	0.70	0.42
$\underline{x}_1(C^*OO)$	0.29	0.26	0.22	0.46
$\underline{x}_2(C^*O_2)$	0.06	0.05	0.07	0.13
$\underline{K}_a$	0.55	0.49	0.58	1

between the value $\underline{K} = 0.58$ and $\underline{K}_a = 0.55(v)$; 0.49(1) suggests the photocatalytic origin of the observed isotope exchange effects. Moreover, that the mass spectrometric intensity ratios (see Table I) for the liquid blank (1) sample are indistinguishable from those observed for the illuminated vapor and liquid Chl $\underline{a}$ samples (v; 1) is indicative of two important facts: (1) Equilibration between the carbon dioxide oxygen with oxygen from the water is readily attainable, without a photocatalyst, via the bicarbonate mechanism. (2) The photocatalyzed equilibration of CO_2 isotope exchange in the vapor phase must necessarily result from a gas-liquid transport mechanism, because the site of photocatalytic activity is located at the immersed Pt/Chl a electrode. The present finding thus corroborates the earlier conclusion of reduction in the carbon dioxide pressure during the Pt/Chl a light reaction.[19] In an ^{18}O-enriched sample the oxygen evolved from reaction 12 is isotopically scrambled.[19] The oxidation of the formic acid by the oxygen in 12, in a direction opposite to that of the photosynthetic process would result in an effective scrambling of the oxygen atoms in the carbon dioxide. In Section IV evidence is provided for the photocatalytic reduction of molecular oxygen to superoxide radicals in water/acetone solutions of Chl a, which suggests a plausible mechanism for the observed isotope exchange effects.

IV. Chlorophyll a catalyzed superoxide formation from molecular oxygen

Free radicals have been proposed by earlier authors as intermediates in biochemical reactions,[35] and have in fact been detected by electron spin resonance (ESR) measurements. The superoxide ion $O_2^-\cdot$ or its protonated form $HO_2\cdot$ has been postulated as an intermediate in many biological oxidation-reduction reactions.[36-38] In particular it has been established that $O_2^-\cdot$ is photogenerated by spinach chloroplasts in the presence of ascorbate or low potential electron acceptors such as methyl viologen.[38] In a recent communication[20] You and Fong reported the use of DMPO (5,5-dimethyl-1-pyrroline-1-oxide) as spin trap in the ESR observation of $O_2^-\cdot$ photogeneration by *in vitro* Chl a in oxygen-saturated water/acetone. The observed ESR signals are identical to those of DMPO adducts obtained from H_2O_2 decomposition, and are in agreement with those reported for the formation of $O_2^-\cdot$ in oxygen-saturated preparations of spinach chloroplasts.

Chlorophyll a was extracted from spinach and purified using the procedures described by Brace *et al.*[4] Absorption spectra were measured using a Cary 17D spectrophotometer immediately after sample preparation and at regular intervals afterwards to monitor any spectral changes.ESR measurements were carried out with a Varian E-109 Century series spectrometer equipped with Hewlett-Packard E-935 data acquisition system. DMPO purchased from Aldrich Chemicals was used as the spin trap agent.

In Fig. 4A are shown the ESR spectra of photogenerated signals in a 2 X 10^{-4} M Chl a solution in 50:50% (v/v) water/acetone. In the absence of oxygen, no ESR signals were observed either in the dark (a) or in the light (b). After the sample was saturated with oxygen, the sequence of observations were made as follows: The dark-adapted (c) sample produced an ESR signal under illumination with visible light (d). This signal had decayed completely five min after the light was turned off (e). The sample was kept in the dark for a period of 16 h, after which it was again illuminated. An ESR signal indistinguishable from that in d was observed (f). The ESR signal photogenerated in the presence of DMPO is shown in g. In addition to the Chl a radical cation signal DMPO spin trap signals (labelled 1,2,3 and 6) were detected. The g-value and peak width of the ESR signal in d and f are 2.0024 ± 0.0002 and 1.5 G, respectively, which readily identifies that signal as being attributable to the radical cation, $(Chl\ a.2H_2O)_n^{+\cdot}$.[14] The absorption spectra of 5 X 10^{-4} M Chl a solutions corresponding to the ESR results in c-g are reproduced in Fig. 4B. The quantitative conversion of the 675-nm band to the remarkably broad 752-nm band took place within 140 min.

Spin trap measurements were also made on a 4.2 X 10^{-4} M Chl a solution in an O_2-saturated 50:50% water/acetone mixture containing 2.5 x 10^{-1} M DMPO (Fig. 5). A dark-adapted sample (a) on illumination produced a multicomponent ESR signal (b) that qualitatively resembles the signal, g, in Fig. 4A. The signal in b persisted under continuous sample illumination for 60 min, although detailed changes were observed (c) indicative of a redistribution of the radical spins as the sample approached steady-state conditions. The Chl a radical cation signal

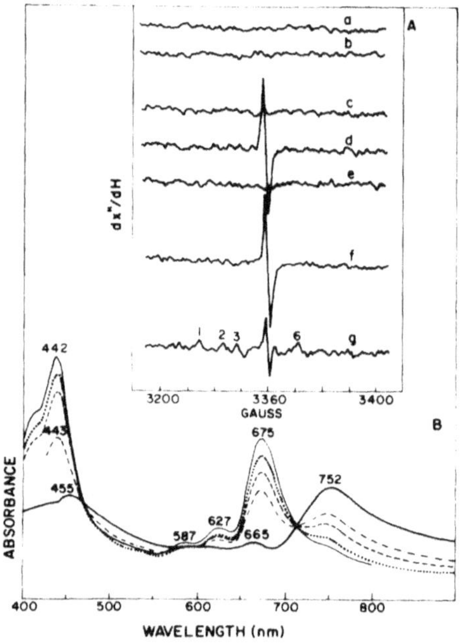

Fig. 4 Absorption and ESR spectra of 2 X 10^{-4} M solutions of Chl a in 50:50 % water/acetone. Effects of oxygen saturation and Chl a aggregate redistribution. Time sequence for conversion of the 675-nm band to the 752-nm: 15, 30, 45, 60 and 140 min after sample preparation.

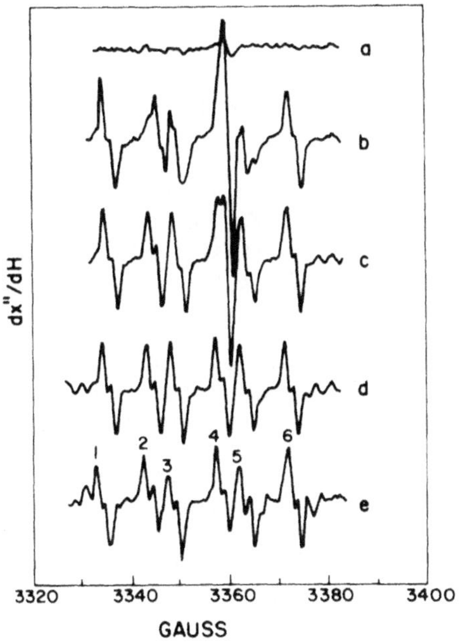

Fig. 5. Comparison of ESR spectra of an illuminated 4.2×10^{-4} M Chl a solution in water/acetone containing 2.5×10^{-1} M DMPO with ESR signals of DMPO spin adducts obtained from hydrogen peroxide decomposition.

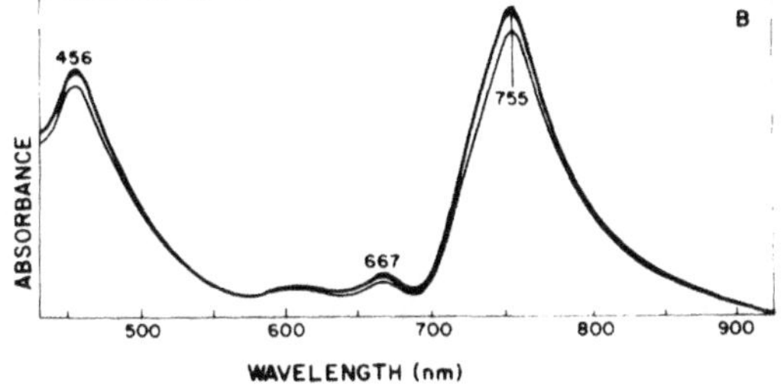

Fig. 6. Lower: absorption spectra of 4.2×10^{-4} M Chl a in water/acetone; decrease in absorbance over a period of 21 h was accompanied by precipitation of polycrystalline (Chl $a.2H_2)_n$. Upper: kinetic response of ESR signal c in Fig. 5.

decayed within 6 min after the light was turned off, leaving the DMPO spin trap signal (d), which is indistinguishable from that (e) obtained from a 2% H_2O_2 solution in 50:50% water/acetone after 4 min of illumination with UV light. The ESR signals in d and e are practically identical to those reported for the formation of O_2^-. in oxygen-saturated preparations of spinach chloroplasts. The typical kinetic response of a light-induced ESR signal from a 5 X 10^{-4} M Chl a solution in O_2-saturated 50:50% water/acetone is illustrated in Fig. 6A. The dark decay has a fast component with a lifetime of about 3 s followed by a slow commponent having a lifetime of several hours. Fig. 6B shows the Cary 17D absorption spectra of the Chl a sample from which the spin trap signal in Fig. 5d was obtained.

The light-dark cycles shown in Figs. 4c-e for the Chl a radical cation ESR signal are indicative of the catalytic origin of the Chl a photoreaction described above. The fact that the amplitude of the ESR signal for $(\text{Chl } a.\ 2H_2O)_n^{2+}$. remains approximately the same in Figs. 4d and f in spite of the observation in Fig. 4B of a quantitative conversion of the 675-nm band to the 752-nm band supports the finding by a fluorescence spectral analysis[39] that the steady-state concentration of $(\text{Chl } a.\ 2H_2O)_n$ remains approximately constant in that conversion.

The labelling (1,2,3 and 6) of the ESR spectrum in Fig. 4g corresponds to that provided for the ESR signals in Figs. 5d and e. All three spectra can be analyzed in terms of the parameters given by Harbour and Bolton:[38] a^N = 14.3 G, a_β^H = 11.7 G, a_γ^H = 1.25 G and g = 2.0061, which were also found earlier to be nearly identical for the HO_2. spin adduct of DMPO.[40] The photogeneration by chloroplasts of ESR signals nearly identical to those shown in Fig. 5d requires the presence of oxygen, as saturating the chloroplast solution with pure oxygen greatly increased the signal amplitude whereas the removal of dissolved oxygen resulted in the elimination of the ESR signal.[36] Likewise, the photogeneration of superoxide by $(\text{Chl } a.2H_2O)_n$ demonstrated in the present investigation was observed only when the in vitro samples were saturated with oxygen, as no ESR signals were detected in the absence of oxygen under otherwise identical conditions (compare Figs. 4b with d,f and g). The close similarities between the experimental behavior of chloroplast preparations and that of the in vitro samples underscore the significance of earlier demonstrations of the photocatalysis of water splitting[18] and carbon dioxide reduction[19] by $(\text{Chl } a.\ 2H_2O)_n$.

The UV radiation of the 2% H_2O_2 solution in Fig. 5e results in splitting the H_2O_2 molecule into two $\cdot OH$ radicals, which then yield $HO_2\cdot/O_2\cdot^-$ according to the reaction sequence:[41]

$$H_2O_2 \longrightarrow 2\cdot OH \tag{21}$$

$$\cdot OH + H_2O_2 \longrightarrow HO_2\cdot + H_2O \tag{22}$$

The observation in Fig. 5d of a Chl a photocatalyzed spin adduct practically identical to that in Fig. 5e suggests a photolytic pathway that generates a nearly identical distribution of radicals:

$$[\text{Chl } \underline{a}.O_2] \xrightarrow{\text{visible light}} [\text{Chl } \underline{a}^{\delta+}\cdot O_2^{\delta-}] \longleftrightarrow \text{Chl } \underline{a}^{+\cdot} + O_2^{-\cdot} \quad (23)$$

$$O_2^{-\cdot} + H_2O \longrightarrow HO_2^{\cdot} + OH^- \quad (24)$$

$$OH^- + [\text{Chl } \underline{a}^{\delta+}.O_2^{\delta-}] \longrightarrow [\text{Chl } \underline{a}\cdot O_2^-] + \cdot OH \quad (25)$$

$$H_2O + H^+ + [\text{Chl } \underline{a}\cdot O_2^-] \longrightarrow \text{Chl } \underline{a} + H_2O_2 + \cdot OH \quad (26)$$

Reactions 23-26, in which (Chl a.$2H_2O)_n$ is written simply as Chl a, account for the collective presence of $\cdot OH$, $HO_2\cdot/O_2\cdot^-$ and H_2O_2 in reactions 21 and 22. The radical cation Chl $a^{+\cdot}$ in reaction 22 accounts for the g = 2.0024 ESR signal in Figs. 4 and 5. The regeneration of Chl a in reaction 26 is compatible with the catalytic Chl a cycle observed in Figs. 4c-e. Reactions 25 and 26 corroborate earlier conclusions by Harbour and Bolton[42] of hydroxyl radical formation by Chl a light reaction in the presence of molecular oxygen. The principal difference between the present results and their findings is that, whereas the Chl a samples were destroyed in the earlier work by the formation of $\cdot OH$ consistent with optical results, the Chl a prepared in the present investigation as (Chl a.$2H_2O)_n$ is photochemically stable in the oxygen photoreductive cycle given by reactions.[23-26]

Behar et al[43] reported the following reaction sequence in formate solutions containing oxygen:

$$\cdot OH + HCOO^- \longrightarrow H_2O + CO_2^- \quad (27)$$

$$CO_2^- + O_2 \longrightarrow O_2^{\bar{\cdot}} + CO_2 \quad (28)$$

The formation of $\cdot OH$ in reactions 23-26 thus provides a plausible mechanism for the isotope exchange effects described in Section III. The involvement of the hydrogen transfer in reaction 27 as a rate determining step could conceivably be the reason for the observation of only deuterated reduction products in Fig. 2. Overcoming the deleterious effects of molecular oxygen thus described appears to hold the key to future work on *in vitro* photosynthesis.

References

1. F. K. Fong, in "Light Reaction Path of Photosynthesis", edited by F. K. Fong, Springer, New York, 1982, Chap. 8.
2. A. S. Holt and E. E. Jacobs, Am. J. Bot. 41, 710 (1954).
3. N. Winograd, A. Shepard, D. H. Karweik. V. Koester, and F. K. Fong, J. Am. Chem. Soc. 98, 2369 (1976).
4. J. G. Brace, F. K. Fong, D. H. Karweik, V. J. Koester, A. Shepard, and N. Winograd, J. Am. Chem. Soc. 100, 5203 (1978).
5. H. Blaha, P. Kis, and H. Springer-Lederer, Anal. Biochem. 112, 282 (1981).
6. F. K. Fong, V. J. Koester, L. Galloway, J. Am. Chem. Soc. 99, 2372 (1977).
7. H. J. Den Blanken and A. J. Hoff, Biochim. Biophys. Acta. 724, 52 (1983).
8. F. K. Fong, Proc. Natl. Acad. Sci. U. S. A. 71, 3692 (1974).
9. F. K. Fong and V. J. Koester, Biochim. Biophys. Acta 423, 52 (1976).
10. M. Hoshino, M. Imamura, K. Koike, K. Kikuchi, and H. Kokubun, Photochem. Photobiol. 38, 255 (1983).
11. M. Hoshino, K. Ikehara, M. Imamura, H. Seki, and Y. Hama, Photochem. Photobiol. 34, 75 (1981).
12. F. K. Fong, M. S. Showell, and A. J. Alfano, J. Am. Chem. Soc. 107, 7231 (1985).
13. H.-C. Chow, R. Serlin, and C. E. Strouse, J. Am. Chem. Soc. 97, 7230 (1975).
14. F. K. Fong, M. Kusunoki, L. Galloway, T. G. Mathews, F. E. Lytle, A. J. Hoff, and F. A. Brinkman, J. Am. Chem. Soc. 104, 2759 (1982).
15. A. J. Alfano and F. K. Fong, J. Am. Chem. Soc. 104, 2767 (1982).
16. M. S. Showell and F. K. Fong, J. Am. Chem. Soc. 104, 2773 (1982).
17. O. Inganas and I. Lundstrom, J. Appl. Phys. 54, 4185 (1983).
18. F. K. Fong and L. Galloway, J. Am. Chem. Soc. 100, 3594 (1978).
19. D. R. Fruge, G. D. Fong, and F. K. Fong, J. Am. Chem. Soc. 101, 3694 (1979).
20. J.-L. You and F. K. Fong, Biochem. Biophys. Res. Commun., in press.
21. H. J. Den Blanken and A. J. Hoff, FEBS Lett. 157, 21 (1983).
22. J.A. Bassham and M. Calvin, "The Path of Carbon in Photosynthesis," Prentice-Hall, Inc., pp. 16-70 (1957).
23. J.A. Bassham and M. Kirk, Biochim. Biophys. Acta, 43, 447 (1960).
24. H.S. Johnson and M.D. Hatch, Biochem. J., 114, 127 (1969).
25. J.M. Galmiche, Plant Physiol., 51, 512 (1973).
26. R. Wurmser, "Oxidations et Reductions", Presses Universitaires de France, Paris, 1930.
27. K. Shibata, "Carbon and Nitrogen Assimilation", translated by H. Gest and R.K. Togasaki, Japan Science Press, Tokyo, in 1975, 1931.
28. C.B. Van Niel, Arch. Mikrobiol., 3, 1(1931).
29. C.B. Van Niel, Cold Spring Harbor Symp. Quant. Biol., 3, 138 (1935).
30. R. Hill and R. Scarisbrick, Nature (London), 146, 61 (1940).

31. R. Emerson and C.M. Lewis, Am. J. Botany. 28, 789 (1941).
32. S. Ruben, M. Randall, M. Kamen and J. Hyde, J. Am. Chem. Soc. 63, 877 (1941).
33. J. G. Bayly, V. B. Kartha, and W. H. Stevens, Infrared Phy. 3, 211 (1963); D. Eisenberg and W. Kauzman, "The Structure and Properties of Water," Clarendon Press, Oxford (1969), pp. 228-246.
34. R. M. Silverstein, G. Clayton Bassler, and T. C. Morrill, "Spectrometric Identification of Organic Compounds," Fourth Ed., John Wiley and Sons, New York (1981), p.96.
35. L. Michaelis, in "Currents in Biochemical Research," D. Green ed., Wiley-Interscience, New York (1946).
36. B.L. Epel and J. Neumann, Biochim. Biophys. Acta 325, 520 (1973).
37. J.F. Allen and D.O. Hall, Biochem. Biophys. Res. Commun. 52, 886 (1973).
38. J.R. Harbour and J.R. Bolton, Biochem. Biophys. Res. Commun. 64, 803 (1975).
39. F.K. Fong, A. Agostiano and K.A. Butcher, Biochim. Biophys. Acta, submitted.
40. J.R. Harbour, V. Chew and J.R. Bolton, Photochem. Photobiol. 20, 263 (1974).
41. L. W. Oberley, Ed., "Superoxide Dismutase," Vol. I, CRC Press, Bocaf, Florida (1982), p.128.
42. J. R. Harbour and J. R. Bolton, Photochem. and Photobiol. 28, 231 (1978).
43. D. Behar, G. Czapski, J. Rabani, L. M. Dorfman, and H. A. Schwarz, J. Phys. Chem. 74, 3209 (1972).

PHOTO- AND ELECTROCHEMICAL REDUCTION OF CARBON DIOXIDE

Raymond ZIESSEL
Institut Le Bel, Université Louis Pasteur,
4, rue Blaise Pascal, 67000 STRASBOURG
France

Abstract. - Efficient photogeneration of either CO and H_2 simultaneously or CO and $HCOO^-$ selectively, occurs on visible light irradiation of an organic solution (solvent + tertiary amine) containing, respectively, a $Ru(bpy)_3^{2+}$, Co^{2+} mixture, fac-$Re(bpy)(CO)_3Cl$ or $Ru(bpy)_3^{2+}$ as homogeneous catalyst. Homogeneous photochemical hydrogen generation and water reduction is mediated by cobaloxime and $Ru(bpy)_3^{2+}$. Selective and sustained electrocatalytic reduction of CO_2 to CO is performed by fac-$Re(bpy)(CO)_3Cl$, at low potential. Results are summarized and discussed with respect to the catalytic systems involved.

Photoinduced splitting of water and reduction of carbon dioxide are the basic reactions of natural photosynthesis (Figure 1). Plants with two photosystems (PS I and PS II) reduce carbon dioxide into chemicals and oxidize water to oxygen. In some cases water or nitrogen reduction takes place in the presence of specific enzymes (hydrogenase or nitrogenase). The electrons needed for these reductions come from the water molecule and travel through a complicated chain of electron carriers. Most of these carriers have been identified and characterized (plastoquinone, cytochrome, plastocyanine, ferredoxin, ...). Nature has resolved the problem of charge separation with this complicated but very efficient system.

Carbon dioxide is an abundant source of raw material in nature and a great deal of attention has been recently given to systems capable of activating it (**1**,**2**). Photochemical activation and transformation of H_2O and CO_2 requires the development of catalytic systems coupled to light absorption and there is much interest in processes which would perform these reactions in vitro. The development of such systems is of interest from three points of view: (i) finding catalysts for water and carbon dioxide activation; (ii) devising means for conversion and storage of solar energy by production of fuels and organic raw materials; (iii) setting up models of natural photosynthesis. Compared to water photolysis (**3**,**4**) little research has been devoted to the photoreduction of carbon dioxide. Few catalytic systems are known, despite the

M. Aresta and G. Forti (eds.), Carbon Dioxide as a Source of Carbon, 113–138.

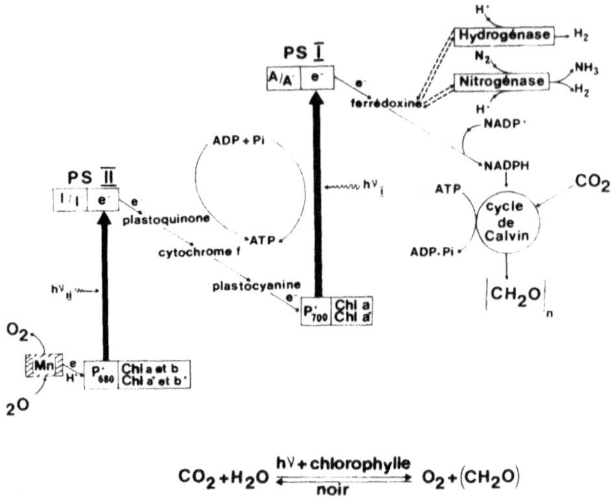

Figure 1. Schematic representation of the redox reactions and electron transfer pathway in natural photosynthesis.

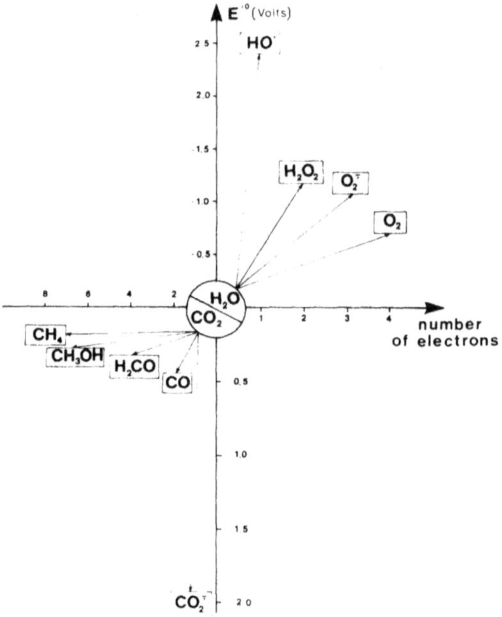

Figure 2. Redox potential of the CO_2 reduction and water oxidation versus the number of electrons involved. Potentials are given versus NHE at pH = 7.

development of a number of different strategies (electro-, photoelectro- and photochemistry). Electrochemical systems use an electrode to either directly reduce carbon dioxide or to reduce a transition metal complex which further reacts with CO_2. By varying the nature of the cathode or the solvent, the following results have been reported: high yields of formic acid (**5-9**), appreciable formation of oxalic acid (**7,9-13**) or carbon monoxide (**7,11,12,14**), low yields of methanol and methane (**15**), and traces of glyoxylate, glycolate and tartrate (**7**). In some electrocatalytic systems, electrodes were modified with an insoluble transition metal complex such as cobalt or nickel phthalocyanine (**16**), with a viologen-based polymer into which metallic palladium is dispersed (**17**), or with a polymerized rhenium complex (**18**). Other materials used as homogeneous catalysts in electrocatalytic systems include soluble phthalocyanine (**19**) or porphyrin complexes (**20,21**), iron-sulfur clusters (**22**), nickel and cobalt complexes of macrocyclic polyamines (**23,24**), as well as rhenium (**25**) or rhodium (**26**) complexes. High current efficiency and long term stability (high overall turnover) were clearly obtained using fac-Re(bpy)$(CO)_3$Cl (**25**), or nickel(II)-cyclam (**23b**). In both cases the average current efficiency is 98 %. Most of the photoelectrochemical systems reported are purely heterogeneous and use single-crystal semi-conductor electrodes (**27**). In only two cases has a semi-conductor been used in association with an electron relay; the first contains a nickel or a cobalt-tetraazamacrocyclic complex (**28**) and in the second, methylviologen is used together with a formate dehydrogenase enzyme (**29**). Numerous studies concern the direct reduction of CO_2 in semi-conductor based photoelectrochemical cells such as: p- or n-GaP (**30,31**), p- or n-GaAs (**15a,15b,32**), p-CdTe (**33**), p-InP (**15a,34**), p-Si (**28,34**). Most of these processes are inefficient and frequently unstable (**31**). Stoichiometric quantities of methane have been obtained by irradiation of a clean single crystal of strontium titanate in the presence of water and carbon dioxide (**35**). The purely photochemical systems may be classified as either heterogeneous, involving the use of semi-conductor suspensions (**36,37**), or as homogeneous, involving the use of aqueous solutions of metal ions (**38**), organic dyes (**39**) or of transition metal complexes (**40-42**).

The task of achieving artificial water photosplitting and carbon dioxide photoreduction is to discover, on the one hand catalytic reactions for their activation and on the other light absorbing substances undergoing photoinduced electron transfer processes. The thermodynamically easiest water oxidation as well as carbon dioxide reduction reactions occur through polyelectronic pathways (Figure 2). It is clearly more difficult to reduce carbon dioxide with a monoelectronic reaction (eq. 1), than with dielectronic reactions (eqs. 2 and 3).

$$CO_2 + 1\,e^- \longrightarrow CO_2^{\overline{\cdot}} \qquad E'^\circ = -2\ V \qquad (1)$$

$$CO_2 + 2\,H^+ + 2\,e^- \longrightarrow CO + H_2O \qquad E'^\circ = -0{,}52\ V \qquad (2)$$

$$CO_2 + 2\,H^+ + 2\,e^- \longrightarrow HCOOH \qquad E'^\circ = -0{,}61\ V \qquad (3)$$

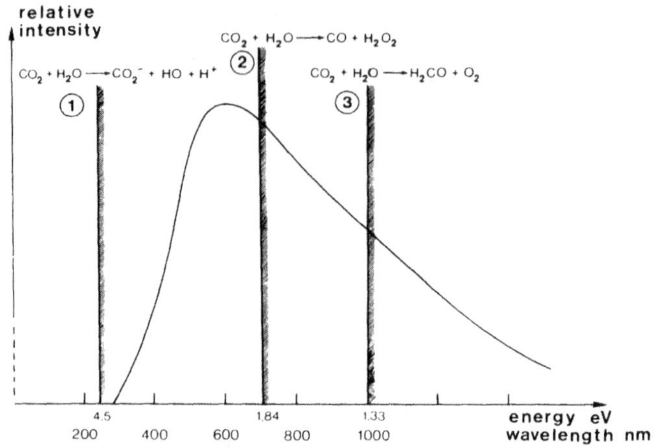

Figure 3. Schematic representation of the relative energy distribution of the solar spectrum, versus energy. Energy and wavelength needed for water oxidation and carbon dioxide reduction: monoelectronic (reaction 1), dielectronic (reaction 2) or tetraelectronic (reaction 3) processes.

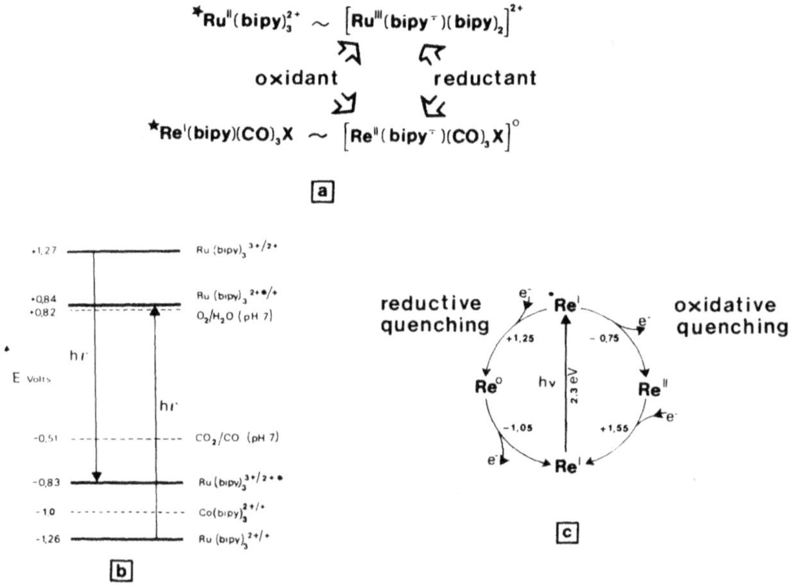

Figure 4. Ground and excited state redox potentials of the $Ru(bpy)_3^{2+}$ and $Re(phen)(CO)_3Cl$ complexes versus NHE, respectively, in H_2O and in CH_3CN. Excitation energy: 2.1 eV for *Ru and 2.3 eV for *Re.

A general behaviour observed is: as the number of electrons involved in the process increases, the energy needed decreases. The monoelectronic reactions need a lot of energy usually absent from the solar spectrum (4.5 eV for the overall reaction 1, Figure 3). This is also the reason why one wishes to perform the photochemical water splitting and carbon dioxide reduction in the visible part of the solar spectrum (1.33 eV for the global reaction 3, Figure 3). 40 to 45 % of the solar energy could be theoretically stored by the use of such polyelectronic reactions (**43,44**). However, water and carbon dioxide do not absorb light above 200 nm. It is therefore necessary to use a photosensitizer and organometallic complexes which are able to transfer more than one electron, e.g. complexes of rhodium(I), cobalt(I), rhenium(0) or ruthenium(I). These species could be oxidized to a higher oxidation state, (III) for Rh, Co and Ru and (II) for Re, by reaction with a small molecule like water or carbon dioxide. This poses problems: (i) of compatibility of redox potentials between the photosensitizers and the catalysts; (ii) of finding mediators and processes capable of performing the polyelectronic reactions concerned and (iii) of setting-up the system in such a way that forward-reactions overcome back-reactions. This led us to take a stepwise approach in which the electron source is unfortunately not water, but a tertiary amine which provides the electron for the reduction. During this process the amine is irreversibly oxidized and thus consumed. This allows the accumulation in solution of the reduced species (vide infra).

We have developed three independant systems for light-driven catalytic reduction of carbon dioxide. In the first system which contains $Ru(bpy)_3^{2+}$ as photosensitizer, (bpy = 2,2'-bipyridine) and, $Co(bpy)_3^{2+}$ or Co(II) ions, as electron mediator and as water and carbon dioxide reduction catalyst, catalytic amounts of both hydrogen and carbon monoxide are produced (**40**). In the second system based on fac-$Re(bpy)(CO)_3Cl$, high yields of CO are obtained photochemically (**41**) and electrochemically (**25**). In the third system which is based on ruthenium complexes, catalytic amounts of formate are generated (**42**).

Properties of the photosensitizers used during these studies (Figure 4).

$Ru(bpy)_3^{2+}$ and fac-$Re(bpy)(CO)_3Cl$ might be considered as a reductant or as an oxidant in the excited state obtained by a metal-to-ligand-charge-transfer. It is interesting to note that $Ru(bpy)_3^{+}$ obtained by reductive quenching of the excited state is a strong reductant capable of reducing $Co(bpy)_3^{2+}$ to the cobalt(I) complex (Figure 4b). Moreover $Ru(bpy)_3^{3+}$ obtained by oxidative quenching of the excited state is a powerful oxidant able to oxidize water into oxygen in the presence of a heterogeneous catalyst like ruthenium dioxide (**46**). These two photosensitizers are interesting for many reasons, three of these being: - their excited state lifetimes are long enough (620 nsec for $^*Ru(bpy)_3^{2+}$ in H_2O at 298 K (**47**) and 600 nsec for $^*Re(bpy)(CO)_3Cl$ in CH_2Cl_2 at 298 K (**48**)) to perform electron transfer to another species before thermal deactivation occurs - their redox potentials could be adjusted by changing the ligand - their chemical and photochemical stabilities are high, so that catalytic systems could be developed.

I PHOTOCHEMICAL REDUCTION OF CO_2 AND H_2O TO CO AND H_2 CATALYZED BY RUTHENIUM AND COBALT COMPLEXES.

Visible light irradiation of an organic solution (solvent + tertiary amine) of $Ru(bpy)_3^{2+}$, $Co(bpy)_3^{2+}$ or Co^{2+} ions and carbon dioxide, led to the simultaneous generation of catalytic amounts of CO and H_2 (analysed by gas-chromatography). No CO is produced in the absence of light, tertiary amine, CO_2 or photosensitizer. When labelled $^{13}CO_2$ was used in some experiments, the results clearly showed, by GC-MS analysis, that all the CO produced was labelled and came from the CO_2 introduced (Figure 5). The peak at mass 29 (^{13}CO) is clearly the major product when $^{13}CO_2$ was used. When unlabelled CO_2 was used, the peaks at mass 29 correspond to ^{15}NN and ^{13}CO present in natural abundance. This is the first photochemical system which catalytically reduces carbon dioxide(**40**).

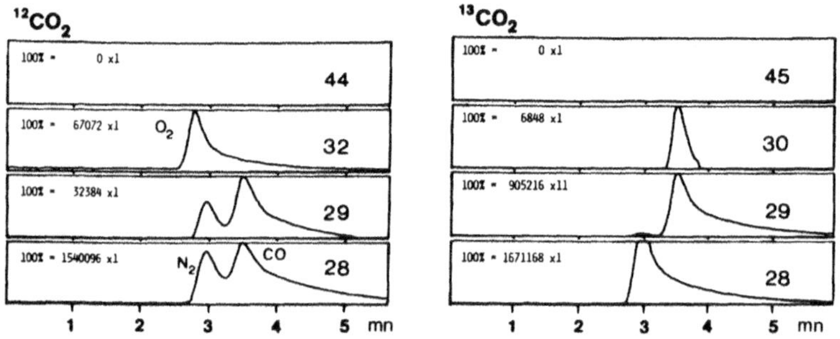

Figure 5. GC/MS analysis of photogenerated CO. Gas obtained from photoreduction of $^{12}CO_2$ (left) and $^{13}CO_2$ (right); the mass peaks are recorded at very different sensitivities, indicated by the figure for 100 % (= height of strongest peak) at left and decreasing from top to bottom (e.g. CO at mass 29 on left is natural abundance ^{13}CO); abscissa GC retention time in min.

Kinetic experiments under the same conditions as described above showed a linear increase in the amounts of H_2 and CO generated, as a function of time (Figure 6). The selectivity (CO/H_2 ratio) was clearly in favour of CO in the absence of 2,2'-bipyridine. Figure 7 shows the dependence of CO and H_2 generation on the number of equivalents of bpy present in the medium. The strongest effect on CO generation was observed on addition of 1 equiv. of bpy, with respect to Co^{2+} ions. Extensive modifications of the process were performed by varying the nature of the tertiary amine (alkylamines or hydroxylated amines), the solvent, the ligand for the Co(II) ions, the photosensitizer and the temperature. These changes led to marked effects on both the efficiency of (CO + H_2) formation and on the selectivity (CO/H_2). Complexing Co^{2+} ions with 2,9-dimethyl-1,10-phenanthroline gave an overall quantum yield of 7.7 % (1.2 % for CO and 6.5 % for H_2)(**40b**).

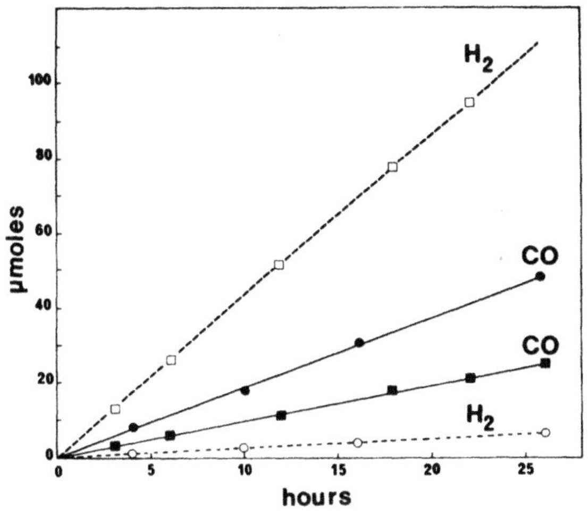

Figure 6. Photogeneration of CO and H_2 as a function of time by photoreduction of CO_2 and H_2O. Soln. contained $MeCN/NEt_3$ or $TEOA/H_2O$ 3:1:1, $Ru(bpy)_3^{2+}$ and $Co(bpy)_3^{2+}$ (■ , □) or Co^{2+} (● , ○).

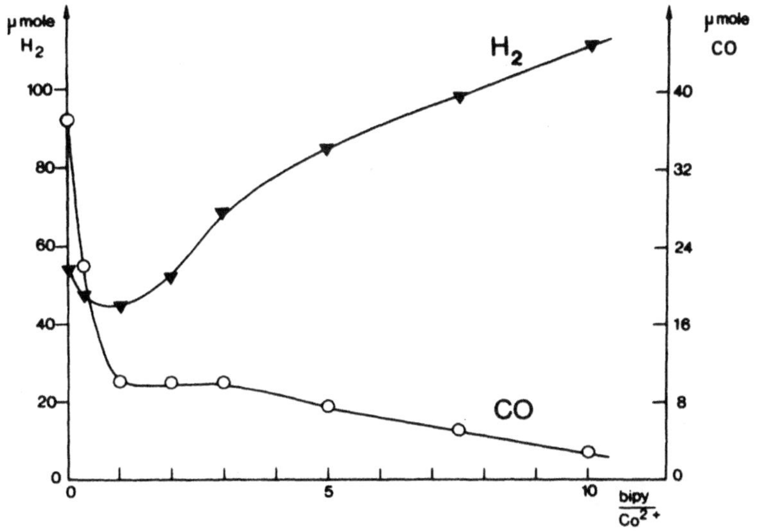

Figure 7. Effect of addition of increasing amounts of bpy (equivalents with respect to Co^{2+} ions) on CO and H_2 photogeneration. Soln. contained $Ru(bpy)_3^{2+}$, $CoCl_2.6\ H_2O$, $MeCN/NEt_3/H_2O$ 3:1:1.

The mechanism of this homogeneous process is schematically represented in Figure 8. Two major cycles could be distinguished: the ruthenium one and the cobalt one.

In the Ru cycle, light is used to generate a reduced cobalt complex. Two processes of quenching of the $Ru(bpy)_3^{2+}$ excited state might be envisaged: (i) reductive quenching by the tertiary amine yielded $Ru(bpy)_3^+$ which, by further thermal reaction, gave $Co(bpy)_n^+$ and the starting photosensitizer; (ii) oxidative quenching by $Co(bpy)_3^{2+}$ formed directly the cobalt(I) species and $Ru(bpy)_3^{3+}$ which was reduced back to $Ru(bpy)_3^{2+}$ by the donor (e.g. triethanolamine; symbolized in the text by TEOA). Conceivably, both mechanisms may participate in the photochemical process (**49**).

In the second cycle, the cobalt(I) complex acts as a homogeneous catalyst for the concurrent reductions of CO_2 or H_2O to CO or H_2 respectively. The Co(I) species could be generated either chemically, electrochemically or photochemically (Figure 9).

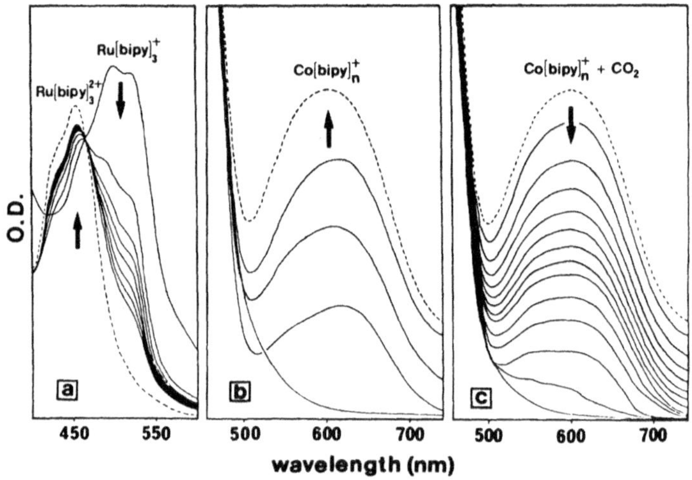

Figure 9. Spectroscopic observation of the redox changes leading to generation of CO by reduction of CO_2. a) Formation of $Ru(bpy)_3^{2+}$ on addition of $Co(bpy)_3^{2+}$ to a soln. of electrochemically generated $Ru(bpy)_3^+$ in DMF; the broken line is the absorption of $Ru(bpy)_3^{2+}$ and was obtained after addition of 1.2 equiv. of $Co(bpy)_3^{2+}$ with respect to $Ru(bpy)_3^+$. b) Observation of $Co(bpy)_n^+$ formation corresponding to the changes shown in a), but at higher concentration and higher sensitivity of the spectrometer. c) Disappearance of the $Co(bpy)_n^+$ band in presence of CO_2; the reduced Co complex was produced photochemically by irradiation of a soln. of $Ru(bpy)_3^{2+}$, $Co(bpy)_3^{2+}$and TEOA in DMF in absence of CO_2.

When $Co(bpy)_3^{2+}$ was gradually added to a solution of $Ru(bpy)_3^+$ prepared electrocnemically, a clean transformation occured. Two new products are formed, $Ru(bpy)_3^{2+}$ (Figure 9a), and $Co(bpy)_n^+$ (Figure 9b).

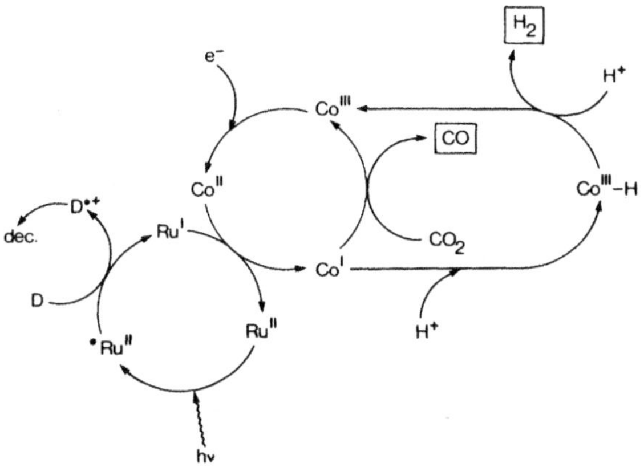

Figure 8. Schematic representation of the processes occuring in the simultaneous and competing generation of CO and H_2 by photoinduced reduction of CO_2 and H_2O using the system $Ru(bpy)_3^{2+}$/Co(II)/donor (D) and following a pathway of reductive quenching of the excited state of the ruthenium photosensitizer. Only heterolytic formation of H_2 by protonation of a cobalt-hydride is shown; see also text; the ligands of the metal ions are not indicated.

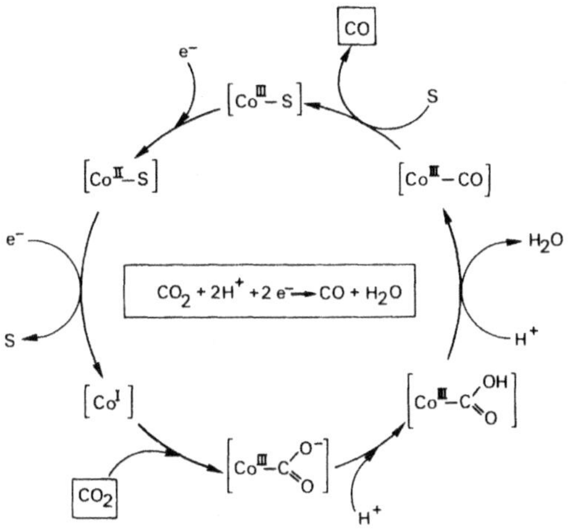

Figure 10. Schematic representation of a possible catalytic Co cycle for reduction of CO_2 to CO via direct reaction of Co(I) with CO_2 to give a Co(III)-carboxylate intermediate. The electrons are provided by the added electron donor via the photochemical Ru cycle; the dielectronic CO_2 to CO reduction shown as insert has a potential of - 0.52 V and is, therefore, thermodynamically feasible via Co(I) complexes; S = solvent or other ligand molecule.

$Co(bpy)_n^+$ could also be generated directly by controlled potential electrolysis of the $Co(bpy)_3^{2+}$ (the same absorption spectrum as in Figure 9b was observed). We also found that photolysis of an organic solution (DMF + TEOA) containing $Ru(bpy)_3^{2+}$ and $Co(bpy)_3^{2+}$, in the absence of CO_2, lead to the formation of the same $Co(bpy)_n^+$ complex (Figure 9c). This reduced species reacted with CO_2 and H_2O to yield 55 % of CO and H_2 with respect to the Ru(I) complex. These experiments clearly demonstrate that light was used only to generate the reduced complex and does not intervene in the CO_2 activation process. The details of the CO_2 reduction process as well as a more precise formulation of the intermediate are not known.

We proposed that the cobalt(I) species reacts as a nucleophile on the electrophilic carbon center of the CO_2 molecule, with formation of a cobalt(III)-carboxylate (Figure 10). Double protonation of this intermediate gave water and a labile cobalt(III)-carbonyl species. Displacement of coordinated CO by a solvent or a ligand molecule gave a Co(III) intermediate, which was reduced back to the Co(II) starting complex with an extra electron (might come from $D^{+\cdot}$, Ru(I) or *Ru(II)). The formation of a Co(III)-carboxylic acid has also been postulated by Bercaw, Goh and Halpern during their studies on the catalytic oxidation of CO catalysed by cobalt(I)-carbonyl complexes (**50**). The dependence of the CO/H_2 ratio on the nature or amount of cobalt ligand and on the nature of the tertiary amine (in the absence of polypyridine ligands), indicates that either, or both, occupation of the coordination sites around the cobalt ion or the change in redox potential due to ligand coordination play an important role in the CO_2 reduction process.

Another example of the importance of the cobalt ligand is illustrated below. When cobalt(II) ions are complexed with dimethylglyoxime, in place of bipyridine or phenanthroline, and photolyzed in the same conditions as before (organic solvent + tertiary amine + dissolved CO_2), carbon dioxide reduction is totally inhibited while H_2 generation is greatly enhanced (**51**). The mechanism of the system is schematically represented in Figure 11. 16 % conversion of light energy into H_2 has been determined by comparaison with an analogous system described in the literature (**52**). This system is of importance for two reasons: first it is homogeneous, no dispersed noble metal was needed for H_2 generation (**3**,**4**); second it is one of the most active photochemical hydrogen generation processes known to date. Addition of triphenylphosphine slowed down the photochemical generation of H_2 by a factor of about 10, while a blue species accumulated in the medium (Figure 12). In the dark, this reduced complex disappeared within minutes, with formation of a high yield of H_2. These data are in good agreement with those of other workers and are consistent with hydride formation (eq. 4)(**53**).

$$[Co^{I}(dmgH)_2]^- + H^+ + S \rightleftharpoons HCo^{III}(dmgH)_2(S) \qquad (4)$$

S = solvent molecule or another axial ligand

The mechanism of hydrogen evolution from hydridocobaloxime has been shown to proceed by parallel heterolytic (eq. 5) and homolytic (eq. 6) cleavage of the hydrogen-cobalt bond (**54**).

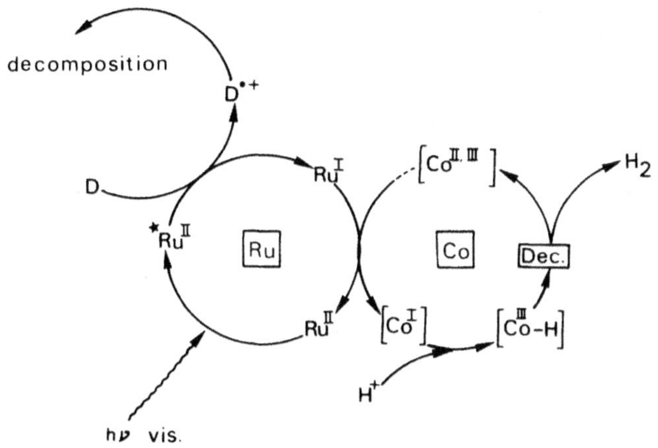

Figure 11. Schematic representation of photochemical hydrogen generation by the system $Ru(bpy)_3^{2+}/Co(dmgH)_2/TEOA$. The ligands of the metal ions have been omitted for clarity; D = TEOA.

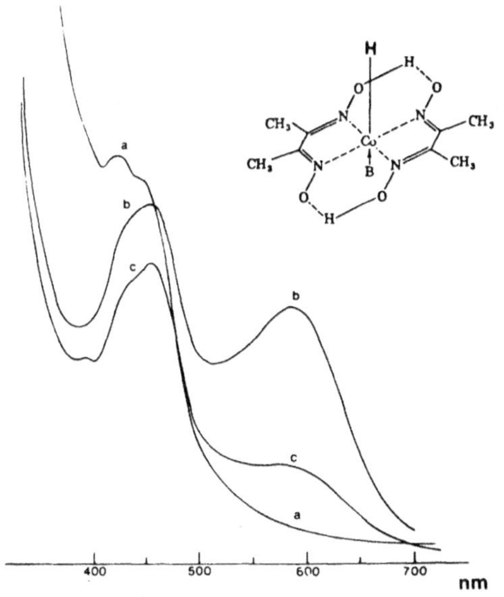

Figure 12. Electronic spectra of a solution (DMF/TEOA 2/1) containing $Ru(bpy)_3^{2+}$ (0.47 mM), $Co(dmgH)_2$ (1.1 mM) and 10 mg of PPh_3 : (a) before irradiation; (b) after irradiation; (c) after 30 min. in the dark. The absorption observed at 585 nm is compatible with cobalt-hydride formation.

$$HCo^{III}(dmgH)_2(S) + H_3O^+ \longrightarrow [(H_2O)Co^{III}(dmgH)_2(S)]^+ + H_2 \quad (5)$$

$$2\ HCo^{III}(dmgH)_2(S) \longrightarrow 2\ Co^{II}(dmgH)_2 + H_2 + S \quad (6)$$

Homogeneous hydrogen generation (dielectronic reaction) occured via the formation and protonation of a soluble metal hydride complex. With TEOA as electron donor, the net light-driven reactions occuring in the systems described here, are given in eqs. (7) and (8). The catalysts are $Ru(bpy)_3^{2+}/Co^{2+}$ for the formation of CO + H_2 and, $Co(dmgH)_2$ for the specific formation of H_2.

$$N(CH_2CH_2OH)_3 + CO_2 \longrightarrow HN(CH_2CH_2OH)_2 + HCO\text{-}CH_2OH + CO \quad (7)$$

$$N(CH_2CH_2OH)_3 + H_2O \longrightarrow HN(CH_2CH_2OH)_2 + HCO\text{-}CH_2OH + H_2 \quad (8)$$

II PHOTO- AND ELECTROREDUCTION OF CO_2 MEDIATED BY Fac-Re(bpy)$(CO)_3$X (X = Br, Cl, OOCH, H) COMPLEXES, AS HOMOGENEOUS CATALYSTS.

A more efficient and selective system (no H_2 was produced) has been discovered by the use of Re(bpy)$(CO)_3$Cl. The process is remarkable by the fact that the complex used acts as both photosensitizer and, as the CO_2 reduction catalyst in its reduced state (**41**). Kinetic studies on the photochemical CO_2 reduction shows (Figure 13) a linear relationship between formation of CO and time when excess chloride is present (14 % quantum yield has been determined (**55,56**)). In the absence of this excess Cl^-, the CO formation decreased (8.7 % conversion of light energy into CO) and was not linear versus time.

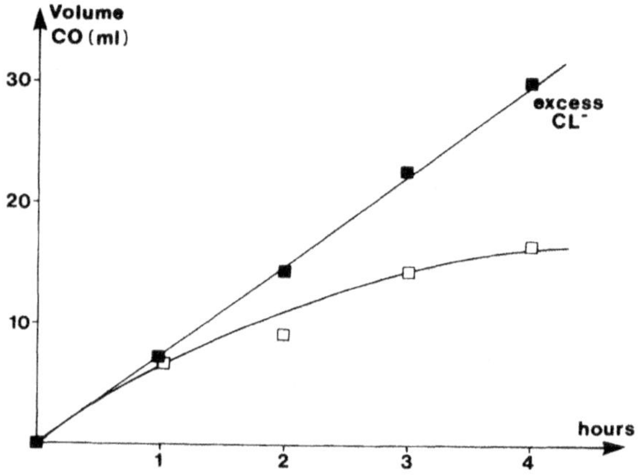

Figure 13. Photogeneration of CO from CO_2, as a function of time, using Re(bpy)$(CO)_3$Cl as catalyst in the absence (□), or presence (■) of excess chloride.

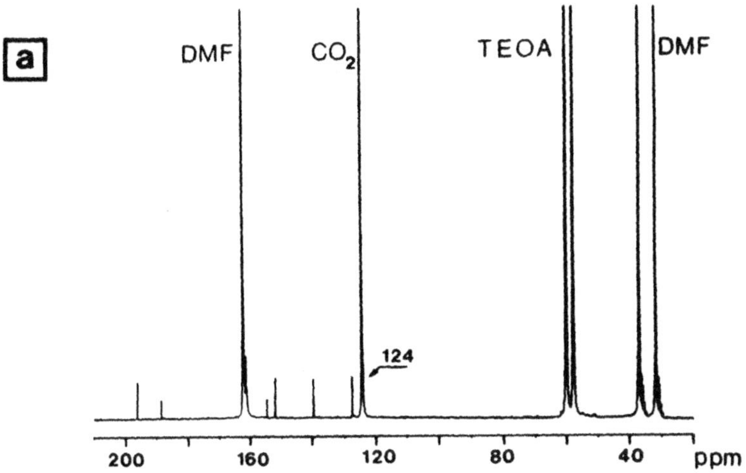

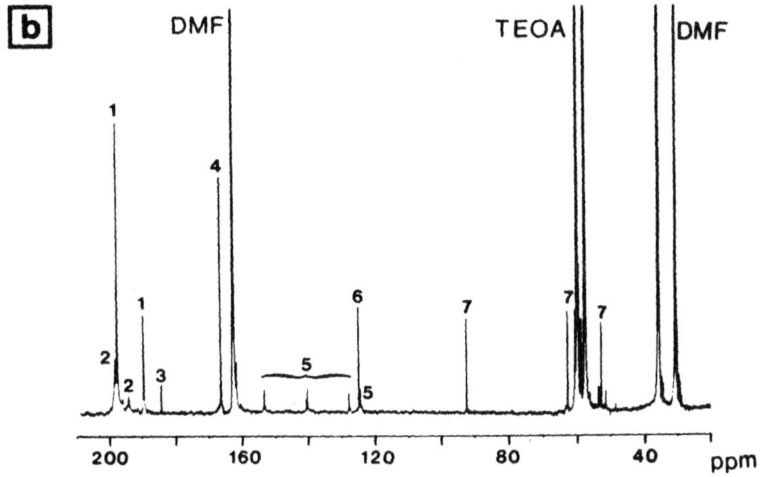

Figure 14. ^{13}C-NMR spectra of a solution containing DMF, TEOA, $^{13}CO_2$ and $Re(bpy)(CO)_3Cl$:
a) in the dark;
b) after 4 hours visible light irradiation.

In these conditions, the photochemical formation of a new complex has been observed in a sealed ^{13}C NMR tube, under an atmosphere of labelled CO_2.Several changes were observed in the spectrum during photolysis (Figure 14). We noted a net decrease in the intensity of peak (n° 6) corresponding to free $^{13}CO_2$ (δ = 125 ppm) and the simultaneous apparition of a new peak (n° 3) due to free ^{13}CO (δ = 184 ppm) which compares with authentic samples. This clearly confirms that the carbon monoxide originates from the photochemical reduction of carbon dioxide. The relative intensities of the rhenium-carbonyl peaks (n° 1 at δ 197 and 189 ppm) are enhanced in comparison with bipyridine signals. All the rhenium carbonyls are exchanged during the photochemical reduction of CO_2. New carbonyl peaks (n° 2 at δ 198 and 194 ppm) and splitting of the peaks (n° 5 at δ 153, 140, 128 and 124 ppm) corresponding to the non-equivalent C-H bonds of the coordinated bpy, demonstrate the formation of a new rhenium complex in solution. The apparition, in the spectrum, of a new intense ^{13}C signal (n° 4 at 167 ppm) which split into a doublet (J CH = 197 Hz) in the absence of proton decoupling, suggested the generation of a formate species from the labelled carbon dioxide. The peaks corresponding to the solvent (d_7-DMF and DMF at δ 162.4 ppm (C=O), δ 35 and 30 ppm (two $-CH_3$) and to the excess TEOA at δ 59 and 57 ppm) are unchanged after irradiation. New peaks due to oxidation of the electron donor (TEOA) are detected after irradiation (at δ 90 and peaks between 50 and 65 ppm, n° 7). These new peaks were assigned to glycolaldehyde (at δ 90 and 62 ppm), diethanolamine and diethanolamine carbamate, and were characterized by comparison with authentic samples and literature data (**57**). Clearly no carbonate (δ = 161 ppm) or bicarbonate (δ = 158 ppm) accumulate in solution (Figure 14b), the final oxygen acceptor from CO_2 being in this case also the proton.

The new complexes formed during the process have been isolated and fully characterized by ^1H, ^{13}C-NMR and IR spectroscopy, and microanalysis. The minor species was a labelled rhenium-formate, the ^1H and ^{13}C-NMR spectra of which are given in Figure 15.

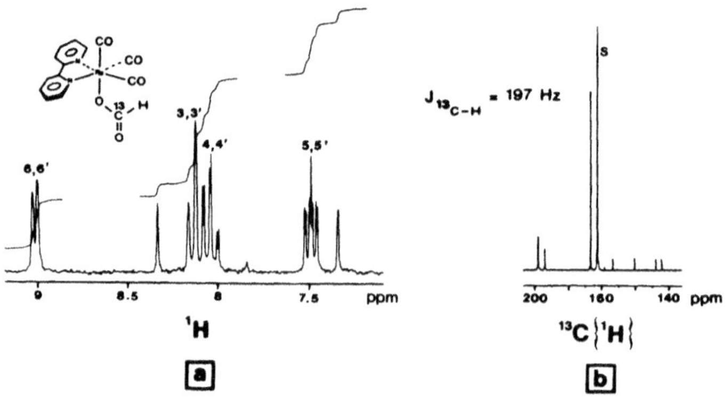

Figure 15. ^1H (in CD_2Cl_2) and ^{13}C (in d_7-DMF + DMF) NMR spectra of fac-Re(bpy)($^{13}CO)_3(OO^{13}CH)$. S = solvent.

The structure of this new complex has been determined (Figure 16), which showed that the configuration of the species was still facial. The formate occupied the coordination site of the chloride anion.

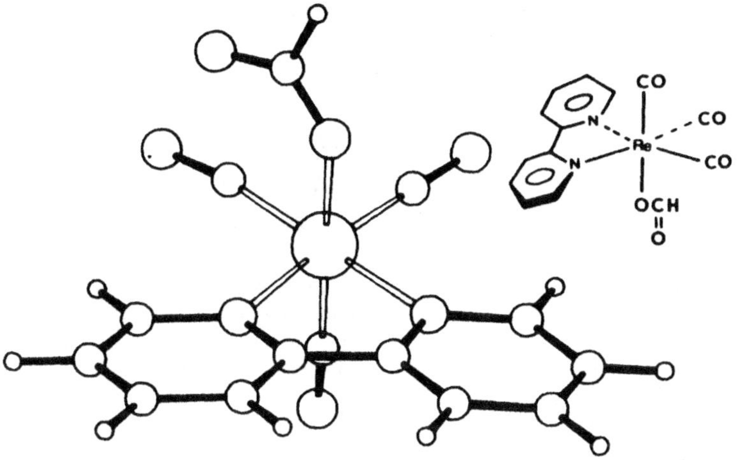

Figure 16. X-Ray crystal structure of fac-Re(bpy)$(CO)_3$(OOCH).

The major new species was the starting rhenium complex with the three carbonyls exchanged fac-Re(bpy)$(^{13}CO)_3Cl$. In order to demonstrate that the rhenium-formate complex was not an intermediate species in the formation of CO, we have studied the photochemical generation of ^{13}CO and Re(bpy)$(^{13}CO)_3$(OO^{13}CH) versus time (Figure 17).

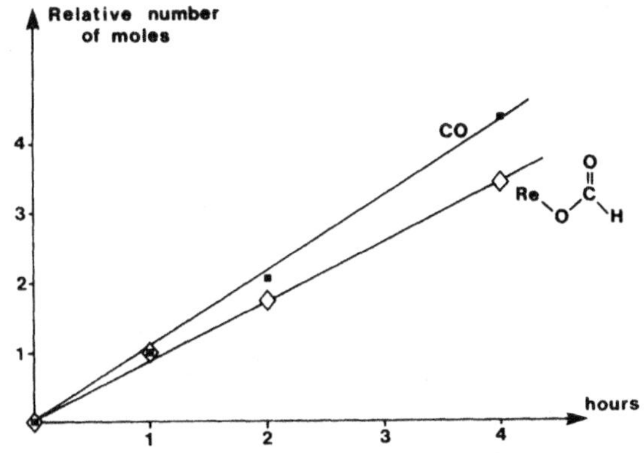

Figure 17. Photogeneration of ^{13}CO and Re(bpy)$(^{13}CO)_3$(OO^{13}CH) as a function of time. The number of moles formed after one hour, is arbitrarly fixed to 1 for each component. In reality, CO generation is 160 times more efficient.

The two reactions are parallel and not consecutive, meaning that the latter was not the intermediate for the generation of the former. The rhenium-formate formation could be inhibited in the presence of excess chloride anion. In this case CO generation increased (vide supra). Other mechanistic aspects have proved that CO and the formate are not produced in the same way (55). A schematic representation of the overall process is shown in Figure 18.

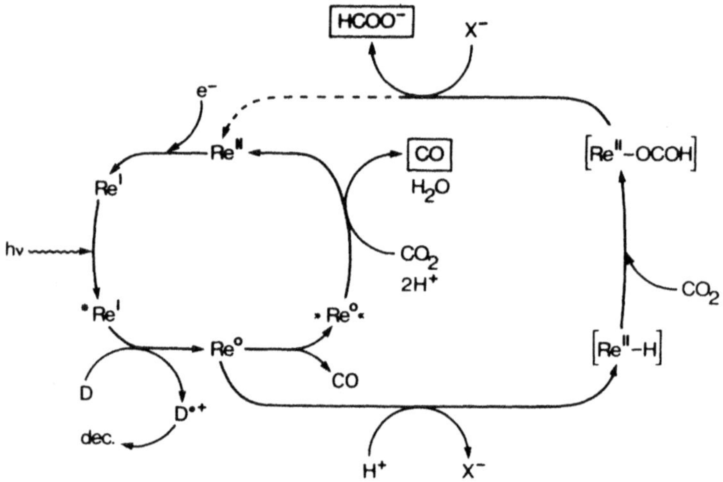

Figure 18 Schematic representation of the photochemical CO and $HCOO^-$ generation process by CO_2 reduction with the system $Re(bpy)(CO)_3X/TEOA/CO_2$ in DMF; the internal and external loops, respectively, lead to CO and $HCOO^-$; the ligands of the rhenium complex are omitted for clarity; Re^0 represents the reduced complex.

The reduced species was generated by reductive quenching of the rhenium-to-bpy-charge-transfer excited state by TEOA. From a Stern-Volmer plot of the luminescence intensity changes, (λ emission = 610 nm) on addition of increasing amounts of quencher (Figure 19), the rate constant for the quenching reaction (eq. 9) was determined to be $kq = 3.4 \times 10^7\ M^{-1} s^{-1}$, calculated from:

$$I/I_o = 1 + kq\tau_o[\text{TEOA}] \qquad K_{SV} = kq\tau_o, \quad \tau_o = 60 \text{ ns for } {}^*\text{Re}$$

$$^*\text{Re(bpy)(CO)}_3\text{Br} + \text{TEOA} \longrightarrow [\text{Re(bpy)(CO)}_3\text{Br}]^- + \text{TEOA}^{+\cdot} \qquad (9)$$

This unstable 19-electron complex is first deactivated by loss of a ligand which then reacts, either directly, with CO_2 to form CO, H_2O and a Re(II) species (major reaction) or, with H_2O and then CO_2 to give a Re(II)-formate complex (minor pathway). These Re(II) complexes were reduced to the oxidation state (I) with an extra electron (originating from $D^{+\cdot}$, *Re or Re°).

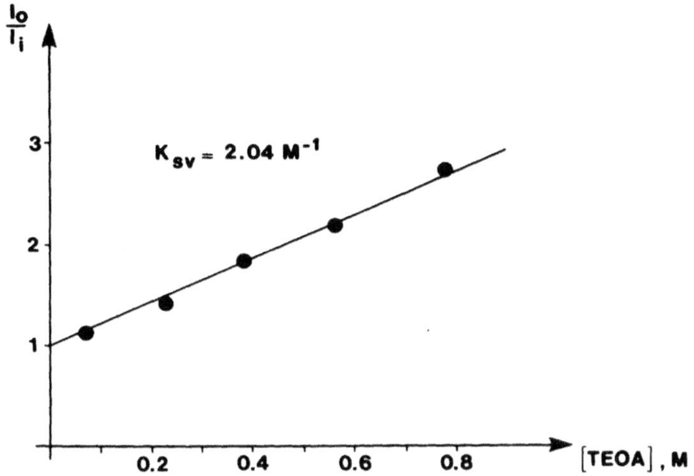

Figure 19. Stern-Volmer plot for the quenching of the luminescence of $Re(bpy)(CO)_3Br$ by TEOA, at 30 °C in DMF, under argon. $K_{SV} = kq\tau_o$, kq = quenching constant and τ_o = 60 ns for excited state lifetime of the rhenium complex in DMF.

If we assume that in the previous photochemical system, light is used only to generate the reduced species, it is also possible to imagine an electrocatalytic system in which the starting rhenium complex is reduced on an electrode. We described such a process and demonstrated that fac-$Re(bpy)(CO)_3Cl$ is a highly efficient (98 % faradic yield) and selective electrocatalyst for the reduction of CO_2 to CO (less than 1 % H_2 was produced)(**25**). Moreover this system displayed high stability (Figure 20). The production of CO depends linearly on time and number of coulombs consumed. The rhenium complex underwent 300 catalytic cycles without loss of activity and remained unchanged, as shown by isolation and characterization by NMR spectroscopy, at the end of the run. The system showed a marked dependence on the amount of water present in solution (Figure 21). When no water was added to the medium, the electrogeneration of CO levelled off and a plateau was observed. At the same time the reduced complex accumulated in solution. It is likely that all the residual water has been consumed. Addition of traces of water to this medium causes the disappearence of the reduced complex and an electroformation of CO, linear versus time, was observed. An optimum efficiency was obtained using 10 % water (see insert in Figure 21). A less efficient system was found when 20 % or more water was present. Formation of a new rhenium(0)-dimer (fac-$Re(bpy)(CO)_3)_2$, hydrogen and traces of the rhenium-formate (same species as in Figure 15) were responsible for the decrease in catalytic activity. When a $nBu_4N^+PF_6^-$ salt was used as supporting electrolyte, in place of a chloride or perchlorate salt (Figure 21), hydrogen and traces of rhenium-formate were additionaly formed. Near quantitative amounts of H_2 (no CO detected) were found when a more acidic salt (NH_4PF_6) was used.

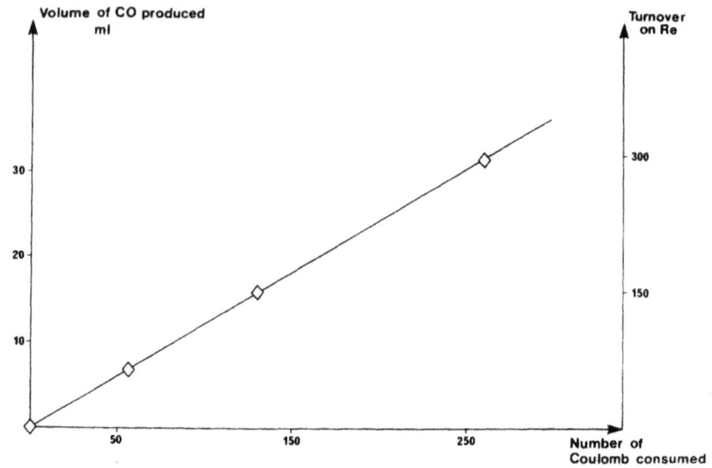

Figure 20. Volume of CO produced and catalyst turnover number as a function of coulombs consumed in the electroreduction of CO_2 to CO by $Re(bpy)(CO)_3Cl$. DMF + H_2O (10 %) + NEt_4Cl as supporting electrolyte.

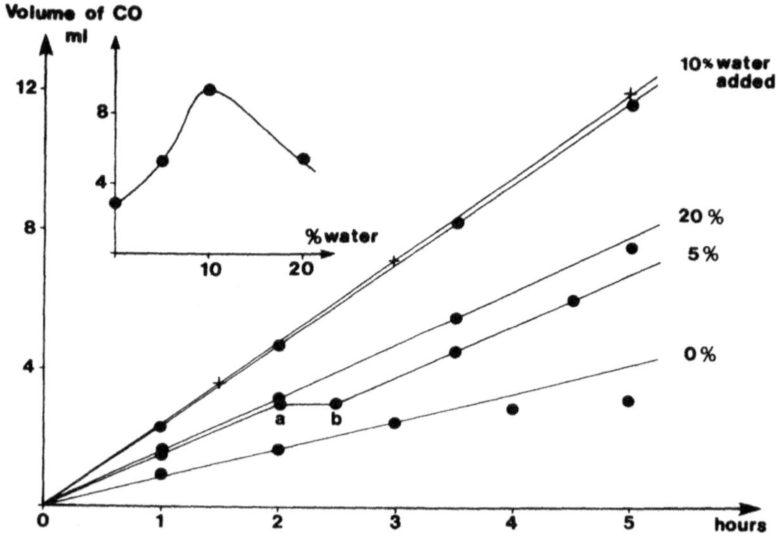

Figure 21. Volume of CO generated as a function of pourcentage water added and electrolysis time; the supporting electrolyte being NEt_4Cl (●) or NBu_4ClO_4 (+). Electrolysis cell was flushed and kept under N_2 during 30 minutes (point a) and then degassed with CO_2 (point b).
Insert: volume of CO produced after 4 hours electrolysis, as a function of the amount of water added.

These electrochemical studies clearly showed that the nature of the supporting electrolyte and in particular whether its anion was ligating (Cl^- or ClO_4^-) or non-ligating (PF_6^-), plays a major role in the electrocatalytic processes observed (CO_2 reduction, dimerization, H_2O reduction).

Thus, we proposed the following mechanisms for the catalytic formation of CO and non-catalytic formation of formate;

a)mechanism of CO formation (Figure 22).
When the vacant site created by anion dissociation was not available (excess ligating anion present), the only process observed was the reduction of CO_2 to CO. The carbonyl ligands of the unstable 19-electron species were labile (fast carbonyl exchange was observed when the species was kept under a ^{13}CO atmosphere). Labilization of a carbonyl gives a 17-electron Re-complex which contains a vacant site available for the reduction of CO_2. The key step is the formation of a metallocarboxylic complex, formed by nucleophilic attack of the formally zero valent rhenium center on the electrophilic carbon center of the CO_2 molecule. Decomposition of the metallocarboxylic intermediate by protonation lead to the formation of water and a triscarbonyl rhenium(II) complex. Further reduction of this species regenerates the starting complex.

b)mechanism of formate formation (Figure 23).
In the absence of excess Cl^- (photochemical experiments) or in the presence of PF_6^- (electrochemical experiments), the vacant site obtained by the rhenium-chloride bond cleavage of the 19-electron species, might allow the formation of a 17-electron Re(II)-hydride complex. This could react with CO_2 to form a coordinated formate. This reaction probably proceeds through a precoordination of CO_2 and formation of a four center intermediate, after loss of a carbonyl ligand. Further reduction of this species gives the isolated rhenium(I)-formate. The Re(II)-hydride species could also react with an acidic proton (NH_4^+), to form molecular hydrogen.

III PHOTOCHEMICAL REDUCTION OF CO_2 TO FORMATE MEDIATED BY POLYPYRIDINE RUTHENIUM COMPLEXES.

In light of the previous systems described, an interesting extension was the search for efficient and catalytic systems for the specific reduction of CO_2 to formate. Literature data show that reduction of CO_2 by H_2 in the presence of various metal catalysts and cocatalysts, yield formate or formamides (**58,59**). Ruthenium cluster anions gave coordinated formate by insertion of CO_2 into a Ru-H bond (**60**). As mentionned previously, ruthenium-polypyridine complexes have remarkable photophysical properties. Thus, combination of these two types of results may represent a potential path for photochemical CO_2 reduction to formate. Visible light irradiation of solutions containing the components $Ru(bpy)_3^{2+}$/TEOA/ $^{13}CO_2$ in DMF yielded mainly formate together with traces of CO and H_2 (**42**). The efficient generation of

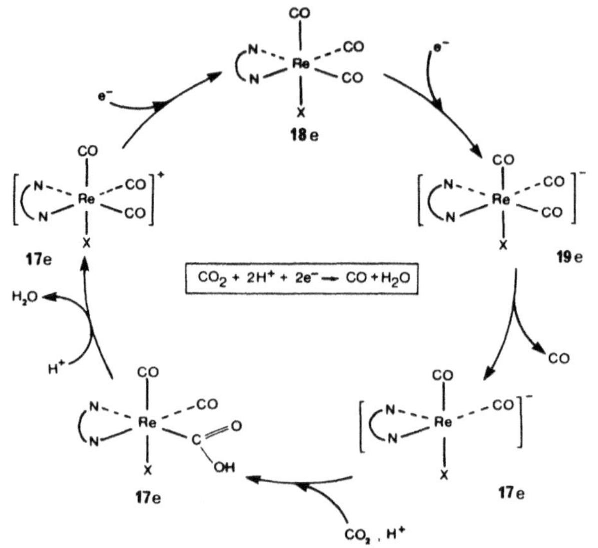

Figure 22. Schematic representation of a possible catalytic rhenium cycle for the reduction of CO_2 to **CO** , in the presence of excess ligating anion.

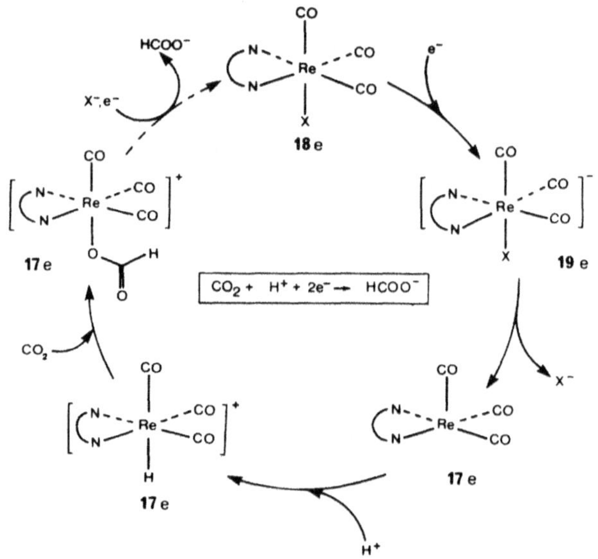

Figure 23. Schematic representation of a possible non-catalytic rhenium cycle for reduction of CO_2 to **HCOO**,in the absence of excess ligating anion.

formate was detected and followed by ^{13}C NMR spectroscopy using labelled CO_2 (Figure 24). The reaction was zero order when the pressure of $^{13}CO_2$ was constant and, first order (in CO_2), after 5 hours irradiation, when the volume of $^{13}CO_2$ was constant (Figure 25). Attempts to isolate the active catalyst, probably a ruthenium-bisbipyridine complex obtained by photolabilization of a bpy ligand from $Ru(bpy)_3^{2+}$, failed. We thus synthesized and tested some ruthenium(II)bisbpy complexes (see Table).

Table. Results on Photoreduction of CO to HCOO⁻.

Catalyst a)	Reagents	Result b)
$Ru(bpy)_3Cl_2$	$^{13}CO_2$, TEOA	catalytic
" + excess bpy	$^{13}CO_2$, TEOA	rate x 2.3
" + excess Cl^-	$^{13}CO_2$, TEOA	rate x 1.4
" + $C_{16}MV^{2+}$	$^{13}CO_2$, TEOA	no formate
cis-$Ru(bpy)_2(CO)(H)^+$	$^{13}CO_2$, $h\nu$ or Δ	no formate
"	$^{13}CO_2$, TEOA	stoich. coord. $HCOO^-$
$Ru(phen)_3Cl_2$	$^{13}CO_2$, TEOA	no formate
" + cis-$Ru(bpy)_2(CO)(H)^+$	$^{13}CO_2$, TEOA	catalytic
" + cis-$Ru(bpy)_2(CO)(Cl)^+$	$^{13}CO_2$, TEOA	"
" + cis-$Ru(bpy)_2Cl_2$	$^{13}CO_2$, TEOA	"
" + cis-$Ru(bpy)_2(S)_2^{2+}$	$^{13}CO_2$, TEOA	"

a) $C_{16}MV^{2+}$ = N-hexadecyl-N'-methyl-4,4'-bipyridinium, phen = 1,10-phenanthroline, S = solvent molecule (DMF or CH_3CN).
b) of formate formation.

Rate enhancing effects of excess bipyridine or chloride salt were observed. Formate formation was totally suppressed in the presence of an electron relay (viologen radical cation species are known to be monoelectronic reducing agents). However ultra-violet irradiation of an aqueous solution of $Ru(bpy)_3^{2+}$, MV^{2+}, TEOA, CO_2 was reported to give formic acid (**39**). Photolysis of $Ru(phen)_3Cl_2$ (less photolabile species), under the same conditions gave no formate. This observation was of special interest because this complex can thus be used to continously generate a reduced complex, allowing the catalytic formation of formate in the presence of a ruthenium-bisbipyridine complex. No catalytic amounts of formate were formed when the $Ru(bpy)_2$-complexes, mentioned in the Table above, were photolyzed in the presence of only TEOA and $^{13}CO_2$. Work is in progress in order to ascertain the nature of the catalytic species and to elucidate the steps of this exciting CO_2 reduction reaction.

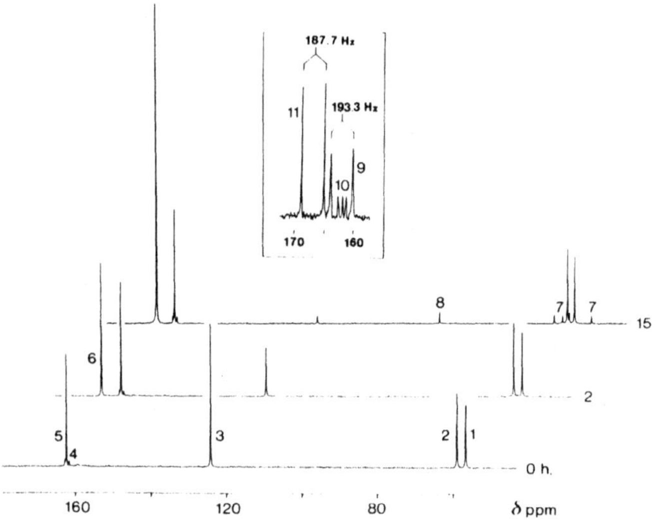

Figure 24. ^{13}C-NMR spectroscopy of $H^{13}COO^-$ generation by visible light irradiation of $Ru(bpy)_3Cl_2/TEOA/^{13}CO_2$, in DMF/d_7DMF.
Insert: carbonyl region without proton decoupling. The CH_3 signals of the solvent are below 40 ppm (not shown).

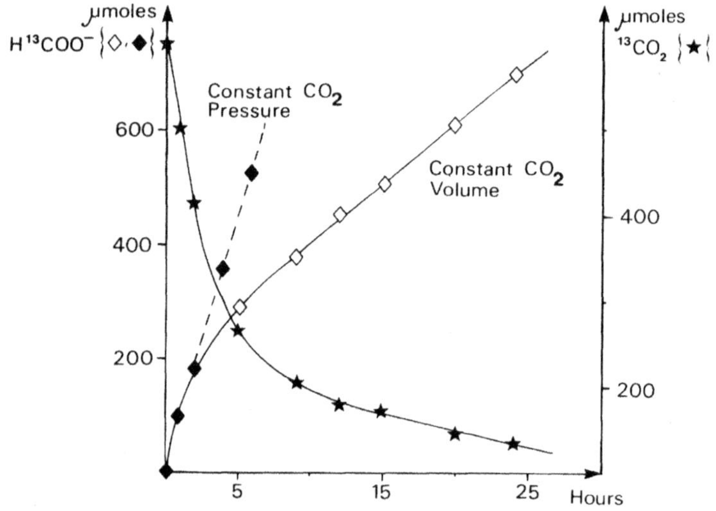

Figure 25. Photogeneration of $H^{13}COO^-$ (◇) and consumption of $^{13}CO_2$ (★) as a function of time. Formate production (◆) by maintaining $^{13}CO_2$ pressure constant.

Conclusion.

The photo- and electrochemical carbon dioxide reduction systems described above represent new CO_2 activation reactions involving organometallic complexes. The rhenium complex displays high efficiency (14 % quantum yield and 98 % current conversion) and remarkable selectivity, both photochemically and electrochemically. Although all the photochemical processes make use of an electron donor (tertiary amine), they represent new processes of artificial photosynthesis and of energy conversion, storing light energy in the form of chemical fuels. The future in this field might be the search for new homogeneous catalysts (e.g. polynuclear species) capable of formally exchanging more than two electrons, so that CO_2 reduction into formaldehyde (4 electrons involved) or to methanol (6 electrons involved) may be envisaged. Another task of artificial photosynthesis and of solar energy conversion is to use the water molecule as electron donor. Water will concomitantly be oxidized to H_2O_2 (a 2 electron process) or to O_2 (a 4 electron process). This poses problems of finding efficient homogeneous catalysts for the water oxidation and, of interactions between the oxidant and reductant simultaneously generated in the same medium.

Acknowledgements.

I would like to thank Dr. Ilesh Bidd and Dr. Romain Ruppert for their kind participation in the preparation of the manuscript.

References.

(1) R. Eisenberg, D.E. Hendricksen, Adv. Catal., **28**, 79, (1979); J.A. Ibers, Chem. Soc. Rev., **11**, 57, (1982); D.J. Darensbourg, R.A. Kudaroski, Adv. Organomet. Chem., **22**, 129, (1983); R. Ziessel, Nouv. J. Chim., **7**, 613, (1983); D.A. Palmer, R. Van Eldik, Chem. Rev., **83**, 651, (1983); D. Walther, E. Dinjus, J. Sieler, Z. Chem., **23**, 237, (1983); A. Behr, in "Catalysis in C1 Chem.", W. Keim ed., D. Reidel Publ. Co., 169, (1983).

(2) "Organic and Bioorganic Chemistry of Carbon Dioxide", S. Inoue and N. Yamazaki eds., Kodansha Ltd., Tokyo, Wiley, New-York (1982).

(3) "Photochemical Conversion and Storage of Solar Energy", J.S. Connolly ed., Academic Press, New-York (1981).

(4) "Photogeneration of Hydrogen", A. Harriman and M.A. West eds., Academic Press, London (1982).

(5) J.L. Roberts, D.T. Soweyer, J. Electroanal. Chem., **9**, 1, (1965).

(6) W. Paik, T. Andersen, H. Eyring, Electrochim. Acta, **16**, 1217 (1969)

(7) U. Kaiser, E. Heitz, Ber. Bunsen Gesell., **77**, 818, (1973).

(8) P.G. Russel, N. Kovac, S. Srinivasan, M. Steinberg, J. Electrochem. Soc., **124**, 1329, (1977); S. Kapusta, N. Hackerman, J. Electrochem. Soc., **130**, 607, (1983).

(9) J.C. Gressin, D. Michelet, L. Nadjo, J.M. Savéant, Nouv. J. Chim., **3**, 545, (1979); C. Amatore, J.M. Savéant, J. Amer. Chem. Soc., **103**, 5021, (1981).

(10) S. Wawzonek, E.W. Blaka, R. Berkey, M.E. Runner, J. Electrochem. Soc., **102**, 235, (1975).

(11) D.A. Tyssee, J.H. Wagenknecht, M.M. Baizer, J.L. Chruma,Tetrahedron Lett., **47**, 4809, (1972).

(12) S. Gambino, G. Silvestri, Tetrahedron Lett., **32**, 3025, (1973).

(13) J. Fischer, Th. Lehmann, E. Heitz, J. Appl. Electrochem., **11**, 743, (1981).

(14) L.V. Haynes, D.T. Sawyer, Anal. Chem., **39**, 332, (1967).

(15) a) D. Canfield, K.W. Jr. Frese, J. Electrochem. Soc., **130**, 1772, (1983); b) K.W. Jr. Frese, D. Canfield, ibid., **131**, 2518, (1984); c) K.W. Jr. Frese, S. Leach, ibid., **132**, 259, (1985).

(16) S. Meshitsuka, M. Ichikawa, K. Tamaru, J. Chem. Soc., Chem. Comm., 158, (1974); S. Kapusta, N. Hackerman, J. Electrochem. Soc., **131**, 1511, (1984); C.M. Lieber, N.S. Lewis, J. Amer. Chem. Soc., **106**, 5033, (1984).

(17) C.J. Stalder, S. Chao, M.S. Wrighton, J. Amer. Chem. Soc., **106**, 3673, (1984); C.J. Stalder, S. Chao, D.P. Summers, M.S. Wrighton, J. Amer. Chem. Soc., **105**, 6318, (1983) and **106**, 2723, (1984).

(18) T.R. O'Toole, L.D. Margerum, T.D. Westmoreland, W.J. Vining, R.W. Murray, T.J. Meyer, J. Chem. Soc., Chem. Comm., 1416, (1985); S. Cosnier, A. Deronzier, J.C. Moutet, J. Electroanal. Chem., **207**, 315, (1986).

(19) K. Hiratsuka, K. Takahashi, H. Sasaki, S. Toshima, Chem. Lett., 1137, (1977).

(20) K. Takahashi, K. Hiratsuka, H. Sasaki, S. Toshima, Chem. Lett., 305 (1979).

(21) J.Y. Becker, B. Vainas, R. Eger, L. Kaufman, J. Chem. Soc., Chem. Comm., 1471, (1985).
(22) a) M. Tezuka, T. Yajima, A. Tsuchiya, Y. Matsumoto, Y. Uchida, M. Hidai, J. Amer. Chem. Soc., **104**, 6835, (1982); b) M. Nakazawa, Y. Mizobe, Y. Matzumoto, Y. Uchida, M. Tezuka, M. Hidai, Bull. Chem Soc. Jpn, **59**, 809, (1986).
(23) a) B. Fischer, R. Eisenberg, J. Amer. Chem. Soc., **102**, 7361, (1980) b) M. Beley, J.P. Collin, R. Ruppert, J.P. Sauvage, J. Chem. Soc., Chem. Comm., 1315, (1984).
(24) A.H.A. Tinnemans, T.P.M. Koster, D.H.M.W. Thewissen, A. Mackor, Recl. Trav. Chim. Pays-Bas, **103**, 288, (1984).
(25) J. Hawecker, J.M. Lehn, R. Ziessel, J. Chem. Soc., Chem. Comm., 328 (1984).
(26) J.H. Wagenknecht, S. Slater, J. Amer. Chem. Soc., **106**, 5367, (1984)
(27) M.S. Wrighton, Pure and Appl. Chem., **57**, 57, (1985).
(28) M.G. Bradley, T. Tysak, J. Electroanal. Chem., **135**, 153, (1982); M.G. Bradley, T. Tysak, D.J. Graves, N.A. Vlachopoulos, J. Chem. Soc., Chem. Comm., 349, (1983); M. Beley, J.P. Collin, J.P. Sauvage J.P. Petit, P. Chartier, J. Electroanal. Chem., **206**, 333, (1986).
(29) B.A. Parkinson, P.F. Weaver, Nature, **309**, 148, (1984).
(30) M. Halmann, Nature, **275**, 115, (1978); M. Halmann, B. Aurian-Blajeni Proceedings of the Second European Community Photovoltaic Solar Energy Conference, (West Berlin, Federal Republic of Germany), 682, (1979); K. Ito, S. Ikeda, M. Yoshida, S. Ohta, T. Iida, Bull. Chem. Soc. Jpn., **57**, 583, (1984); Y. Taniguchi,H. Yoneyama H. Tamura, Bull. Chem. Soc. Jpn., **55**, 2034, (1982).
(31) S. Ikeda, M. Yoshida, K. Ito, Bull. Chem. Soc. Jpn., **58**, 1353, (1985).
(32) M. Zafrir, M. Ulman, Y. Zuckerman, M. Halmann, J. Electroanal. Chem., **159**, 373, (1983); W.M. Sears, S.R. Morrison, J. Phys. Chem., **89**, 3295, (1985).
(33) T. Taniguchi, B. Aurian-Blajeni, J.O'M. Bockris, J. Electroanal. Chem., **157**, 179, (1983).
(34) B. Aurian-Blajeni, I. Taniguchi, J.O'M. Bockris, J. Electroanal. Chem., **149**, 291, (1983).
(35) J.C. Hemminger, R. Carr, G.A. Somorjai, Chem. Phys. Letters, **57**, 100, (1978).
(36) T. Inoue, A. Fujishima, S. Konishi, K. Honda, Nature, **277**, 637, (1979); B. Akermark, U. Eklund-Westlin, P. Baeckstrom, R. Lof, Acta Chem. Scand. Ber. Ser. B, **B34**, 27, (1980); B. Aurian-Blajeni, M. Halmann, J. Manassen, Sol. Energy, **25**, 165, (1980); M. Ulman, A.H.A. Tinnemans, A. Mackor, B. Aurian-Blajeni, M. Halmann, Int. J. Sol. Energy, **1**, 213, (1982).
(37) A. Henglein, M. Gutierrez, Ber. Bunsenges. Phys. Chem., **87**, 852, (1983).
(38) B. Akermark, in Solar Energy Photochemical Conversion and Storage, eds. S. Claesson and L. Engstrom (National Swedish Board for Energy Source Development, Stockholm), chap. **6**, 1, (1977); N. Getoff, Z. Naturforsch., **B17**, 87, (1962) and **B18**, 169, (1963).
(39) S. Tazuke, N. Kitamura, Nature, **275**, 301, (1978)

(40) a) J.M. Lehn, R. Ziessel, Proc. Natl. Acad. Sci. USA, **79**, 701, (1982); b) R. Ziessel, J. Hawecker, J.M. Lehn, Helv. Chim. Acta, **69**, 1065, (1986).
(41) J. Hawecker, J.M. Lehn, R. Ziessel, J. Chem. Soc., Chem. Comm., 536 (1983).
(42) J. Hawecker, J.M. Lehn, R. Ziessel, J. Chem. Soc., Chem. Comm., 56, (1985).
(43) V. Balzani, L. Moggi, M.F. Manfrin, F. Bolleta, M. Gleria, Science, **189**, 852, (1975).
(44) J.R. Bolton, Science, **202**, 705, (1978).
(45) C.V. Krishnam, N. Sutin, J. Amer. Chem. Soc., **103**, 2141, (1981).
(46) J.M. Lehn, J.P. Sauvage, R. Ziessel, Nouv. J. Chim., **3**, 423, (1979)
(47) K. Kalyanasundaram, Coord. Chem. Rev., **46**, 159, (1982).
(48) M. Wrighton, D.L. Morse, J. Amer. Chem. Soc., **96**, 998, (1974).
(49) a) F.R. Keene, C. Creutz, N. Sutin, Coord. Chem. Rev., **64**, 247, (1985); b) C. Creutz, N. Sutin, ibid., **64**, 321, (1985).
(50) J.E. Bercaw, L.Y. Goh, J. Halpern, J. Amer. Chem. Soc., **94**, 6534, (1972).
(51) J. Hawecker, J.M. Lehn, R. Ziessel, Nouv. J. Chim., **7**, 271, (1983).
(52) a) C.V. Krishnan, N. Sutin, J. Amer. Chem. Soc., **103**, 2141, (1981); b) C.V. Krishnan, C. Creutz, D. Mahajan, H.A. Schwarz, N. Sutin, Israel J. Chem., **22**, 98, (1982).
(53) G.N. Schrauzer, R.J. Holland, J. Amer. Chem. Soc., **93**, 1505, (1971)
(54) T.H. Chao, J.H. Espenson, J. Amer. Chem. Soc., **100**, 129, (1978).
(55) J. Hawecker, J.M. Lehn, R. Ziessel, Helv. Chim. Acta, in press.
(56) C. Kutal, M.A. Weber, G. Ferraudi, D. Geiger, Organometal., **4**, 2161 (1985).
(57) D. Barth, P. Rubini, J.J. Delpuech, Nouv. J. Chim., **7**, 563, (1983).
(58) P. Haynes, L.H. Slaugh, J.F. Kohnle, Tetrahedron Lett., **5**, 365, (1970).
(59) Y. Inoue, Y. Sasaki, H. Hashimoto, J. Chem. Soc., Chem. Comm., 718, (1975); Y. Inoue, H. Izumido, Y. Sasaki, H. Hashimoto, Chem. Lett., 863, (1976).
(60) D.J. Darensbourg, M. Pala, J. Waller, Organometal., **2**, 1285, (1983) D.J. Darensbourg, C. Ovalles, M. Pala, J. Amer. Chem. Soc., **105**, 5937, (1983).

THE ENZYME CARBONIC ANHYDRASE

I. Bertini, C. Luchinat and R. Monnanni
Department of Chemistry, University of Florence
Via G. Capponi 7, 50121 Florence
Italy

ABSTRACT. The properties of Carbonic Anhydrase, the enzyme which catalyzes the CO_2 hydration and HCO_3^- dehydration, are reviewed. The mechanism is discussed in details in terms of Zn-OH attacking CO_2 to give rise to HCO_3^- which is then released after H_2O uptake. H^+ is then released. The inhibition mechanism is also discussed. The native zinc can be substituted by cobalt with retention of activity. The information gained through spectroscopic investigations of the cobalt enzyme are presented. Some discussion is also devoted to manganese, copper and cadmium derivatives.

1. INTRODUCTION

Carbonic Anhydrase (CA) is a ubiquitous protein which contains a zinc ion per molecule[1]. Its physiological role is that of catalyzing the hydration of carbon dioxide and the dehydration of bicarbonate[2] according to the following scheme:

$$CO_2 + H_2O \rightleftharpoons HCO_3^- + H^+ \quad (1)$$

such reaction is slow if uncatalyzed at pH 7[3]. The reaction has $k = 3.5 \times 10^{-2}\ s^{-1}$; at pH 10 bicarbonate formation becomes fast since OH^- attacks CO_2:

$$CO_2 + OH^- \rightleftharpoons HCO_3^- \quad (2)$$

the kinetic constant is $8.5 \times 10^3\ M^{-1}\ cm^{-1}$[4]. The reaction catalyzed by high activity CA isoenzymes is one of the fastest known with k_{cat} of the order of 10^6 mols of CO_2 converted per mol of enzyme per second at pH 7[5].

M. Aresta and G. Forti (eds.), Carbon Dioxide as a Source of Carbon, 139–167.

Several CA isoenzymes from different sources have been studied. The X-ray structure at 2.0 A nominal resolution is known for the low activity human CAI (HCAI)[6] (formerly HCAB) and high activity HCAII (formerly HCAC)[7]. A further refinement of the X-ray data for the latter enzyme is being completed[8].

The shape of the enzyme is that of a rugby ball with a 15 A deep cavity in the middle; at the bottom of the cavity there is the zinc atom which is coordinated to three histidine residues. The zinc atom is thus solvent accessible; at pH 8.7 in ammonium sulfate[6,7] a fourth ligand is present in the coordination sphere which is proposed to be a solvent molecule, possibly OH^-. A hydrogen bond net occurs between the latter solvent molecule, a threonine residue, and a histidine residue[8]. These two residues are conserved in the aforementioned enzymes as well as in the high activity bovine isoenzyme (BCAII) and probably are means of proton release according to reaction (1).

Another isoenzyme has now been characterized, BCAIII (from muscle) which has relatively low catalytic activity and a sensibly different aminoacid composition without free histidine in the active cavity[9-11].

Since its discovery the enzyme has been investigated with the aim of understanding the catalytic mechanism. The enzyme is stable under a large variety of conditions and therefore is easy to study; however, the native enzyme has essentially been studied through kinetic measurements. As it will be shown, the enzyme needs buffer to explicate its full activity; however the small anions of the buffer interact at the metal site[12-14]; only large buffer species such as HEPES, MES, etc. have been found to be relatively inert[15-17].

Zinc is essential for catalysis but it cannot be monitored through any spectroscopy because it is white, diamagnetic and the sensitivity of ^{67}Zn is too low to use NMR at the enzyme concentration[18]. Zinc however can be removed and substituted with other metal ions (see later) some of which do maintain partial or full activity. Many information which have added to the present knowledge on the structure and function of the enzyme have been obtained from artificial enzymes. Here we present the whole picture and then the spectroscopic details of the derivatives. Several ^1H NMR studies have been performed with the enzyme[17,19-22] and some histidine signals have been assigned; this has allowed researchers to have direct information on histidines pK_a's.

2. KINETIC FACTS

CA's obey Michaelis-Menten kinetics over a wide range of conditions[5,23-25]. The kinetic parameters obtained for hydration of CO_2

and dehydration of HCO_3^- are collected in Table I for the three isoenzymes[24].

TABLE I

Michaelis-Menten Parameters for Carbonic Anhydrase Isoenzymes toward the Hydration of CO_2 (A) and the Dehydration of HCO_3^- (B). References in Parenthesis.

		HCAI	HCAII	BCAII	BCAIII
A	K_m (mM)	4(26)	9(26)	12(27)	10(28)
	k_{cat} (s^{-1})	1.5×10^5(29)	1.0×10^6(29)	1.0×10^6(26)	3.0×10^3(29)
B	K_m (mM)	12(30)	60(31)	26(23)	-
	K_{eff} (mM)	6.3(30)	32(31)	500(32)	50(33)
	k_{cat} (s^{-1})	3.6×10^4(30)	6.0×10^5(34)	4.0×10^5(23)	-

We summarize here the salient features of kinetic experiments. 1) The hydration activity, expressed as k_{cat}/K_m, increases with increasing pH. The pH dependence is essentially contained in k_{cat}, while K_m is roughly pH independent. 2) The pH profile does not follow a single sygmoidal behavior: this has been realized recently[17], and ascribed to the presence of two mutually interacting acid-base groups in the active site cavity. 3) The dehydration profiles of HCO_3^- show the opposite pH dependence; this is rationalized if one considers that H^+ is a cosubstrate for the dehydration reaction. It follows that the only active form of the enzyme is the high pH species, i.e. one with the catalytic acid base group in the deprotonated form. 4) With no added buffer or at low ($<$ 1 mM) buffer concentration a proton transfer from the enzyme to the solution is the rate-limiting step[35-37]. 5) At high buffer concentration the rate-limiting step becomes an internal proton transfer, presumably between the two active site acid-base groups[9,10,34,38,39]; in the low activity isoenzyme I bicarbonate dissociation partially contributes as a rate-limiting step[26,30,40]. 6) Anionic inhibitors are competitive with respect to bicarbonate[25]; as far as CO_2 is concerned, inhibition is non competitive at low pH[23-25] and uncompetitive at high pH[41,42]. At high pH imidazole and phenol are competitive with respect to CO_2, while 1,2,4 triazole and trichloroacetaldehyde are non competitive[26,28]. 7) Equilibrium kinetic measurements show that the presence of buffer is no longer needed to ensure maximal catalytic efficiency[30,32,43,44]. Interestingly, $K_{eff}(CO_2)$ is probably[33] much larger than the corresponding K_m values.

A minimal reaction scheme consistent with all kinetic data is the following[29]

$$E + CO_2 + B \left(\underset{k_{-1}}{\overset{k_1}{\rightleftharpoons}} E\text{-}CO_2 + B \right) \underset{k_{-2}}{\overset{k_2}{\rightleftharpoons}} EH^+\text{-}HCO_3^- + B \underset{k_{-3}}{\overset{k_3}{\rightleftharpoons}} EH^+ + HCO_3^- + B$$

$$\underset{k_{-4}}{\overset{k_4}{\rightleftharpoons}} HE^+ + HCO_3^- + B \underset{k_{-5}}{\overset{k_5}{\rightleftharpoons}} E + HCO_3^- + BH^+$$

where B and BH^+ stand for the basic and acidic forms of the buffer, E is the deprotonated form of the enzyme, EH^+ is the form with the protonated catalytic group and HE^+ is the form with the protonated proton-transfer group.

3. THE ENZYME

It has been pointed out that the cavity allows two ligands to bind zinc ion; therefore four or five coordination can occur[12,45,46]. It is possible that at low pH zinc binds one or two water molecules in an equilibrium between coordination number four and five. The position of the equilibrium depends on the particular isoenzyme[47]. The chromophore is eventually transformed in a tetracoordinated species containing an OH^-; the pK_a of this equilibrium depends on the properties of the solvent inside the cavity, which are determined by the nature and by the reciprocal position of the residues which determine the cavity surface. As already mentioned many isoenzymes of type I and II contain at least one histidine inside the cavity. When this histidine is protonated, starting from high pH, protons will distribute between the basic OH^- group and the basic histidine residue, and the protonation state of one group will affect the apparent pK_a of the other. The general treatment of two interacting groups requires four microconstants and the general scheme in the particular case of carbonic anhydrase is shown in Scheme I,[17,48] where OH_2 is a metal-coordinated water molecule and N is His-64 for human and bovine II, and His-200 for human I isoenzymes.

If we study the electronic spectra of cobalt(II) substituted enzyme (Figure 1) two apparent pK_a's can be determined from the spectral intensity (Figure 2)[48]. In general it is not possible to obtain the four microconstants; however, assuming that the two water-containing species, as well as that the hydroxo-containing species, have the same spectra, the four microconstants can be evaluated[48]. They are reported in Table II.

Scheme I

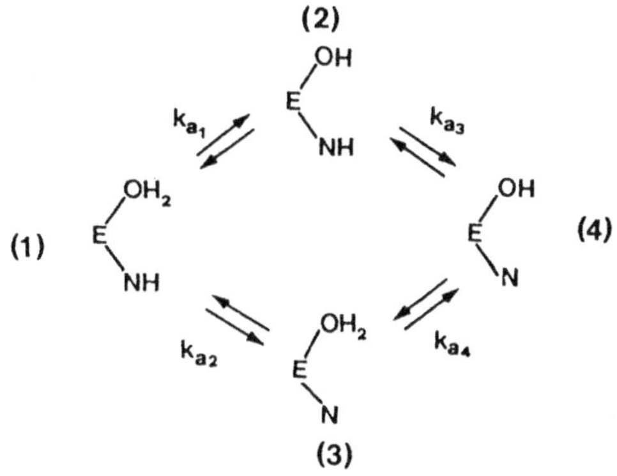

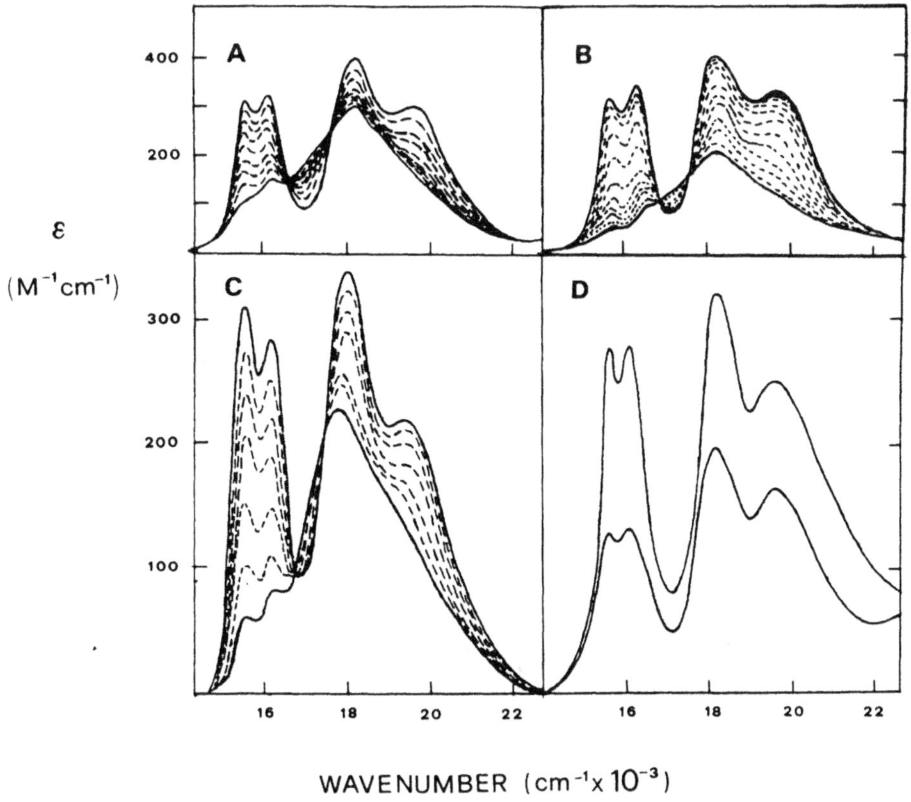

Figure 1. Electronic spectra at various pH (increasing with increasing absorption at 15.6×10^{3} cm^{-1}) of Co(II)BCAII (A), Co(II)HCAII (B), Co(II)HCAI (C) and Co(II)BCAIII (D) (from ref. 49 with permission).

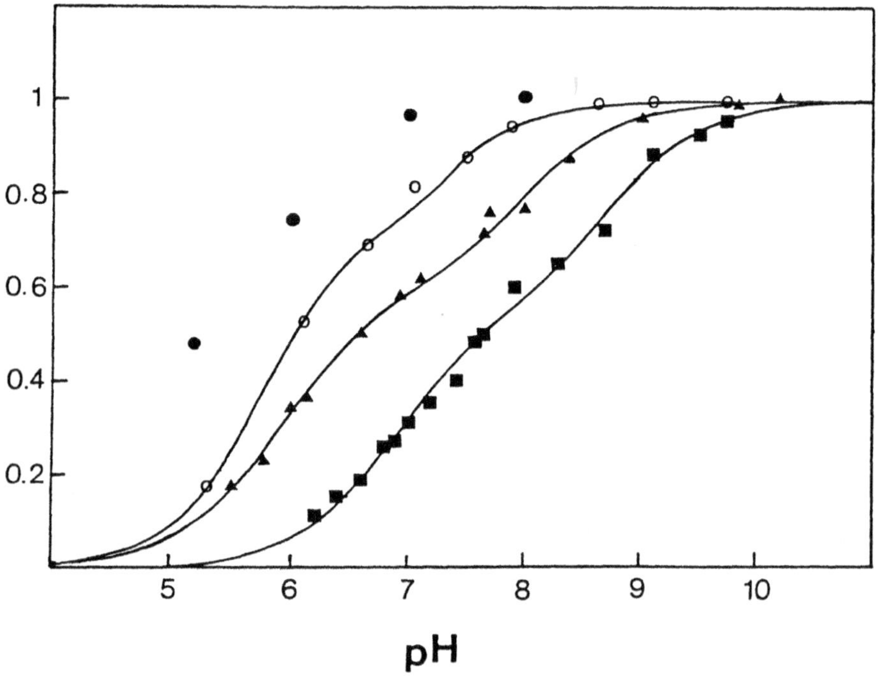

Figure 2. pH dependence of the molar absorbance at 15.6×10^{3} cm^{-1} for Co(II)HCAI (■), Co(II)BCAII (▲), Co(II)HCAII (○), and Co(II)BCAIII (●). The curves are calculated using the pk_a values of Table II. The absorbance values are normalized to a limit high pH value of 1 for each isoenzyme (from ref. 50 with permission).

TABLE II

Microconstant Values Associated with Proton Transfer Equilibria in Cobalt(II) Substituted Carbonic Anhydrases[48].

	pk_{a1}	pk_{a2}	pk_{a3}	pk_{a4}
CoHCAI	7.14	7.21	8.45	8.38
CoHCAII	5.95	5.62	6.62	6.95
CoBCAII	6.12	6.28	7.75	7.59

Analogous values are obtained for the native enzyme from the pH dependence of the activity[17]. By inspection of Table II we learn that the pK_a's of water in the diprotonated enzymes are very low, lower in general than found in simple model complexes. In Table III the pK_a's of water coordinated to several cobalt(II) complexes are reported[51-57]. The lowering of the pK_a from the pure water pK_a is due to: i) the binding to a tetracoordinated zinc(II) without anionic ligand in the coordination sphere; ii) the presence of the histidinium ion inside the active cavity. The pK_a of water in presence of neutral histidine is about 8, as found in model complexes; this is a further proof that the properties of the enzyme are not due to an entatic state of the metal but simply to the ability of various residues to tune the chemical properties of the cavity. The pK_a values of Table II indicate that at

TABLE III

pK_a's of Coordinated Water for Cobalt(II) Complexes

	pK_a		pK_a
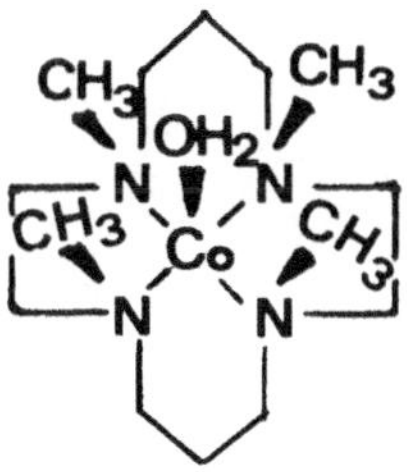	8.4 [51]	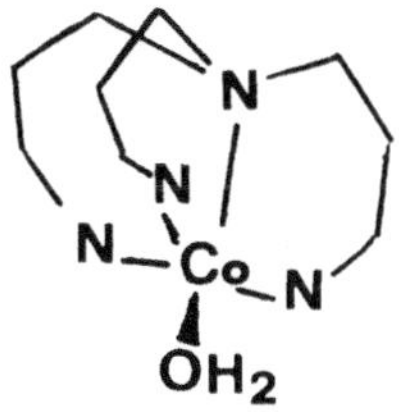	10.8 [54]
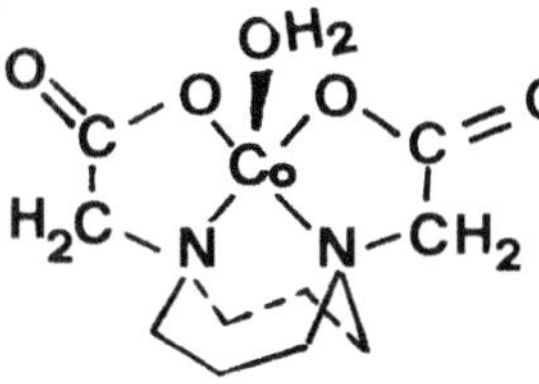	9.4 [52]	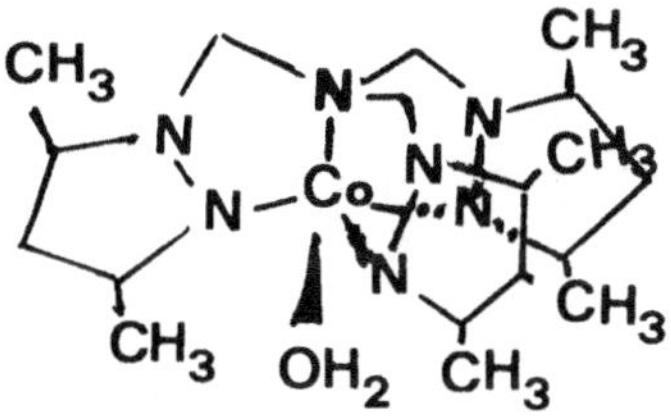	9.0 [55]
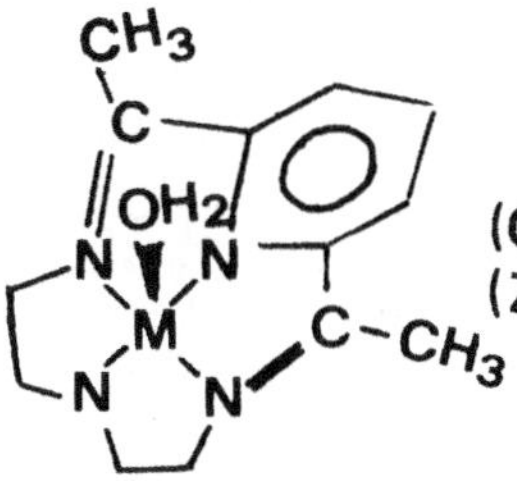	(Co) 8 [53] (Zn) 8.7	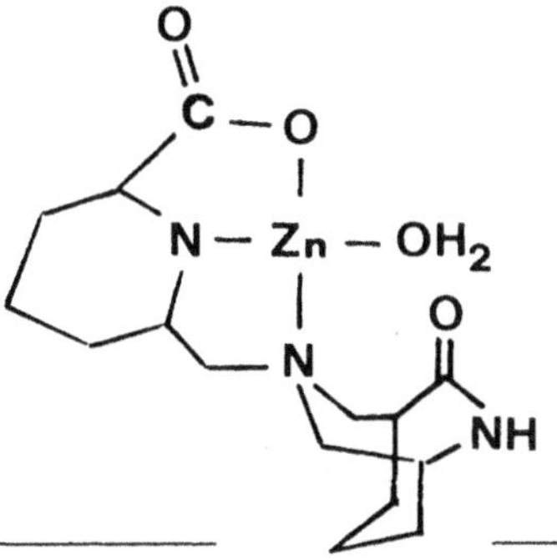	7 [56]

physiological pH the enzyme is completely in the hydroxo-form, which is the active form of the enzyme[45]. Carbon dioxide is believed to bind to a specific site of the enzyme. Although it is not clear which is its binding site some estimates indicate a metal-carbon distance of about 5 to 7 A[58-60], based on the assumption that the affinity constant of CO_2 for the enzyme is larger than 50 M^{-1}. Probably this binding site is not productive in the sense that at this distance the carbon is not available by nucleophilic attack of the coordinated hydroxo group; therefore it may be that the Zn-OH group performs the nucleophilic attack without any particular role of the enzyme. There are theoretical estimates that a metal coordinated hydroxo group is a better nucleophile than free H_2O, which in turn is a better nucleophile than a metal-coordinated water[61]. The result of the attack is the formation of bicarbonate, which is however coordinated through the O-H group.

Scheme II

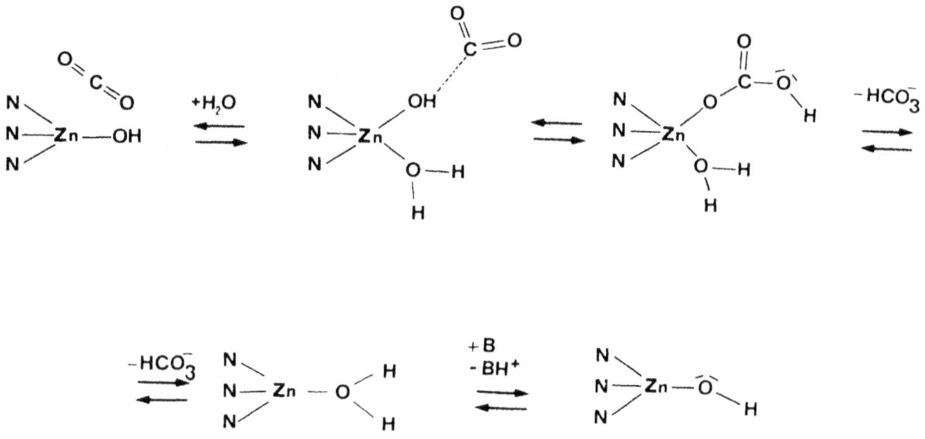

Somehow the proton has to move in such way that bicarbonate binds through a terminal oxygen; this may occur through the hydrogen bond net or through bicarbonate acting as bidentate in an intermediate state. At this point a water molecule may enter the coordination sphere giving rise to a five coordinated intermediate. There are independent evidences that bicarbonate added to Co(II)CA gives rise to an equilibrium between four and five coordinated species[12]. The affinity constant of bicarbonate for Co(II)CA is 10 M^{-1} at pH 8.2[62], and the affinity constant for the native isoenzyme II is 40 M^{-1} at pH 7.5[23]. A low affinity constant is necessary for any enzyme to be active. The water molecule can enter the coordination sphere also at the stage of the hydroxo phase before nucleophilic attack; this transition state of which we have no evidence would act on the nucleophilicity of the hydroxo group. The next step is the detachment of the bicarbonate followed by the deprotonation of the water form of the enzyme. In the high activity form of the enzyme the rate limiting step is the proton release that goes with a diffusion coefficient of 10^{11} s^{-1}; therefore the turnover cannot exceed $10^4 s^{-1}$ at pH 7. In the presence of buffer the actual turnover is enhanced through buffer assisted proton release; under these circumstances the rate limiting step becomes an internal proton transfer; a path from coordinated water oxygen to buffer is proposed that involves proton transfer from oxygen to histidine. In the low activity form internal proton transfer is still the rate limiting step, although detachment of bicarbonate starts contributing to it. In carbonic anhydrase III where no histidines are present in the cavity the rate limiting step continues to be the internal proton transfer. The problem of where CO_2 binds is one of major relevance; if K_m or even K_{eff} for CO_2 are taken as thermodynamic dissociation constants they point to a binding site for CO_2; however there is not a clear indication of a pocket for CO_2 nor of a group which can interact with the oxygen atom in order to activate CO_2. There are suggestions that CO_2 binds the metal ion in one of the known ways[63,64]. Particularly the following binding would suit the mechanism in terms of activation of CO_2[65]:

Scheme III

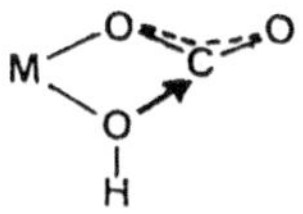

In order to obtain information on the binding site of CO_2 ^{13}C NMR studies have been performed in presence of substituted enzymes containing paramagnetic metal ions[59,60,66]. In presence of Co(II)CAII the interconversion rate of CO_2 in HCO_3^- is so fast that a single ^{13}C signal is obtained which contains an average information on both CO_2 and HCO_3^- interacting with the protein (Figure 3).

In the low activity species, CoHCAI, under appropriate circumstances two signals are obtained, one for CO_2 and one for HCO_3^-, but their linewidths and T_1's are equal: the interconversion rate is slow with respect to the shift difference but is fast with respect to the difference in nuclear relaxation rates; therefore the latter cannot yet be used to obtain information on each substrate with respect to the enzyme[67]. Kinetic information are obtained on the detachment of HCO_3^-, which has been found to be close to the rate limiting step[66]. If on the other hand an inactive copper(II) enzyme is used the two signals of CO_2 and HCO_3^- are essentially in the diamagnetic position owing to the presence of large excess of substrate, and show different enhancements of both longitudinal (T_{1M}^{-1}) and transverse (T_{2M}^{-1}) nuclear relaxation rates. From the analysis of these rates it is learned that[58,59]: i) the line broadening is not due to residual exchange rate but only to the interaction of each substrate with copper; ii) that HCO_3^- binds to the metal with a dissociation constant of $5 \times 10^3\ s^{-1}$ for bovine isoenzyme II and $4.3 \times 10^2\ s^{-1}$ for human isoenzyme I, and that these values account by themselves for the low activity of the derivative; iii) CO_2 binds at a copper-carbon distance $\lesssim 7$ A. Investigation of the same system in presence of the manganese enzyme (done under the condition in which the effect of interconversion is negligible) provides the same kind of information[60]. Provided that CO_2 does not bind the metal ion, and that we have not identified a site of binding which would permit the identification of activating groups, if any, the CO_2 binding site cannot be far away from the HCO_3^- binding site. Since it is known that

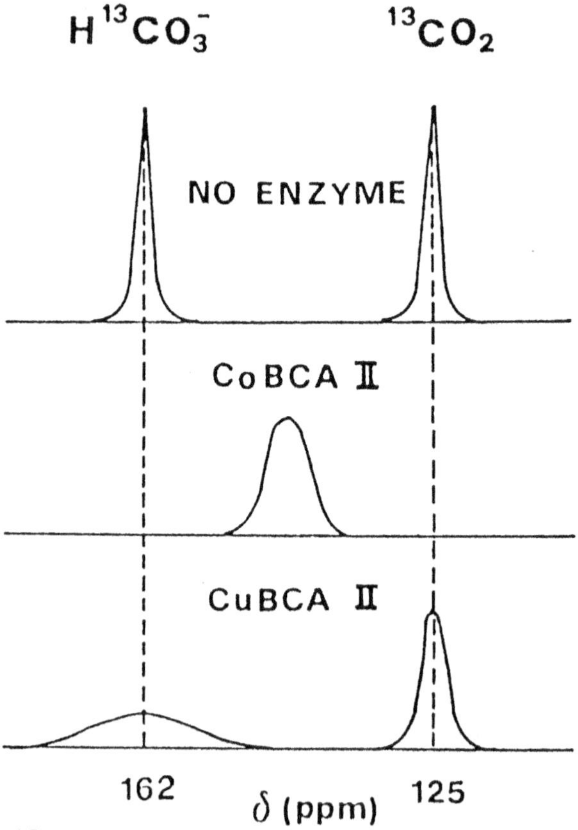

Figure 3. ^{13}C NMR signals for the system CO_2/HCO_3^- under various conditions.

the latter interacts with the metal ion, a computer graphics analysis has provided the following picture[29]:

Scheme IV

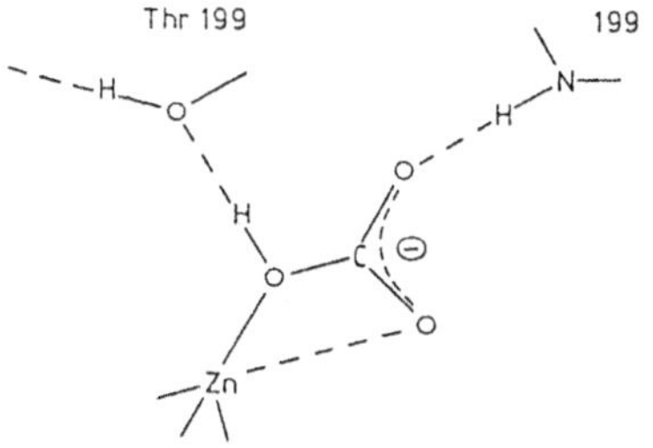

The metal ion active cavity of CA is very reactive, in the sense that it binds most of the molecules which in coordination chemistry are known as ligands, particularly anions of the type: N_3^-, NCO^-, NCS^-, CN^-, carboxylates, etc. either substituting the water molecule and giving rise to four coordination or providing a five coordinate adduct containing a water molecule[12].

Scheme V

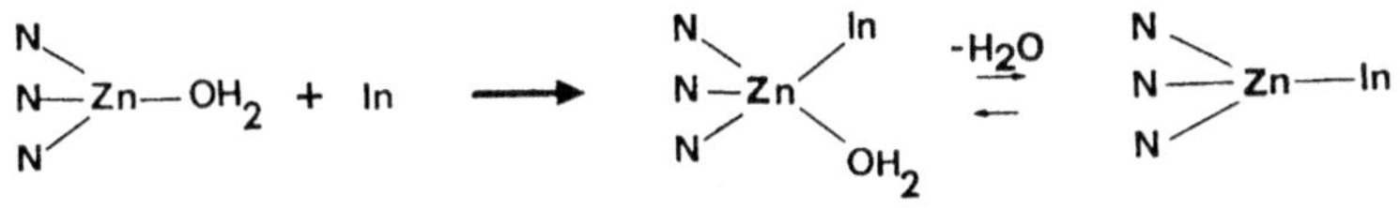

Often an equilibrium between the two species occurs. The kind of coordination number provided is not easily predictable on the basis of inhibitors shape and of their donor atoms. Since these ligands bind with relatively high affinity, they are competitive inhibitors of HCO_3^- binding and therefore of the whole mechanism. Carboxylates generally give rise to largely five coordinated species, whereas sulfonamides give rise to essentially tetracoordinated species, although one of the oxygens may also interact as a fifth, distal ligand[68]. The azide derivative is tetracoordinated whereas NCS^- is five coordinated. It is possible that both the interactions with the cavity and the donor strength at the metal atom determine the overall binding capability[69,70].

As mentioned, anions bind the metal in the low pH species; therefore the apparent affinity constant is pH dependent[71] (Figure 4). Furthermore, by varying the pH, the intrinsic capability of anion binding changes as a result of the protonation of many accessible groups; therefore a detailed understanding of a pK_{app} versus pH profile may be hard to be understood. When anions are relatively strong bases like CN^- and HS^- then the pK_{app} versus pH profile is bell shaped[69,71]. Sulfonamides which are the strongest inhibitors[72] of the enzyme have the same behavior[69], thus indicating that they bind like sulfonamidato anions. From a thermodynamic point of view it is equivalent to say that sulfonamidates bind the acidic form of the enzyme or neutral sulfonamides the alkaline form. Finally imidazole[73] and triazole[74] have an essentially pH independent profile; imidazole is able to bind both the acidic and basic forms of the enzyme, the resulting adducts being formed with neutral imidazole and imidazolate, respectively. The attack of the neutral molecule to the alkaline form is kinetically favored accounting for the competitive behavior of the latter. Triazoles may

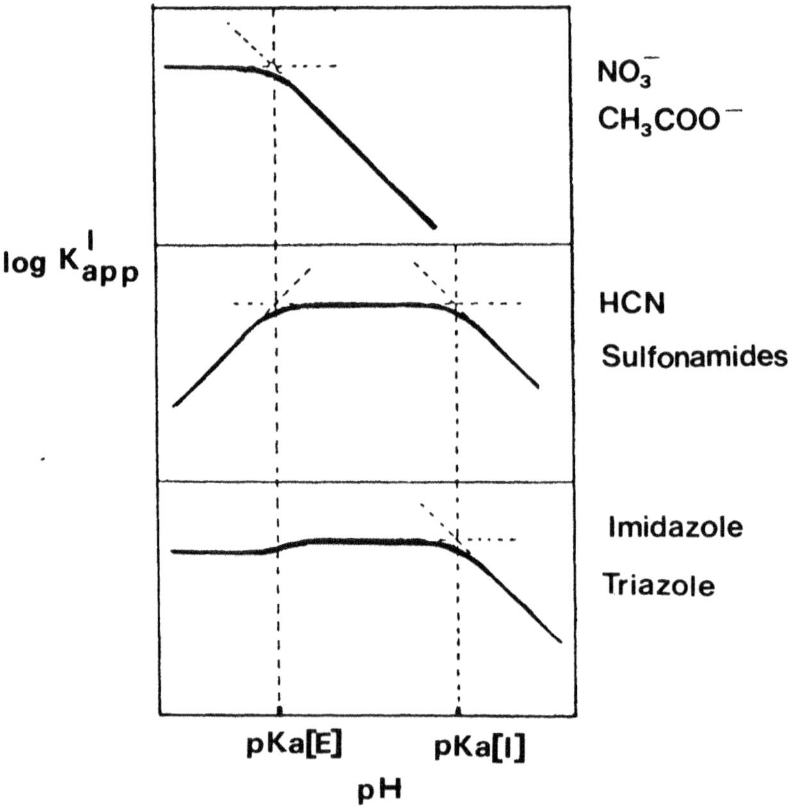

Figure 4. Experimental pH dependence of the apparent affinity constants of different kinds of inhibitors for carbonic anhydrases (from ref. 45, with permission).

bind in a similar way[74], but kinetic measurements[26,41] suggest that binding as anions may be favored[28].

In order to compare the affinity of anions for the various isoenzymes a full profile has been obtained for NO_3^- with three isoenzymes[50]. If reference is made to Scheme I it comes out that the data are consistent with the binding of the anion NO_3^- to the diprotonated form of the enzyme with high activity and with significantly lower activity to the monoprotonated species EH^+ (Figure 5, Figure 6, Table IV). It is assumed that the affinity for the

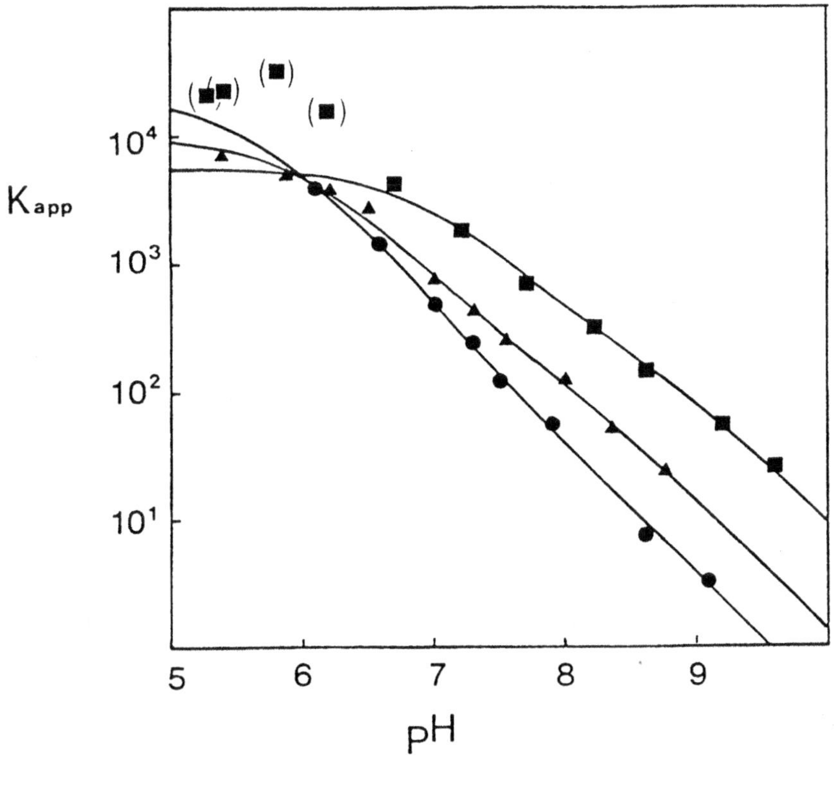

Figure 5. pH dependence of the apparent affinity constants of nitrate ion for Co(II)HCAI (■), Co(II)BCAII (▲), and Co(II)HCAII (●). The data are fitted using the pk_a values of Table II and K_1 and K_3 values of Table IV. The points in parentheses have been excluded from the fitting procedure due to the possible formation of a different nitrate adduct at low pH (from ref. 50, with permission).

TABLE IV

Best Fit Affinity Constants of Nitrate for Species 1 and 3 (Scheme I) of Co(II) Carbonic Anhydrases[50].

	HCAI	BCAII	HCAII
log K_1	3.74+0.04	4.01+0.02	4.34+0.04
log K_3	2.62+0.06	2.56+0.04	2.61+0.05

fully deprotonated species is about zero. It comes out also that the pattern of affinity for the isoenzymes is contrary to the pattern of activity; this is probably related to the fact that the larger the affinity of anions, HCO_3^- included, the lower the catalytic activity of the enzyme.

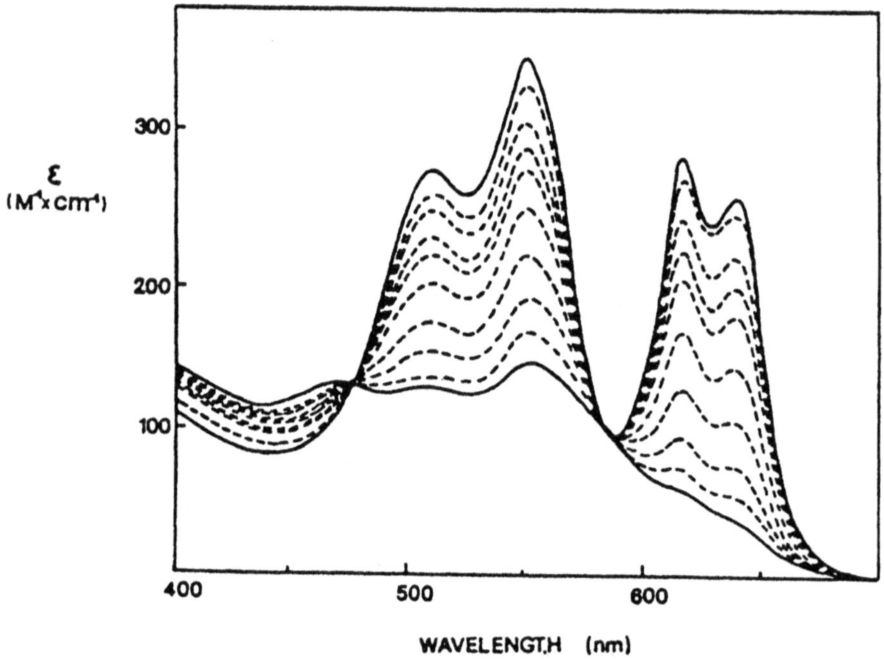

Figure 6. Spectrophotometric titration of Co(II)HCAII with nitrate ion at pH 8.1. Molar absorbance decreases in intensity with increasing nitrate concentration (from ref. 50, with permission).

4. METAL DERIVATIVES

The activity of metal-substituted carbonic anhydrases as compared to that of the native enzyme is reported in Table V. It appears that cobalt (II) preserves the activity and in one case it gives even more activity than the native enzyme[75,78]; manganese (II) is only slightly active[24,76], whereas copper (II)[24,76] nickel (II)[24,76] and oxovanadium(IV)[24,76] do not have appreciable activity. In the case of cadmium(II) the activity is maintained at high pH[70,79,80]; this is due to the high pK_a of the coordinated water. The investigation of this derivative has added precious information on the ligands rigidity in

TABLE V

Specific Activities for CO_2 and *p*-nitrophenylacetate Hydrolysis of Metal Substituted Carbonic Anhydrases Referred to those of the Native Enzymes.

	CO_2	ref.	*p*-NPA	ref.
ZnHCAI	100	75	100	75
CuHCAI	0	75	18	75
NiHCAI	5	75	12	75
CoHCAI	60	75	322	75
FeHCAI	5	76	-	
MnHCAI	4	75	14	75
CdHCAI	2	75	7	75
ZnHCAII	100	40	-	
CoHCAII	50	76	-	
CdHCAII	2	41	-	
ZnBCAII	100	24	100	71
CuBCAII	0	24	0	71
NiBCAII	2	24	1	71
CoBCAII	50	24	97	71
FeBCAII	4	76	1	71
MnBCAII	8	24	4	71
VOBCAII	0	76	-	
CdBCAII	2	24	0	71
ZnBCAIII	100	77	-	
CoBCAIII	50	77	-	

the active cavity and on the interaction of substrate with the enzyme; these information have contributed to the understanding of the enzyme as already discussed.

Here some properties of Co(II)CA which support the biochemical behavior of the enzyme discussed in the previous sections will be reviewed. Also copper and manganese will be briefly discussed. For the details regarding Cd(II)CA and Ni(II)CA, former review articles should be looked at[18,24,81]. Here it can be mentioned that the coordination numbers of cadmium parallel those of cobalt, at least from what can be inferred from ^{113}Cd NMR spectroscopy[79,80,82].

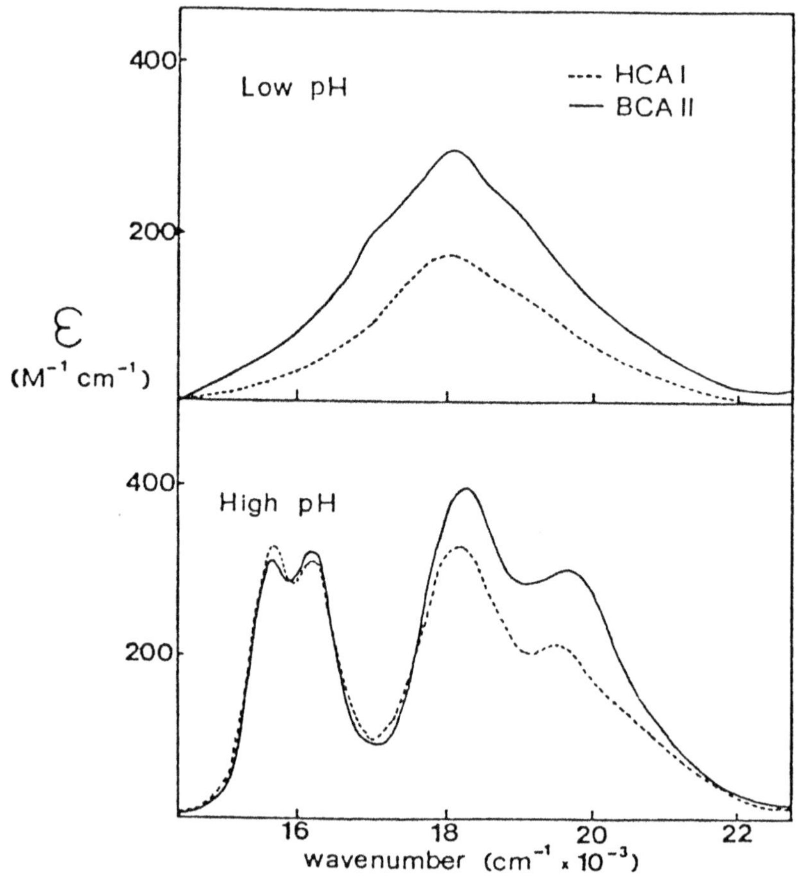

Figure 7. Limit electronic spectra of Co(II)-substituted carbonic anhydrases (from ref. 45, with permission).

4.1. Cobalt(II)

Cobalt(II)CA is an S=3/2 high spin system with beautiful violet-blue color[83]. The electronic spectra and their pH dependence are shown in Figure 1[81]; the limit high and low pH spectra are reported in Figure 7[45]. The many electronic spectra obtained with inhibitors (Figure 8) have provided a variety of spectral features which have been

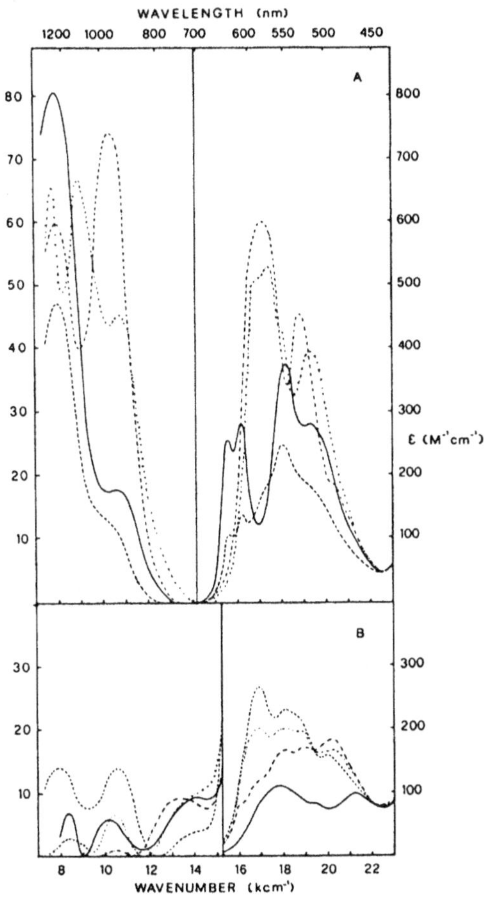

Figure 8. A) Electronic spectra of Co(II)BCAII at pH 5.9 (– – –) and pH 8.0 (——) and limit electronic spectra of the derivatives with acetazolamide, pH 8.0 (•••••••) and cyanate, pH 5.9 (–•–•). B) Chloride (– – –), bromide (.....), iodide (–.–.–), and acetate (——) adducts, pH 5.9 (from ref. 12, with permission).

interpreted in terms of five coordinated or four coordinated derivatives or in terms of equilibria between the two species[12,84]. Five coordinated compounds have been associated with low intensity spectra (ε_{max} < 200 M^{-1} cm^{-1}) whereas four coordinated have been associated with high intensity spectra (ε_{max} > 300 M^{-1} cm^{-1})[12,45,85]. If the electronic spectra are extended to the near-infrared region then five coordinated spectra show a low absorption in the 13-15 cm^{-1} $x10^{-3}$ region which is absent in four coordinated compounds[12,84,86]. Even the CD spectra (Figure 9) are different for the two kinds of derivatives

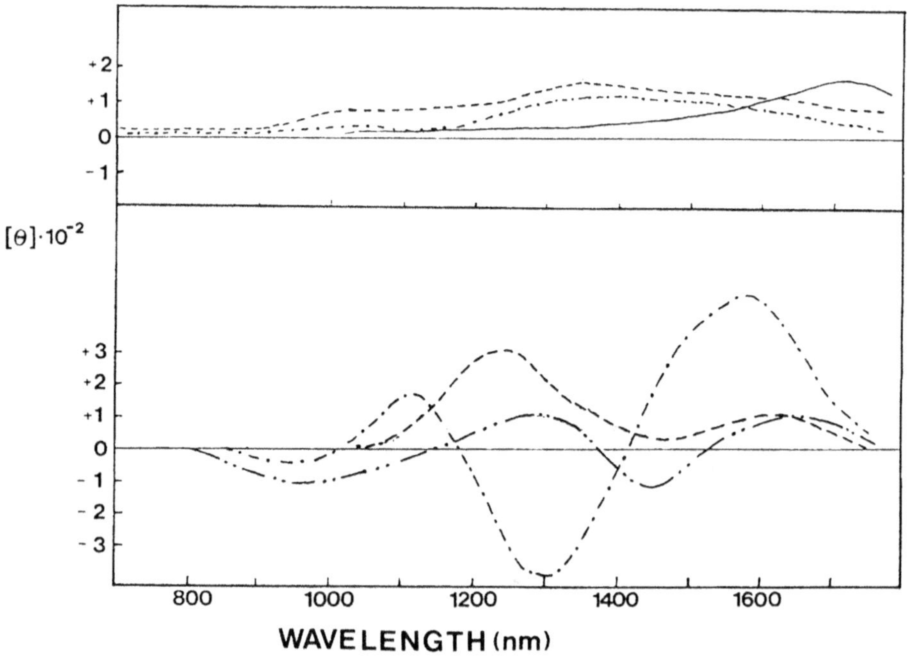

Figure 9. Near infrared CD spectra of: top, Co(II)BCAII NCS (– – –), Co(II)BCAII acetate (–·–·–), Co(II)HCAI at high pH (——); bottom, Co(II)BCAII p-toluenesulfonamide (–·–·–), Co(II)BCAII NCO^- (–··–··), Co(II)BCAII at high pH (– – –)[88].

reflecting the difference in both intensity and number of transitions[87,88]. The MCD spectra (Figure 10) also provide a criterion to distinguish the two different coordination numbers, particularly because four coordinated compounds have a week positive absorption on the high energy side of the bands in the visible region[89,90].

The spectroscopic data of the cobalt enzyme have been interpreted as due to a tetrahedral complex at high pH and to an equilibrium with

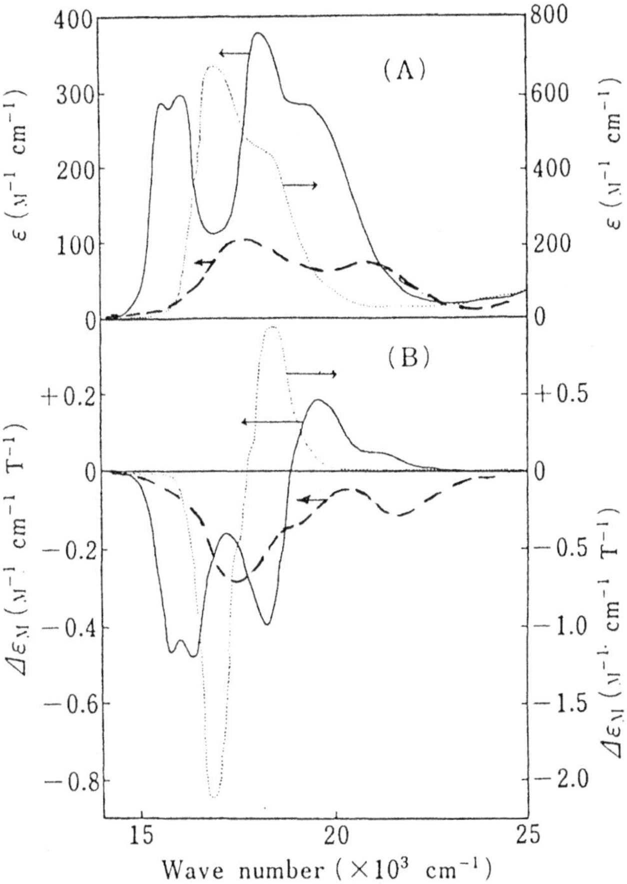

Figure 10. Absorption (A) and MCD (B) spectra of Co(II)BCAII (——)[89,90] and its cyanide (······)[90] and acetate (- - - -)[90] derivatives in 0.1 M Hepes buffer (pH 8.0). Arrows indicate the scale.

one or two water molecules at low pH (Scheme VI). The human isoenzyme I has a large share of five coordination, consistent with a low intensity of the spectra.

Scheme VI

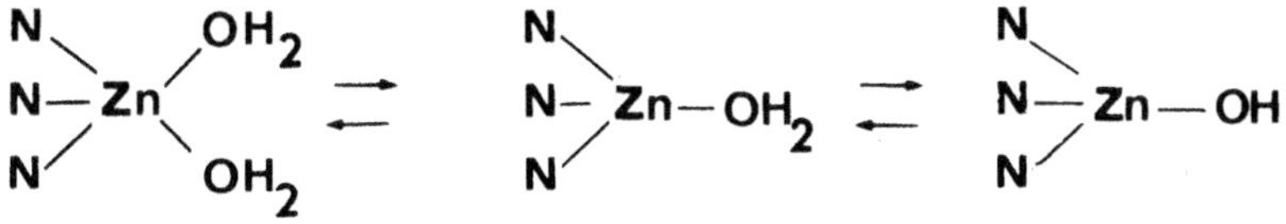

The cobalt carbonic anhydrase derivatives have been widely investigated through ^{1}H NMR which showed the protons of the histidines coordinated to the metal ion[91]; this has shown that three histidines are coordinated under any circumstance (Figure 11). The presence of exchangeable protons bound to one oxygen directly coordinated to the metal ion permit their detection through water solvent ^{1}H NMRD[92]. The knowledge of whether coordinated water is attached to the metal ion is an important piece of information regarding the overall coordination number and should fit within the other independent data. If the water ^{1}H NMRD experiments are performed in a large range of magnetic fields[93,94], then an estimate of the electronic relaxation time can be obtained. This has been achieved by reworking out the theory of the dipolar electron-nucleus coupling[95] (Figure 12).

Even this approach has provided a further criterion to assign the coordination number of Co(II)CA derivatives. In fact, electronic

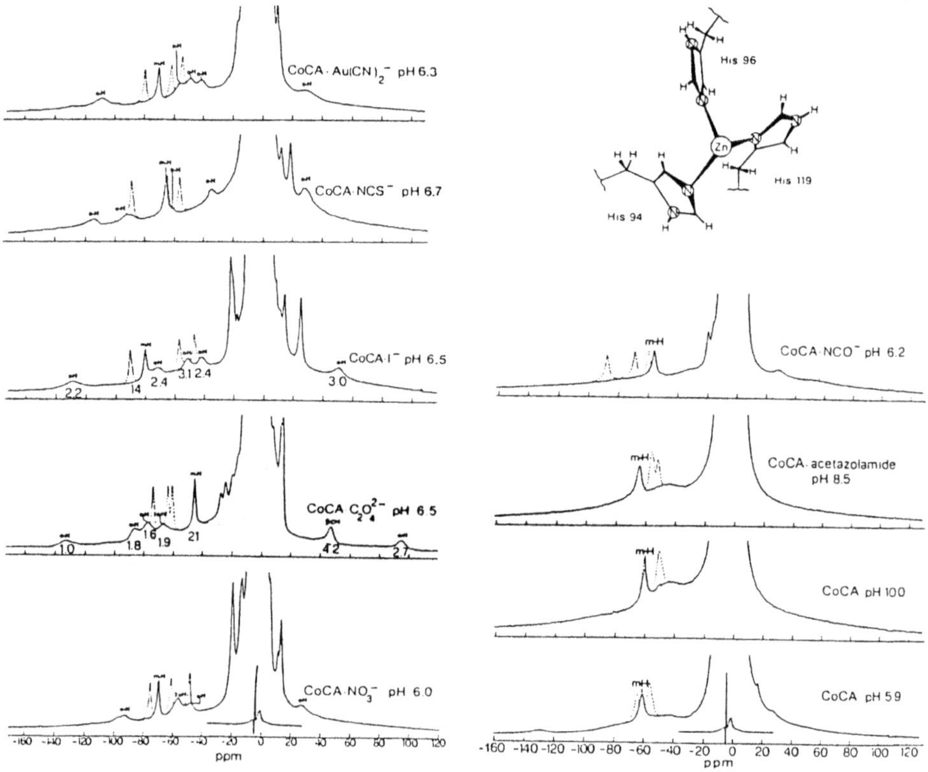

Figure 11. 60 MHz ^{1}H NMR spectra of Co(II)BCAII and its inhibitor derivatives[91] and T_1 values (ms) for iodide and oxalate adducts.

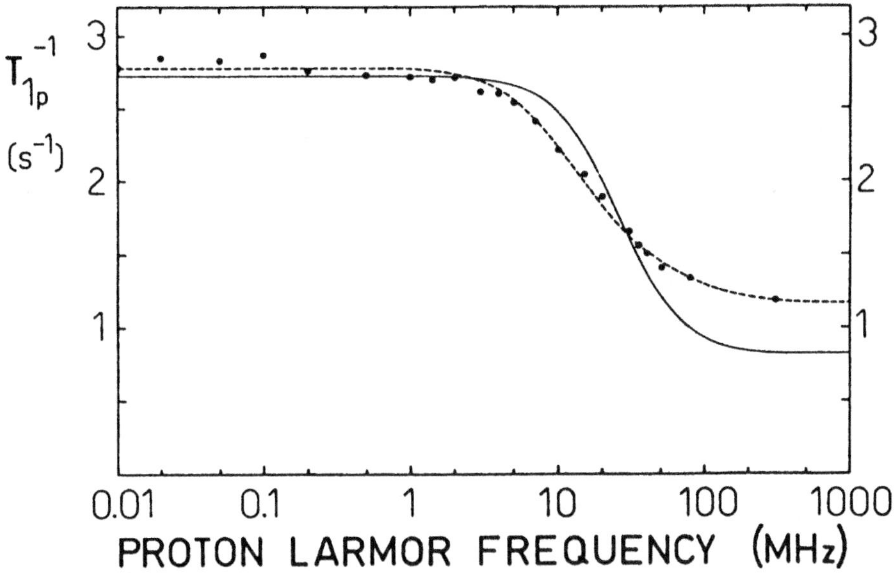

Figure 12. Field dependence of the paramagnetic contribution to the water proton longitudinal relaxation rate, T_{1p}^{-1}, in 3 mM solution of Co(II)HCAI at pH 9.9[94] as fitted by Solomon's equation (——)[96] and by a reworked theory (- - - -)[95]. (From ref. 95 with permission).

relaxation times of five coordinated chromophores are shorter than those of tetrahedral species[97] (typically 1.5 10^{-12} s against 10^{-11} s if allowance is made for some systematic errors due to neglect of particular relaxation mechanisms)[98,99]. Shorter electronic relaxation times cause a small broadening of the ^{1}H NMR signals; therefore the ^{1}H NMR spectra of five coordinated chromophores are definitely of better quality than those of tetrahedral species[91] (Figure 11, left and right hand side). These properties of five-coordinated derivatives have been exploited with ^{13}C enriched carboxymetylated histidine 64 of HCAI; a full characterization of this chemical modified enzyme including its behavior with inhibitors has been achieved[100-104]. A detailed assignment of the ^{1}H NMR spectra of five coordinated chromophores can also provide information on the geometry of the ligands; particularly, the high field broad signal assigned to one ortho-like histidine proton indicates that the chromophore is essentially tetragonal with one apical histidine, the other two histidines and the water molecule defining the basal plane (Figure 13).

The variation of the electronic relaxation time with coordination number can be monitored through the T_1 of any observed NMR signal,

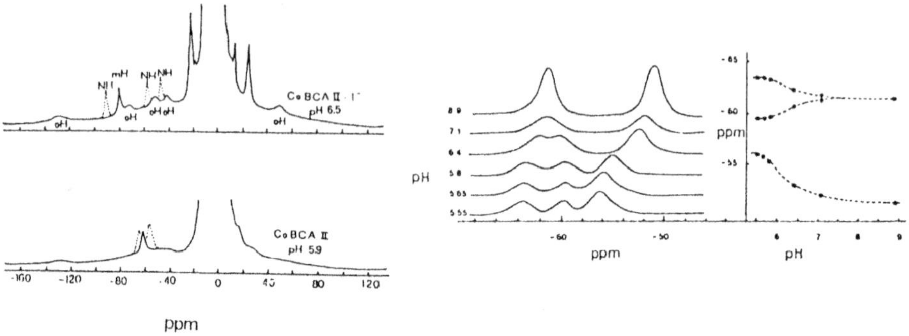

Figure 13. Left: 60 MHz ^{1}H NMR spectra of Co(II)BCAII at pH 5.9 and of its adduct with iodide. Right: pH dependence of the chemical shifts for 300 MHz ^{1}H NMR signals in H_2O solution of histidines bound to the Co(II) ion in Co(II)BCAII[91].

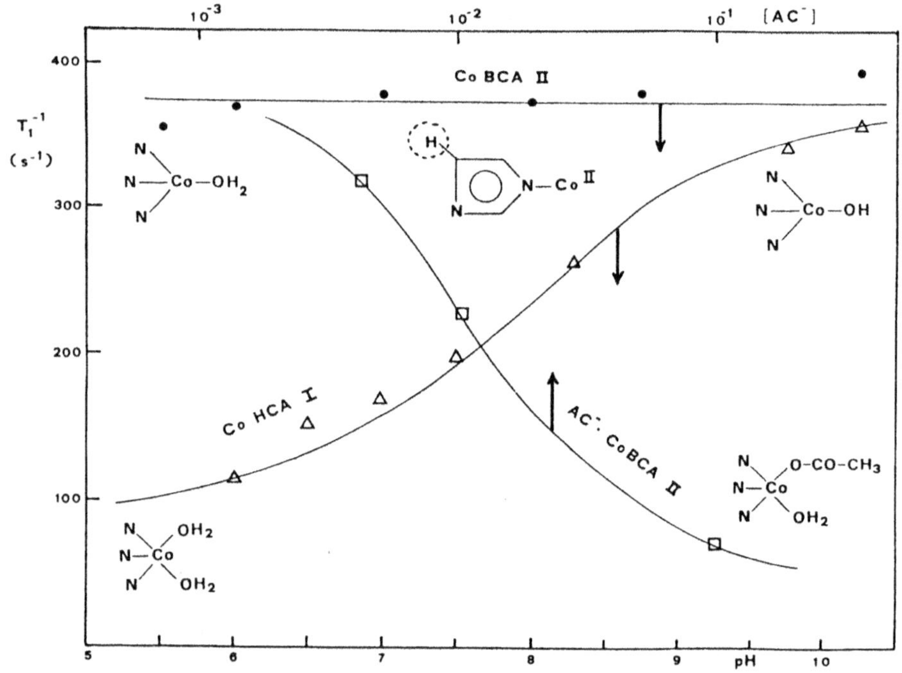

Figure 14. Longitudinal relaxation rates for the 4H His-119 signal in cobalt(II) derivatives of bovine and human carbonic anhydrases under various conditions[105]. Arrows indicate the scale.

particularly the meta-like proton of histidine 119. The T_1 of such proton has a value of 2.5 ms in the essentially tetrahedral form of bovine isoenzyme II and moves to 15 ms upon formation of the five coordinate acetate derivative[105] (Figure 14). Conversely the essentially five coordinate human isoenzyme I displays a T_1 of 8.7 ms for the same signal, which goes to 2.7 ms at high pH when the enzyme becomes tetrahedral[105].

The EPR spectra of Co(II)CA derivatives have been also investigated with relative success in terms of understanding the electronic structure[106,107]. More promising appears the determination of zero field splitting values through saturation studies of the EPR signals; such values may be related to the coordination number[108]. CN^- at low temperature gives rise to a $Co(II)CA(CN)_2$ derivative which is five coordinated low spin with a square pyramidal geometry[109].

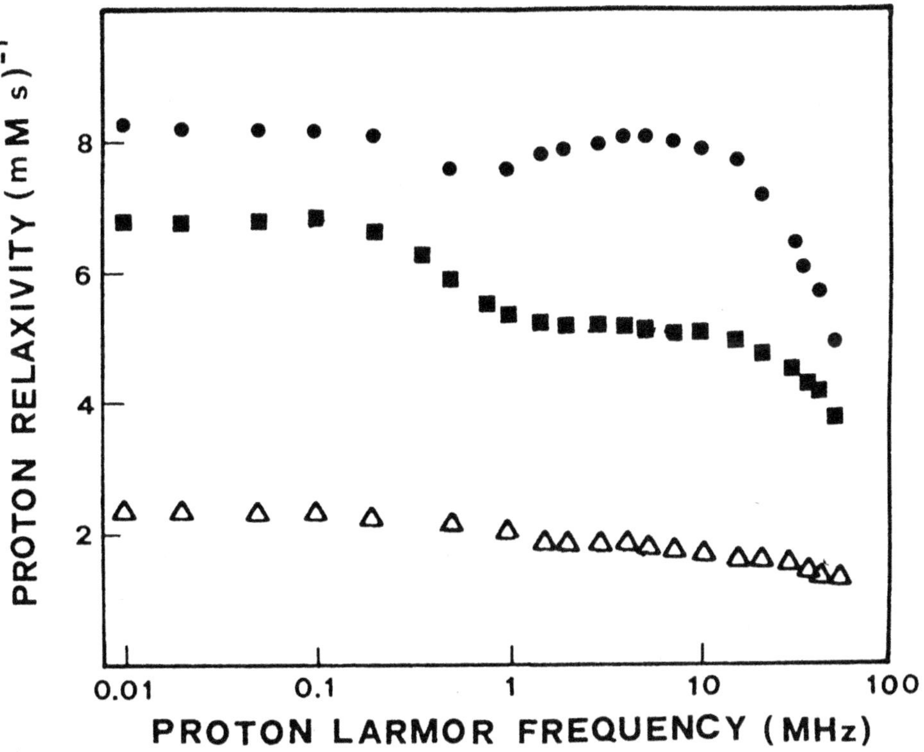

Figure 15. Water proton NMRD for Cu(II)BCAII (■), Cu(II)BCAII N_3^- (●) and Cu(II)BCAII HCO_3^- (▲) at 25 °C (from ref. 59 with permission).

4.2. Copper(II)

Cu(II)CA has been deeply investigated in the late seventies. Electronic spectra[110,111], EPR spectra[110-113] and later ^1H NMRD measurements at various magnetic fields[59,99,111,114] have shown that the electronic relaxation times of copper are essentially constant from a derivative to another and that variations in water ^1H T_1 are due to different copper-water interactions. Cu(II)CA HCO_3^- has a water ^1H T_1 profile different from that of other anions (Figure 15)[59]. This may be due to the particular shape which has to suit the CA cavity. The independent knowledge of the electronic relaxation times is a precious aid in the interpretation of ^{13}C NMR spectra.

4.3. Manganese(II)

After the early papers by Navon[115,116] no much has been done on this system owing of the lack of efficient investigation tools. The difficulties in understanding the water ^1H NMR T_1 values have been recently reviewed[117,118]. Recent ^{13}C NMR studies on CO_2 and HCO_3^- interacting with Mn(II)CA are consistent with the data obtained on the copper derivatives[60].

5. REFERENCES

1) Keilin, D., and Mann, T., Biochem. J., **34**, 1163 (1940).
2) Meldrum, N. U., and Roughton, F. J. W., J. Physiol. (London), **80**, 113 (1933).
3) Faurholt, C., J. Chim. Phys., **21**, 400 (1924).
4) Edsall, J. T. and Wyman, J., Biophys. Chem., Acad. Press, New York, **1**, 550 (1958).
5) Lindskog, S., in Zinc Enzymes, Metal Ions in Biology, **5**, Spiro, T. G., Ed., Wiley-Interscience, New York, p. 78, (1983).
6) Kannan, K. K., Notstrand, B., Fridborg, K., Lövgren, S., Ohlsson, A., and Petef, M., Proc. Natl. Acad. Sci. USA, **72**, 51 (1975).
7) Liljas, A., Kannan, K. K., Bergsten, P. C., Waara, I., Fridborg, K., Strandberg, B., Carlbom, U., Järup, L., Lövgren, S., and Petef, M., Nature, **235**, 131 (1972).
8) Eriksson, E. A., Jones, T. A. and Liljas, A., in Zinc Enzymes, Bertini, I., Luchinat, C., Maret, W., and Zeppezauer, M., Eds., Birkhäuser, Boston, p. 317 (1986).
9) Tu, C. K., Sanyal, G., Wynns, G. C., and Silverman, D. N., J. Biol. Chem., **258**, 8867 (1983).
10) Kararli, T., and Silverman, D. N., J. Biol. Chem., **260**, 3484 (1985).

11) Engberg, P., Millqvist, E., Pohl, G., and Lindskog, S., Arch. Biochem. Biophys., **241**, 628 (1985).
12) Bertini, I., Canti, G., Luchinat, C., and Scozzafava, A., J. Am. Chem. Soc., **100**, 4873 (1978).
13) Bertini, I., Canti, G., Luchinat, C., and Scozzafava, A., Biochem. Biophys. Res. Commun., **78**, 158 (1977).
14) Koenig, S.H., Brown, R.D., and Jakob, G.S., in Biophysics and Physiology of Carbon Dioxide, Bauer, C., Gros, G., and Bartels, M., Eds., Springer Verlag, Berlin, p. 238 (1980).
15) Barzi, D., Bertini, I., Luchinat, C., and Scozzafava, A., Inorg. Chim. Acta, **36**, L431-L432 (1979).
16) Bertini, I., Luchinat, C., and Scozzafava, A., Inorg. Chim. Acta, **46**, 85 (1980).
17) Simonsson, I., and Lindskog, S., Eur. J. Biochem., **123**, 29 (1982).
18) Bertini, I., and Luchinat, C., in Metal Ions in Biological Systems, Sigel, H., Ed., M. Dekker Inc., New York, **15**, p. 101 (1983).
19) Campbell, I. D., Lindskog, S., and White, A. I., J. Mol. Biol., **90**, 469 (1974).
20) Campbell, I. D., Lindskog, S., and White, A. I., J. Mol. Biol., **98**, 597 (1975).
21) Campbell, I. D., Lindskog, S., and White, A. I., Biochim. Biophys. Acta, **484**, 443 (1977).
22) Pesando, J. M., Biochemistry, **14**, 675 (1975).
23) Lindskog, S., Henderson, L. E., Kannan, K. K., Liljas, A., Nyman, P. O., and Strandberg, B., in The Enzymes, Boyer, P. D., Ed., Academic press, New York, **5**, 587 (1971).
24) Lindskog, S., in Advances in Inorganic Biochemistry Eichhorn, G. L., and Marzilli, L. G., Eds., Elsevier, New York, **4**, 115 (1982).
25) Pocker, Y., and Sarkanen, S., Adv. Enzymol., **47**, 149 (1978).
26) Khalifah, R. G., J. Biol. Chem., **246**, 2561 (1971).
27) Kernohan, J.C., Biochim. Biophys. Acta, **96**, 304 (1965).
28) Tibell, L., Thesis, University of Umea, Umea, Sweden, (1984).
29) Lindskog, S., in Zinc Enzymes, Bertini, I., Luchinat, C., Maret, W., and Zeppezauer, M., Eds., Birkhäuser, Boston, p. 307 (1986).
30) Simonsson, I., Jonsson, B.-H., and Lindskog, S., Eur. J. Biochem., **129**, 165 (1982).
31) Steiner, H., Jonsson, B.-H., and Lindskog, S., FEBS Lett., **62**, 16 (1976).
32) Simonsson, I., Jonsson, B.-H., and Lindskog, S., Eur. J. Biochem., **93**, 409 (1979).
33) Sen, A.C., Tu, C.K., Thomas, H., Wynns, G.C., and Silverman, D.N., in Zinc Enzymes, Bertini, I., Luchinat, C., Maret, W., and Zeppezauer, M., Eds., Birkhäuser, Boston, p. 329 (1986).

34) Steiner, H., Jonsson, B.-H., and Lindskog, S., Eur. J. Biochem., **59**, 253 (1975).
35) Silverman, D. N., and Tu, C. K., J. Am. Chem. Soc., **97**, 2263 (1975).
36) Jonsson, B. H., Steiner, H., and Lindskog, S., FEBS Lett., **64**, 310 (1976).
37) Rowlett, R. S., and Silverman, D. N., J. Am. Chem. Soc., **104**, 6737 (1982).
38) Pocker, Y., and Bjorquist, Biochemistry, **16**, 5698 (1977).
39) Venkatasubban, K. S., and Silverman, D. N., Biochemistry, **19**, 4984 (1980).
40) Lindskog, S., Engberg, P., Forsman, C., Ibrahim, S. A., Jonsson, B.-H., Simonsson, I., and Tibell, L., Ann. N. Y. Acad. Sci., **429**, 61 (1984).
41) Tibell, L., Forsman, C., Simonsson, I., and Lindskog, S., Biochim. Biophys. Acta, **789**, 302 (1984).
42) Pocker, Y., and Deits, R. L., J. Am. Chem. Soc., **104**, 2424 (1982).
43) Jacob, G. S., Brown, R. D., and Koenig, S. H., Biochem. Biophys. Res. Commun., **82**, 203 (1978).
44) Silverman, D. N., Tu, C. K., Lindskog, S., and Wynns, G. C., J. Am. Chem. Soc., **101**, 6734 (1979).
45) Bertini, I., and Luchinat, C., Acc. Chem. Res., **16**, 272 (1983).
46) Clementi, E., Corongiu, G., Jönsson, B., and Romano, S., FEBS Lett., **100**, 313 (1979).
47) Bertini, I., Lanini, G., and Luchinat, C., J. Am. Chem. Soc., **105**, 5116 (1983).
48) Bertini, I., Dei, A., Luchinat, C., and Monnanni, R., Inorg. Chem., **24**, 301 (1985).
49) Bertini, I., Luchinat, C., and Monnanni, R., J. Chem. Ed., **62**, 924 (1985).
50) Bertini, I., Dei, A., Luchinat, C., and Monnanni, R., in Zinc Enzymes, Bertini, I., Luchinat, C., Maret, W., and Zeppezauer, M., Eds., Birkhäuser, Boston, p. 371 (1986).
51) Burki, S., Thesis, University of Basel, Basel, Switzerland (1977).
52) Billo, E. J., Inorg. Nucl. Chem. Lett., **11**, 491 (1975).
53) Wolley, P., Nature, **258**, 677 (1975).
54) Dei, A., Paoletti, P., and Vacca, A., Inorg. Chem., **7**, 865 (1968).
55) Bertini, I., Canti, G., Luchinat, C. and Mani, F., Inorg. Chem, **20**, 1670 (1981).
56) Groves, T.J., and Olson, J.R., Inorg. Chem., **24**, 2715 (1985).
57) Bertini, I., Luchinat, C., and Messori, L., Inorg. Chem., **21**, 3426 (1982).
58) Bertini, I., Borghi, E., and Luchinat, C., J. Am. Chem. Soc., **101**, 7069 (1979).

59) Bertini, I., Borghi, E., Canti, G., and Luchinat, C., J. Inorg. Biochem., **18**, 221 (1983).

60) Led, J.J., Nesgaard, E., and Johansen, J.T., FEBS Lett., **147**, 74 (1982).

61) Pullman, A., Ann. N.Y. Acad. Sci. U.S.A., **367**, 340 (1981).

62) Pocker, Y., and Stone, T., Biochemistry, **7**, 2936 (1968).

63) Palmer, D.A., and Van Eldik, R., Chem. Rev., **83**, 651 (1983).

64) Bianchini, C. and Meli, A., J. Am. Chem. Soc., **106**, 2698 (1984).

65) Lindskog, S., Ibrahim, S.A., Jonsson, B.-H. and Simonsson, I, in The Coordination Chemistry of Metalloenzymes, Bertini, I., Drago, R. S., and Luchinat, C., Eds., Reidel, D., Dordrecht, Holland, p. 49 (1983).

66) Williams, T.J., and Henkens, R.W., Biochemistry, **24**, 2459 (1985).

67) Stein, P.J., Merrill, S.T., and Henkens, R.W., J. Am. Chem. Soc., **99**, 3194 (1977).

68) Vedani, A., and Dunitz, J. D., J. Am. Chem. Soc., **107**, 7653 (1985).

69) Taylor, P.W., King, R.W., and Burgen, A.S.V., Biochemistry, **9**, 3894 (1970).

70) Coleman, J.E., Nature, **214**, 193 (1967).

71) Thorslund, A. and Lindskog, S., Eur. J. Biochem., **3**, 117 (1967).

72) Mann, T., and Keilin, D., Nature, **146**, 164 (1940).

73) Bauer, R., Limkilde, P., and Johansen, J.T., Carlsberg Res. Commun., **42**, 325 (1977).

74) Alberti, G., Bertini, I., Luchinat, C., and Scozzafava, A., Biochim. Biophys. Acta, **668**, 16 (1981).

75) Coleman, J.E., Biochemistry, **4**, 2644 (1965).

76) Lindskog, S., and Nyman, P.O., Biochim. Biophys. Acta, **85**, 462 (1964).

77) Engberg, O., and Lindskog, S., FEBS Lett., **170**, 326 (1984).

78) Duff, T.A., and Coleman, J.E., Biochemistry, **5**, 2009 (1966).

79) Coleman, J.E., Ann. N.Y. Acad. Sci. USA, **429**, 61 (1984).

80) Coleman, J.E., and Gettins, P., in Zinc Enzymes, Bertini, I., Luchinat, C., Maret, W., and Zeppezauer, M., Eds., Birkhäuser, Boston, p. 77 (1986).

81) Bertini, I., Luchinat, C., and Scozzafava, A., Struct. Bonding, **48**, 46 (1982).

82) Evelhoch, J.L., Bocian, D.F., and Sudmeier, J.L., Biochemistry, **20**, 4951 (1981).

83) Lindskog, S., and Nyman, P.O., Biochim. Biophys. Acta, **85**, 141 (1964).

84) Bertini, I., Luchinat, C., and Scozzafava, A., Inorg. Chim. Acta, **22**, L-23 L-24 (1977).

85) Rosenberg, R.C., Root, C.A., Wang, R.-H., Cerdonio, M., and Gray, H.B., Proc. Natl. Acad. Sci. U.S.A., **70**, 161 (1973).
86) Bertini, I., Canti, G., Luchinat, C., and Romanelli, P., Inorg. Chim. Acta, **46**, 211 (1980).
87) Vallee, B.L. and Holmquist, B., in Advances in Inorganic Biochemistry, Elsevier North Holland, New York, **2**, 27 (1980).
88) Unpublished data from our laboratory.
89) Holmquist, B., Kaden, T.A., and Vallee, B.L., Biochemistry, **14**, 1454 (1975).
90) Hirose, J., Noji, M., and Kidani, Y., Chem. Farm. Bull., **32**, 2481 (1984).
91) Bertini, I., Canti, G., Luchinat, C., and Mani, F., J. Am. Chem. Soc., **103**, 7784 (1981).
92) Bertini, I., and Luchinat, C., NMR of Paramagnetic Molecules in Biological Systems, Benjamin/Cummungs, Boston, Ch. 9, (1986).
93) Fabry (Riepe), M.E., Koenig, S.H., and Shillinger, W.E., J. Biol. Chem., **245**, 4256 (1970).
94) Bertini, I., Brown III, R.D., Koenig, S.H., and Luchinat, C., Biophys. J., **41**, 179 (1983).
95) Bertini, I., Luchinat, C., Mancini, M., and Spina, G., J. Magn. Reson., **59**, 213 (1984).
96) Solomon, I., Phys. Rev., **99**, 559 (1955).
97) Bertini, I., Canti, G., and Luchinat, C., Inorg. Chim. Acta, **56**, 99 (1981).
98) Bertini, I., Lanini, G., and Luchinat, C., Inorg. Chim. Acta, **80**, 123 (1983).
99) Banci, L., Bertini, I., and Luchinat, C., Inorg. Chim. Acta, **100**, 173 (1985).
100) Khalifah, R.G., Rogers, J.I., Mukherjee, J., and Morley, P.J., Ann. N.Y. Acad. Sci. U.S.A., **429**, 114 (1984).
101) Khalifah, R.G., Rogers, J.I., Morley, P.J., and Carroll, S.B., Biochemistry, **23**, 3129 (1984).
102) Morley, P.J., Carroll, S.B., Harmon, P., and Khalifah, R.G., J. Am. Chem. Soc., **106**, 1830 (1984).
103) Jeffers, P.K., Sutherland, W.M., and Khalifah, R.G., Biochemistry, **17**, 1305 (1978).
104) Khalifah, R.G., Rogers, J.I., and Mukherjee, J., in Zinc Enzymes, Bertini, I., Luchinat, C., Maret, W. and Zeppezauer, M Eds., Birkhäuser, Boston, p. 357 (1986).
105) Bertini, I., Lanini, G. and Luchinat, C., J. Am. Chem. Soc., **105**, 5116 (1983).
106) Desideri, A., Morpurgo, L., Raynor, J.B., and Rotilio, G., Biophys. Chem., **8**, 267 (1978).
107) Bencini, A., Bertini, I., Canti, G., Gatteschi, D., and Luchinat, C., J. Inorg. Biochem., **14**, 81 (1981).

108) Makinen, M.W., Kuo, L.C., Yim, M.B., Wells, G.B., Fukuyama, J.M., and Kim, J.E., J. Am. Chem. Soc., **107**, 5245 (1985).
109) Haffner, P.H., and Coleman, J.E., J. Biol. Chem., **248**, 6630 (1973).
110) Morpurgo, L., Rotilio, G., Finazzi Agrò, A., and Mondovì, B., Arch. Biochem. Biophys., **170**, 360 (1975).
111) Bertini, I., Canti, G., Luchinat, C., and Scozzafava, A., J. C. S. Dalton, 1269 (1978).
112) Taylor, J.S., and Coleman, J.E., J. Biol. Chem., **246**, 7058 (1971).
113) Haffner, P.H., and Coleman, J.E., J. Biol. Chem., **250**, 995 (1975).
114) Bertini, I., and Luchinat, C., in Biochemical and Inorganic Aspects of Copper Coordination Chemistry, Karlin, K.D., and Zubieta, J., Eds., Adenine Press, p. 23 (1986).
115) Lanir, A., and Navon, G., Biochemistry, **11**, 3536 (1972).
116) Lanir, A., Gradstajn, S., and Navon, G., Biochemistry, **14**, 242 (1975).
117) Kushnir, T., and Navon, G., J. Magn. Reson., **56**, 373 (1984).
118) Banci, L., Bertini, I., Briganti, F., and Luchinat, C., J. Magn. Reson., **66**, 58 (1986).

EVOLUTION OF MODELS FOR THE STRUCTURAL, SPECTROSCOPIC, AND CATALYTIC PROPERTIES OF CARBONIC ANHYDRASE

R.S. Brown
Department of Chemistry
University of Alberta
Edmonton, Alberta, Canada
T6G 2G2

ABSTRACT. The evolution of simple, low molecular weight models for the active site of carbonic anhydrase (CA) is presented. The models have generally followed two paths: the first as structural models for the active site coordination in the enzyme as well as models for the spectroscopic properties of CoCA and the second as models for the catalytic properties of the enzyme. With respect to the first category, a presentation of the structural models follows a chronological order, finally ending with the tripod ligands which have three M^{2+}-chelating groups emanating from a central pivitol anchor. These models to some extent as their Co^{2+}-complexes have spectral properties similar to that of CoCA. In addition, such models provide evidence that the ionization of a Co^{2+}-OH_2 unit occurs at relatively low pH and produces changes in the visible absorption spectrum reminiscent of those seen for the cobalt enzyme.

Models which catalyse the interconversion of $CO_2 \rightleftarrows HCO_3^-$, this being the natural function of the enzyme, are not as well evolved as the structural models. Nevertheless, many systems are presented in which a M-coordinated OH_2 has a low pK_a but the M-OH^- unit retains much of the nucleophilicity of free OH^-. Possible mechanisms by which these models function are presented and discussed. Finally, conclusions concerning the state of the art of the models and future considerations are given.

1. Introduction

Carbonic anhydrase (CA) is a widely occurring Zn^{2+}-metalloenzyme found in both plants and animals.[1] The complete structures of the human B and C isozymes (HCAB and HCAC respectively) have been elucidated by x-ray crystallographic analysis[2] (Fig. 1) which indicates that the essential Zn^{2+} is bound in a tetrahedral fashion by three histidine imidazole units; a fourth and possibly fifth ligand position is occupied by H_2O or OH^- depending upon the pH. Although the only known function of CA is to facilitate the interconversion of CO_2 and HCO_3^- as in eq. 1, it plays a key role in such diverse processes as photosynthesis, physiological pH control, gas balance, and calcification.[1] It will however, catalyse

M. Aresta and G. Forti (eds.), Carbon Dioxide as a Source of Carbon, 169–197.

$$1. \quad CO_2 + H_2O \underset{}{\overset{CA}{\rightleftharpoons}} H^+ + HCO_3^-$$

other processes involving nucleophilic attack of OH^- on electrophilic centres including hydration of aldehydes,[3a] pyruvate,[3b] and alkyl pyruvates[3c] as well as the hydrolysis of carboxylic, sulfonic, carbonic and phosphoric esters.[4]

Figure 1 A stylized portion of the active site of Human CAB showing the Zn^{2+}-environment and hydrogen-bonding network. (Redrawn from ref. 2b)

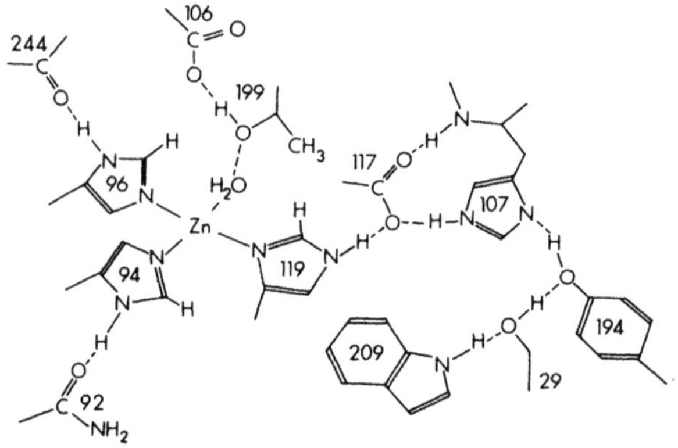

The Zn^{2+} ion of native CA can be removed by dialysis of the enzyme against 1,10-phenanthroline in acetate buffer (pH 5)[5a] or more rapidly by EDTA.[5b] Other divalent metals can be inserted into the apoenzyme,[6] but of these, only Zn^{2+} and Co^{2+} result in significant restoration of activity, (CoCA ~ 45% of native enzyme[5a]). CoCA is a reddish blue species and displays a visible absorption spectrum characteristic of a four and/or five-coordinate environment about the metal.[6,7] Analysis of the pH dependence of the Co^{2+} visible absorption spectrum indicates that the alkaline form of bovine CoCA (which is also catalytically active in CO_2 hydration) is predominantly distorted tetrahedral while the acid form apparently consists of an equilibrium distribution of 4 and 5 coordinate forms, (eq. 2).[7]

2.

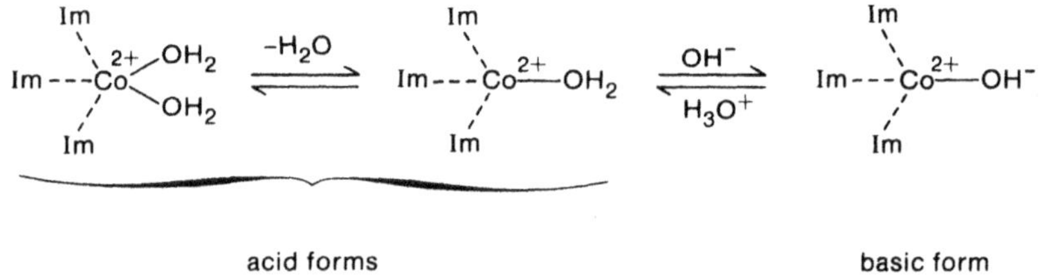

Although the pK_a of the group responsible for the spectroscopic changes in CoCA was originally reported to be ~ 7,[6] more recent work in the presence of non-interacting buffers or ion-free solutions indicates the absorption vs pH profile is not as simple as that expected for a single dissociating group.[7b] Also, the spectrophotometric pK_a is markedly dependent upon the presence of inhibitory mono-anions, and, for example increases from 5.6 to 7.9 as $[Cl^-]$ increases from 0-0.6 M.[6c] Finally, the Co absorption spectrum is markedly dependent upon the presence of added anions, a fact that indicates these anions are in some way coordinated to the metal.[6,7] Presented in Table I are the spectral properties of some anionic derivatives of CoCA.

Table I Characteristic Coordination Environments for Various Complexes of CoCA.[a]

4 Coordinate	4-5 Coordinate	5 Coordinate
(ε_{max}>300 $M^{-1}cm^{-1}$)	(200 $M^{-1}cm^{-1}$<ε_{max} <300 $M^{-1}cm^{-1}$	(ε_{max}<200 $M^{-1}cm^{-1}$)
Bovine CoCAB (high pH)	Bovine CoCAB low pH	Human CoCAB low pH
Human CoCAB (high pH)	HCO_3-	SCN^-
CN^-	F^-	HSO_3-
NCO^-	Cl^-	I^-
SH^-	Br^-	acetate
aniline	N_3^-	formate
imidazole (high pH)	imidazole (low pH)	oxalate
sulfonamides	benzoate	succinate

a Bertini, I.; Luchinat, C., Accts. Chem. Res. 1983, **16**, 272.

Considering the physiological importance of the enzyme, and the fact that the active site is relatively well structurally characterized, it is not surprising that several attempts have been made to model one or more features of the complex enzyme using simpler systems.[8] A large number of the model studies have centred on the design of chelating agents that approximate the distorted 4,5-coordinate active site. Of particular interest has been the spectroscopic properties of the Co^{2+}-chelates and the pK_a values metal-associated groups. A smaller number of studies have been directed toward modelling the catalytic properties of the enzyme, and in particular the ability to hydrate CO_2 or aldehydes. The following will deal first with the evolution of the models for the spectroscopic properties of CoCA and then in turn with catalytic models. Particular emphasis for the latter will be placed upon the $CO_2 \rightleftarrows HCO_3^-$ equilibrium although a number of models have been studied as to their ability to effect other catalytic tasks.

2. Spectroscopic and Structural Models

Octahedrally coordinated Co^{2+} gives rise to pink solutions having molar absorbancies of ~ 10 $M^{-1}cm^{-1}$ around 490-500 nm.[9] On the other hand, solutions of tetrahedral Co^{2+} are bright blue and have extinction coefficients of > 300 $M^{-1}cm^{-1}$ at 550-650 nm. Five coordinate Co^{2+} is typified by lighter blue solutions (ε_{max} < 200 $M^{-1}cm^{-1}$ at 530-630 nm) as well as a diagnostic weak transition (ε_{max} < 50 $M^{-1}cm^{-1}$) between 700 and 800 nm. As can be seen from Fig. 2, the spectrum of bovine CoCAB in unbuffered medium changes as a function of pH. At low pH (5.8) the spectrum is characteristic of an equilibrium 4-5 coordinate environment which changes to 4-coordinate as the pH is raised. Although originally it appeared that the spectrophotometric change adhered to a process having a single pK_a ~ 7,[6a] more recent work in the absence of buffers and anions incidates the profile can be reproduced assuming ionization of two active site groups, one having a pK_a < 6, and the other > 7.[7b] Possibly these are a metal coordinated H_2O and the active site histidine (His 64).

Figure 2 The absorption spectrum of Bovine CoCAB (top) and Human CoCAB (bottom) in unbuffered medium as a function pH. (Redrawn from ref. 7b)

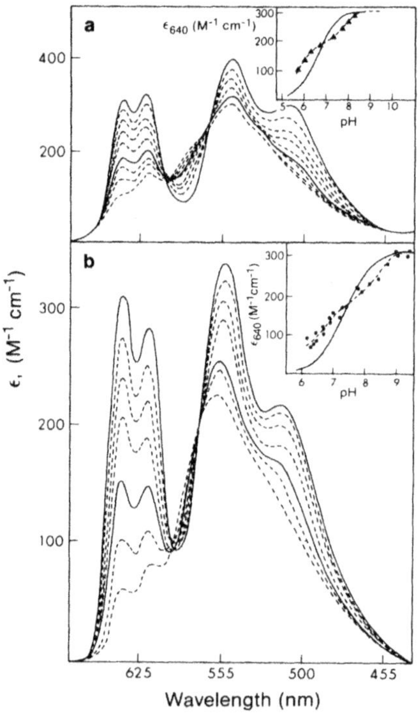

Several attempts to model the absorption/pH dependence for CoCA have been made. Much work has centred on the nature of the ionizing group responsible for the spectrophotometric pK_a. The two principal candidates for this group are M^{2+}-coordinated imidazole or H_2O.

Early spectroscopic measurements by Morris and Martin[10] with 2:1 complexes of L-histidine and Co^{2+} indicated that at pH 9, the absorption spectrum was characteristic of octahedral coordination (eq. 3).

3.

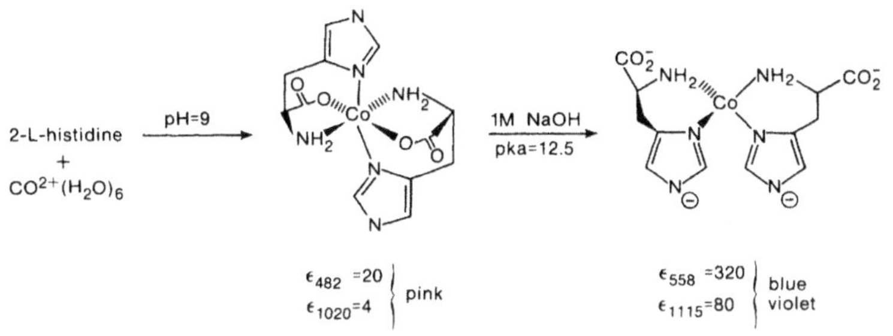

However, as the pH was raised to that of 1N NaOH, a spectroscopic transition occurred in which a tetrahedral environment was formed. In this case,[10] ionization of the coordinated histidine pyrrole N-H (pK_a = 12.5) produced a ligand sufficiently anionic that it decreased the Co^{2+} coordination requirements. That the ionization of the Co^{2+}-coordinated imidazole ring NH was responsible for the spectroscopic change was implied by the observation that histidine, histamine and imidazole produced similar changes, but N-methylimidazole and 1, or 3-methylhistidine did not. The major shortcoming of these as spectroscopic models for CoCA stems from their inability to approximate the known coordination site in the enzyme (3 protein based imidazoles). Also, the pK_a of 12.5, attributable to ionization of the pyrrole NH seems too high to be accepted as an analogue for the origin of the ionization in the enzyme, (although it is difficult to exclude a hydrophobic environment in the active site region as perturbing the pK_a of the imidazole: Zn^{2+} unit downward.) Subsequent titration studies of some Zn^{2+} imidazole complexes were interpreted[11] in terms of ionization of Zn-coordinated imidazole having a pK_a near 7. However, these conclusions were criticized[12] on the basis of lack of precedence and the fact that they were drawn from titrations performed under nonequilibrium conditions. It appears that coordinated pyrrole N-H pK_a values in Zn^{2+} complexes should be closer to ~ 13, similar to the corresponding Co^{2+} complexes,[10] and that M^{2+}-coordinated H_2O should deprotonate at lower values. NMR data for histamine-Zn^{2+} complexes are in agreement in that ionization of Zn^{2+}-OH_2 commences at ~ pH 8.[13]

In response to the approximating requisite coordination geometry of the enzyme, several groups have investigated the complexes of the so-called "tripod" ligands such as **1-9.**[14-21] The unifying feature of these complexes is a central anchor atom X, (which may or may not have coordinating capabilities) to which is attached three coordinating ligands. Tris-imidazolyl or tris-pyridyl carbinols **1-4** are easily synthesized and, in aqueous solutions bind divalent metal ions rather strongly.[14-16] However, none of these ligands, when bound to Co^{2+} gives evidence for low coordination numbers and spectra typical of octahedral Co^{2+} are observed.[14,15] They are too small and when bound do not extend far enough over the M^{2+} to prevent access of a second ligand or three H_2O molecules. This is particularly evident from ^{1}H-NMR studies of the ligands **1-3** as a function of $[Zn^{2+}]$[15] which show the formation of symmetric 2:1 complexes at ligand:Zn^{2+} ratios of 2 which give way to 1:1 complexes upon the addition of excess Zn^{2+}.

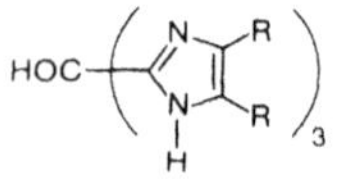

1[14,15]

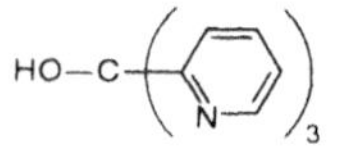

2[15,16]

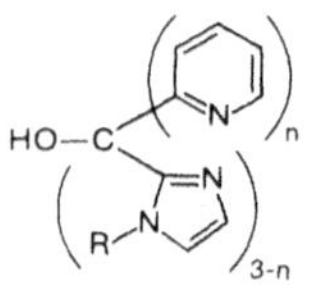

3[15]

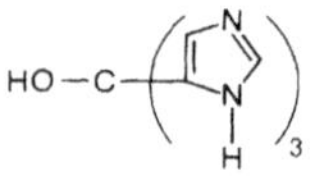

4[14]

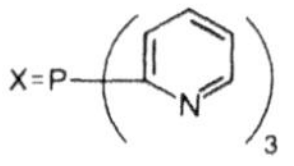

5 a X=:[17]
b X=O

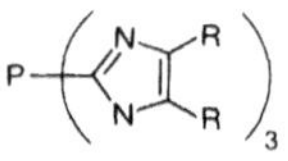

6 R=H, CH_3, iPr[18]

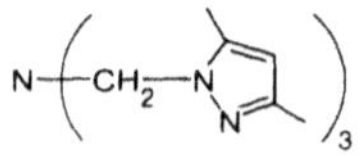

7[19]

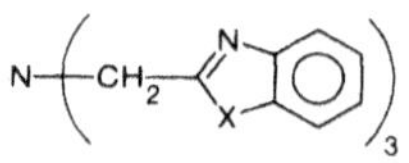

8 a X=NH[20]
b X=S[20]

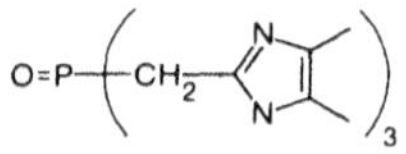

9[21]

An important clue about increasing the affinity of the ligand for M^{2+} comes from comparison of the stability constants of ligands containing N-methylimidazole vs. N-H imidazole. From the data above it seems clear that each N-methylation reduces the metal affinity by ~ 1-1.5 pK units. This may be related to a better hydrogen-bonding by solvent H_2O of the N-H complexes relative to the N-CH_3 complexes.[15]

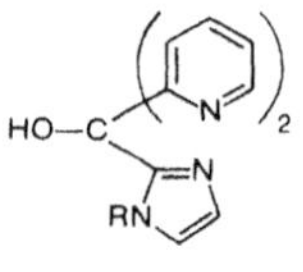

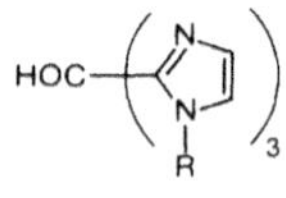

$R=CH_3$[15]	R=H[15]		$R=CH_3$[15]	R=H[14]
5.9	7.5	$pk_{Zn^{2+}}$ (M)	3.15	6.77
6.6	8.7	$pk_{Co^{2+}}$ (M)		

pk_M values defined as -log (dissociation constants) of 1:1 complexes
T=25°C, μ=0.16M KNO_3[14,15].

In an effort to reduce the 2:1 complexation and thereby better approximate the binding cavity of CA, ligand **1** was modified by the introduction of bulky groups such as CH_3 or $CH(CH_3)_2$ at the 4 and 5 positions of the imidazole.[15] While 1:1 ligand: M^{2+} complexes of the 4,5-diisopropyl ligand were formed exclusively, these were unstable in solution and tended to form intensely colored species even in the presence of Zn^{2+}. This was a result of deleterious dehydration of the complexes to form the extensively conjugated fulvene-like derivatives **10.**[15] Related dehydration of tris (2-phenylimidazole-5-yl) carbinol has been reported[22] and clearly stems from the extreme compression of the groups about the SP^3 hybridized carbinol. It is therefore apparent that the utility of such symmetric tris carbinols as CA models will be limited, particularly when the binding groups contain substituents which impede the access of more than one or two small ligands such as H_2O to the surface of the metal.

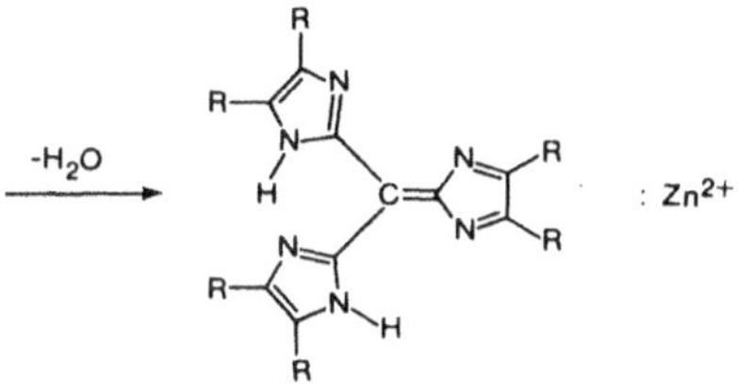

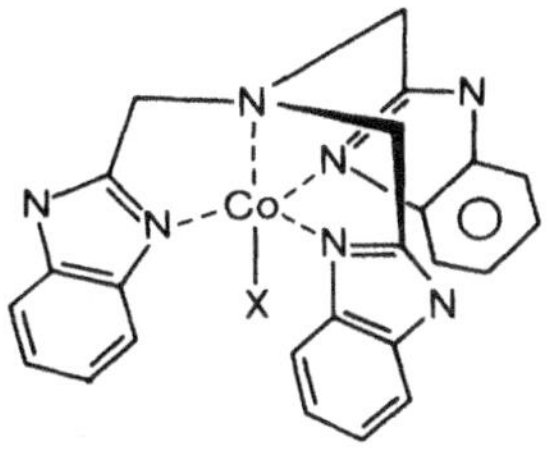

11 X=Cl, Br, NCS

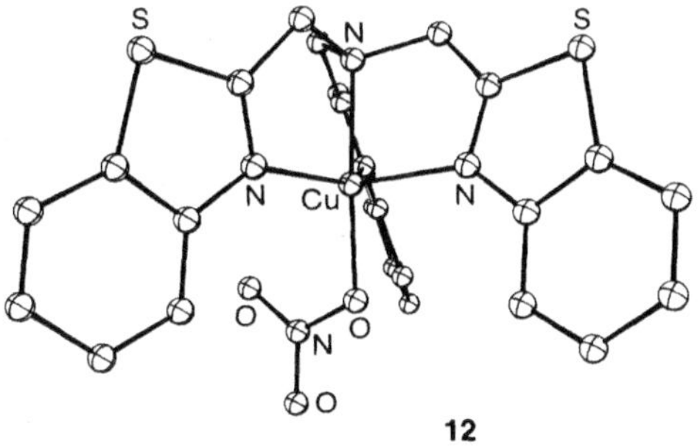

12

Conceptually one of the ways to circumvent the deleterious dehydration of the central carbinol carbon would be to replace it entirely. At the same time, it is advantageous to extend the ligands outward from the pivotal point so that they encapsulate the metal more completely and hence reduce the coordination number. The earliest of these were the Co^{2+} and Zn^{2+} complexes of tris(2-benzimidazolylmethyl) amine (**8a**[20]). In the presence of one equivalent of the non-coordinating counterion tetraphenylborate, a series of purple 5-coordinate Co^{2+}complexes (**11**) were prepared and characterized.[20]

A related Cu^{2+} complex of the trisbenzthiazole derivative **8b** was prepared and characterized by x-ray crystallography.[20b] In that example (**12**) the apical N acts as a fifth ligand with a Cu^{2+}-N distance of 2.059Å, roughly the same as the other Cu^{2+}-N distances of 2.048-2.120Å.

In a more recent report, Bertini and coworkers reported[19c] the visible absorption spectrum of the Co^{2+} complex aquo[tris[(3,5-dimethyl-1-pyrazolyl) methyl]amine] Co^{2+} perchlorate (**13** in eq. 4). This complex is also 5-coordinate and the spectral data as a function of pH indicate the presence of two ionizing groups associated with the complex (Fig. 3). Overall, the process appears to be that in eq. 4, and the high pH spectrum is reminiscent in general appearance of that observed for CoCA, although the maximal extinction coefficient for the complex is ~ ¼ that of the enzyme (~ 70 $M^{-1}cm^{-1}$ vs > 300 $M^{-1}cm^{-1}$).[19c]

Figure 3 The absorption spectrum of aquo [tris[(3,5-dimethyl-1-pyrazolyl)methyl]amine]Co^{2+} as a function of pH (redrawn from ref. 19c).

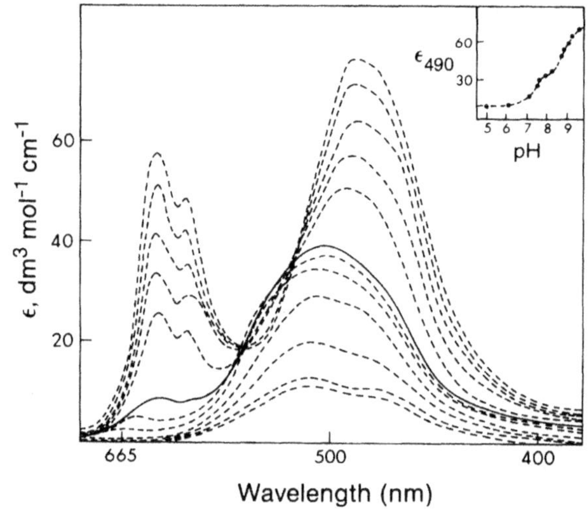

4.

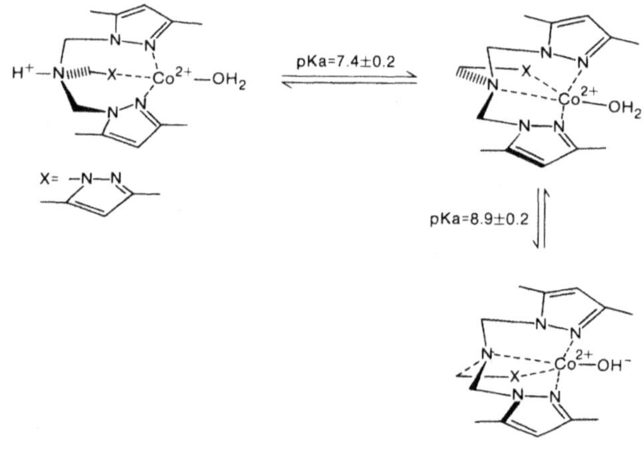

13

The important conclusion from the above model is that the deprotonation of a metal-bound H_2O successfully mimics the spectral variations of the Co enzyme: the authors concluded that this makes less speculative the same process in the enzyme. However, it should be mentioned that the pK_a of 8.9 reported for the $Co^{2+}-OH_2$ ionization in the complex exceeds that seen for CoCA by at least 2 units and possibly as much as 3 or more.[7] Despite the similarities between the complex and enzyme, no catalytic activity toward CO_2 hydration is observed. Also, at high pH, precipitation of some hydroxo species ($Co^{2+}(OH)_x^-$) is observed. This is a fundamental problem for essentially all models studied to date: they do not bind metal ions strongly enough to prevent OH^- sequestering of the ion away from the complex at high pH.

Recently, the x-ray crystallographically determined structure of aquo[tris[3,5-dimethyl-1-pyrazolyl]methyl]amine Co^{2+} perchlorate was determined (Fig. 4).[19d] That structure confirmed the 5-coordinate geometry of the metal and also indicated the presence of both an apically bonded N and OH_2, the bond lengths being Co-N, 2.293Å and Co-OH_2, 2.024Å respectively.

Figure 4 The x-ray crystallographically determined structure of aquo [tris[(3,5-dimethyl-1-pyrazolyl)methyl]amine]Co^{2+} perchlorate (redrawn from ref. 19d).

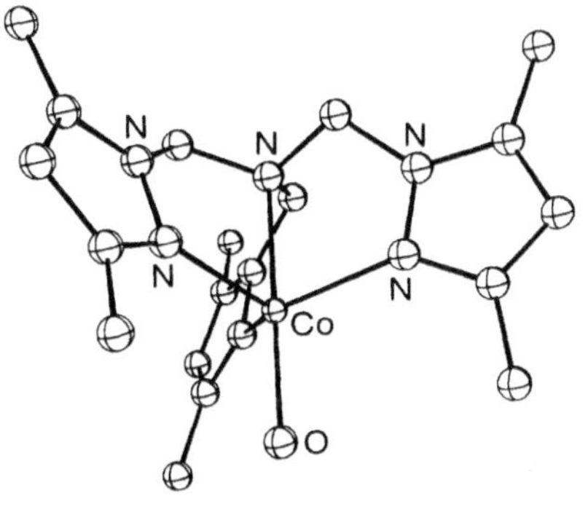

The bulk of the tripod ligands discussed so far suffer from obvious deficiencies when viewed in the context of models for the active site of CA. The most obvious one is the dissimilarity in number and nature of the ligating groups relative to the histidine imidazoles in the naturally occurring system. The tris-imidazole carbinols, while attempting to approximate the three chelating groups of the enzyme also have the problems of instability when bound to M^{2+}. This is most easily visualized by considering in **14** that the M^{2+}-N bond lengths should be 2.0-2.1Å, while the HO-C-C bond lengths are ~ 1.4-1.5Å. Hence the binding of the M^{2+} has the net effect of inducing strain into the complex by pushing the imidazoles outwards from each other thereby flattening the carbinol unit. Ideally one could rectify the strain

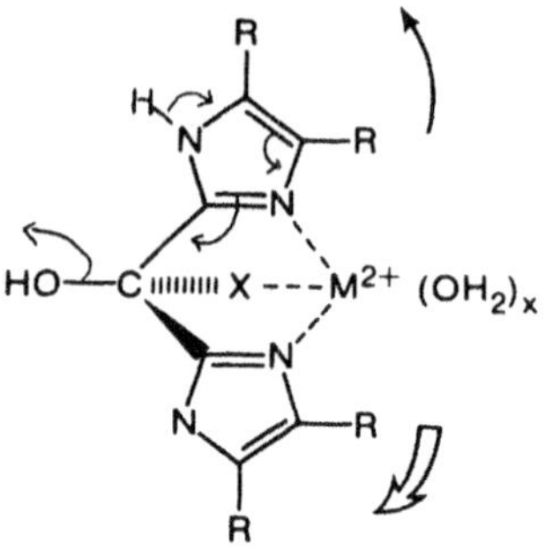

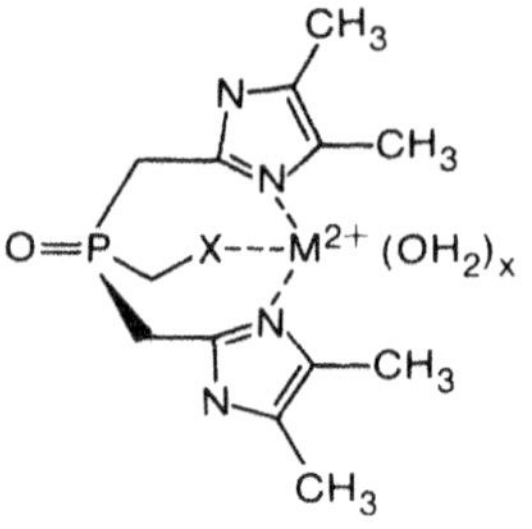

15

problem by increasing the bond length from the central anchor atom to the ring carbon. One such attempt at this approach was that of Brown and Huguet[18a] who showed that tris(4,5-diisopropylimidazol-2-yl)phosphine (6 iPr) when bound to Zn^{2+} or Co^{2+} displayed several features in common with the enzyme including the ability to catalyze the interconversion of HCO_3^- and CO_2. The x-ray structure of the Zn^{2+}-Cl^- complex (Fig. 5)[23] confirmed the postulated mode of binding and in

Figure 5 The x-ray crystallographically determined structure of [tris(4,5-diisopropylimidazol-2-yl)phosphine]dichlorozinc (11) bis[N,N-dimethylformamide], redrawn from ref. 23).

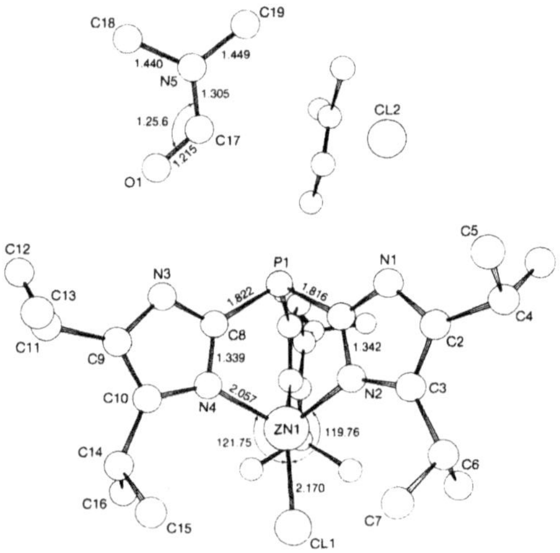

addition revealed features (perhaps fortuitously) in common with the active site binding in the enzyme (Fig. 1). Firstly, the isopropyl groups at the imidazole 4 and 5 positions can be seen to encapsulate the metal sufficiently to allow access of only one or possibly two small ligands. In that distorted tetrahedral complex, the N-Zn^{2+}-N angles are 94-96°. Secondly, the crystals were isolated from a solution containing dimethyl formamide (DMF) and two of these can be seen to be hydrogen bonded to the distal N-H of two of the chelating imidazoles. The third imidazole is hydrogen bonded to the second anionic chloride required for neutrality of the complex. For the enzyme, two of the three imidazole units are hydrogen bonded to amide linkages of Gln-92 and Asn 244; the third is hydrogen bonded to the anionic carboxylate of Glu 117.

That the symmetric tridentate chelation is maintained in solution is shown by ^{1}H-NMR studies with **6** iPr in the absence and presence of equimolar $ZnBr_2$ (Fig. 6). The presence of two equal intensity sets of isopropyl resonances in the complex rules out all modes of binding other than symmetric-tridentate and furthermore indicates that the ligand is not undergoing any exchange processes on the NMR time scale.

Figure 6 The 100 MHz ^{1}H-NMR spectrum of tris-(4,5-diisopropylimidazole-2-yl)phosphine in the absence and presence of 1 eq $ZnBr_2$, (d_4MeOH/D_2O redrawn from ref. 18b).

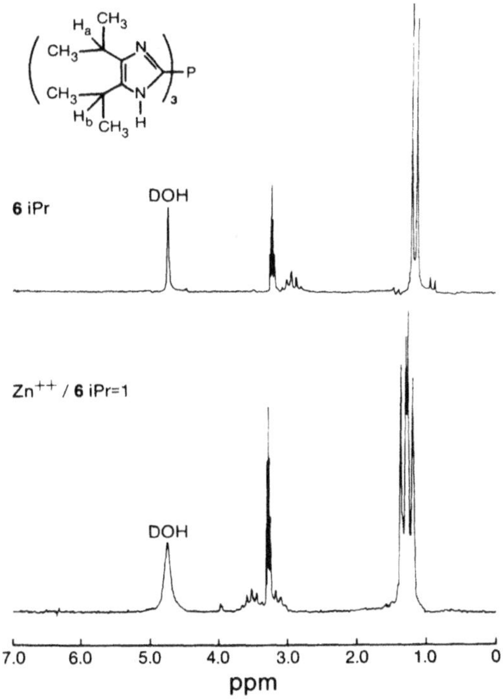

Cobalt CA exhibits a visible absorption spectrum which is markedly dependent upon the presence of anions.[6] So too is the Co^{2+} absorption spectrum of **6** iPr:Co^{2+}. As is illustrated in Fig. 7 the spectra primarily are indicative of 4-coordination.

Figure 7 Anion-dependent visible absorption spectra of 1 x 10^{-3} M **6** iPr:Co(II) determined in 80% ethanol/H_2O solution saturated with NaX (X = I, Br, Cl, F) compared with that of CoCA (redrawn from ref. 18b, Co(II) spectra from ref. 9a).

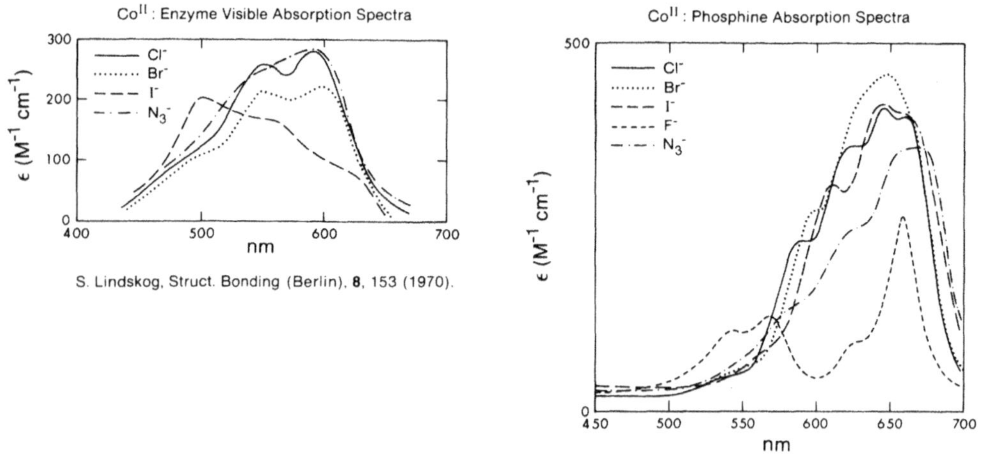

As stated before, the Co^{2+} absorption spectrum of CoCA is pH dependent and appears, depending upon the presence of monovalent anions to depend upon the presence of some active site associated group(s) having pK_a values between 5.6 and 7.9.[6,7] The absorption spectrum of Co^{2+}:**6**iPr:Cl^- is also pH dependent as is shown in Fig. 8: a comparison spectrum for CoCA reported by Lindskog[6a] is also given in the figure.

Figure 8 pH dependence of absorption spectrum of 1 x 10^{-3} M **6** iPr:Co(II) in the presence of 10^{-1}M Cl^- (80% EtOH/H_2O) compared with pH dependence of CoCA (redrawn from ref. 18b, CoCA spectrum from ref. 6a).

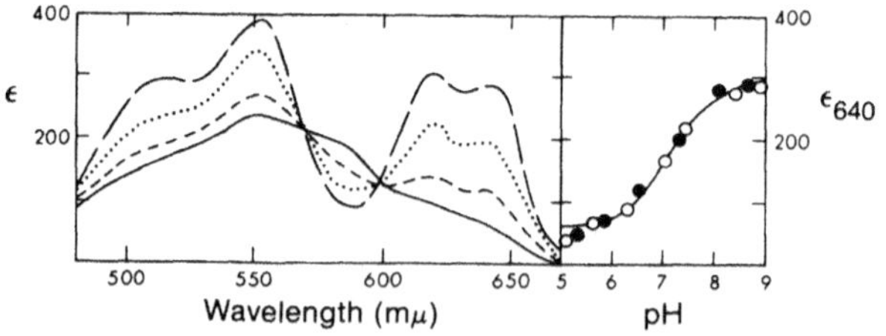

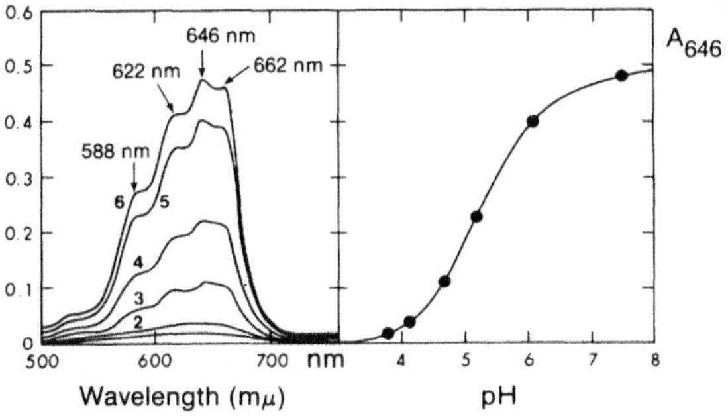

The development of the blue color adheres to the ionization of a complex associated group with a pK_a of ~ 5.6; at low pH values only the spectrum of octahedral Co^{2+} is observed. Importantly however, the spectrum does not develop in the presence of ClO_4^-, a presumably non-coordinating counterion. Therefore it is not likely that the titrating group is Co^{2+}-OH_2. The process responsible for the development of the tetrahedral complex is best described as in eq. 5 wherein at low pH, the imidazole units become protonated and therefore no longer bound to Co^{2+} in a way which would ensure a low coordinate environment.

5. 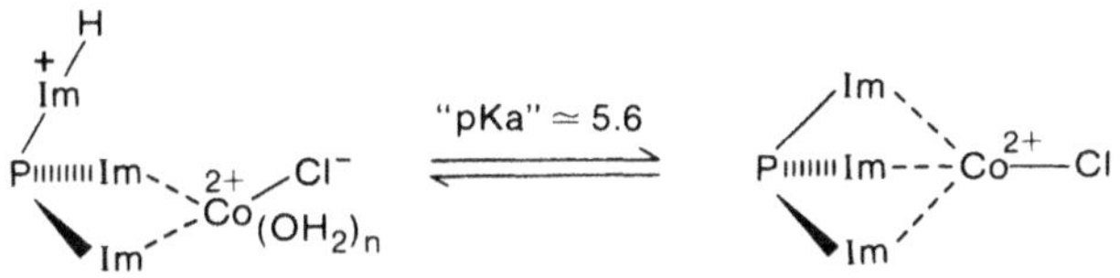

low pH form, octahedral high pH form, tetrahedral

6 iPr + $Co^{2+}Cl^-$

The metal complexes of trisimidazolyl phosphines, while providing some features in common with CA, suffer from certain serious shortcomings. Firstly they do not provide evidence for the ionization of a metal coordinated H_2O. Secondly the x-ray structure of **6** iPr:$Zn^{2+}Cl^-$ shows that the coordinating environment about the Zn^{2+} is severely distorted from tetrahedral. This fact must lead to a decreased metal binding affinity and perhaps is the reason why the 4-coordinate complexes can only be found in the presence of a small monovalent anion.

In order to circumvent these problems, the approach of extending the imidazoles outward from the central phosphorus anchor was adopted.[21]

[Tris[(4,5-dimethyl-2-imidazolyl)methyl]phosphine oxide] (**9**) was synthesized with the expectation that when bound to M^{2+}, the imidazole units could optimize their chelation by adjustment of the "pitch" of the propeller-shaped ligand. In part this appears true since the metal dissociation constants for **9**:M^{2+} are less than for **6** iPr:M^{2+} (eq. **9**:M^{2+}; $pK_{Zn^{2+}}$ = 6.67, $pK_{Co^{2+}}$ = 3.70; **6** iPr:M^{2+}; $pK_{Zn^{2+}}$ = 6.0, $pK_{Co^{2+}}$ = 3.5).[21] However, the affinity for metal is considerably less than one would have hoped and probably stems from an unfavourable O=P---M^{2+} dipole-ion interaction as in **15**. Unfortunately the corresponding phosphine (which would not suffer from this problem) could not be prepared since it spontaneously oxidized to the phosphine oxide on purification.

Nevertheless, study of **9**:M^{2+} led to some interesting observations of direct relevance to the enzyme. Quantitative data for titration of the ligand:$2HNO_3$ in the absence and presence of 1.0 eq $Zn(ClO_4)_2$ or $Co(ClO_4)_2$ are presented in Fig. 9. In the absence of M^{2+}, 2 eq. of OH^- are consumed only after the third pK_a for the ligand is passed (pK_a^3 = 7.18). In the presence of Zn^{2+}, the analogous titration shows the

Figure 9 Quantitative titration data for solutions of 5 x 10^{-3} M **9**:$2H^+$ in the presence of no added M^{2+} (Δ), 1.0 eq. Zn^{2+} (O); 1.0 eq. Co^{2+} (□); 80% ethanol/H_2O; μ = 0.2 M $NaClO_4$) (redrawn from ref. 21).

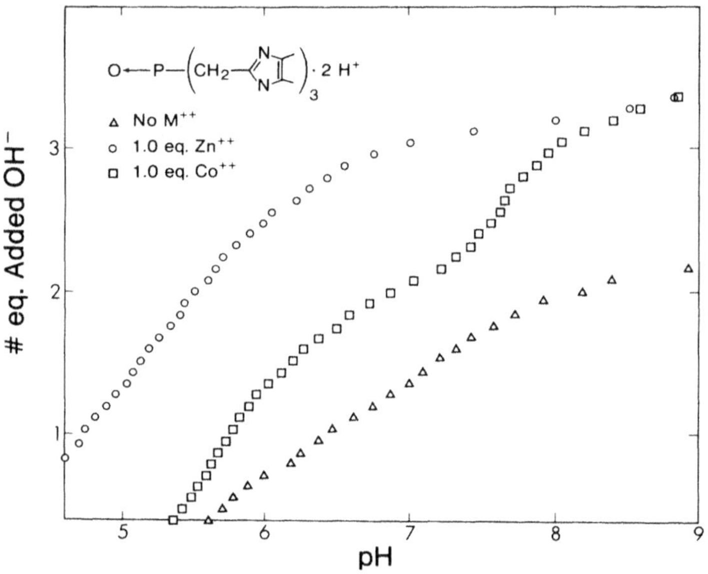

consumption of 3 eq. of OH^- before pH 7 indicating a formal titration of one additional group associated with the 1:1 complex having a pK_a ~ 6. Additionally, with 1 eq Co^{2+}, two OH^- are consumed before pH 7 (indicating complete deprotonation of the ligand by that point and hence complete complex formation) followed by consumption of a third OH^- in some process having an apparent pK_a of ~ 7.8. In these cases, the titrations are completely reversible and no sign of precipitation is seen up to at least pH 9.

The visible absorption spectrum of an equimolar solution of **9** and Co^{2+} as a function of pH (Fig. 10) also shows that the titrimetric pK_a

is also tied to the formation of an intensely blue colored species; spectrophotometric $pK_a \simeq 7.8$. Thus, while a complex is formed by pH 7, it is octahedral and requires the ionization of a metal coordinated group (probably H_2O) to further reduce the coordination number to 4 or 5.

Figure 10 Visible absorption spectrum of 8.3×10^{-4} M **9**:Co(II) in 80% ethanol/H_2O as a function of pH in the presence of added ClO_4^- (Δ) or Cl^- (O), (0.2 M added as Na salts) (redrawn from ref. 21).

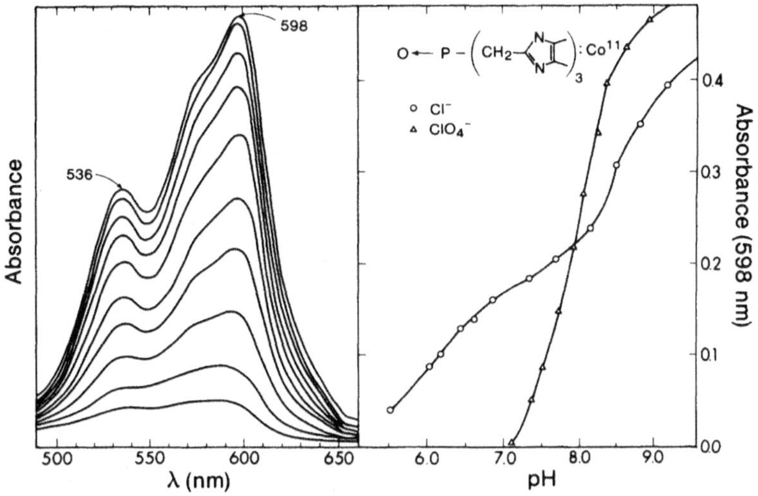

The **9**:Co^{2+} spectrum is markedly anion dependent which indicates the association of monovalent anions such as Cl^-, I^-, Br^-, AcO^- with the metal. No unique spectrum indicative of a new species is observed in the presence of ClO_4^- which suggests that it is only weakly if at all associated with the metal. However, spectrophotometric determinations as a function of pH in the presence of 0.2 M Cl^- clearly show the formation of a unique 4(5) coordinate species which subsequently gives way on addition of further OH^- to the same spectrum as was observed with only ClO_4^- as a counterion. The appearance of the absorbance vs pH profile at 598 nm in the presence of 0.2 M Cl^- is shown in Fig. 10 and suggests that the spectral change is somewhat more complicated than expected on the basis of a single ionizing group. Indeed, as shown in Fig. 11, the spectrophotometric "pK_a" of **9**:Co^{2+} has general trend to increasing values in the presence of associating anions. A similar phenomenon is observed with the enzyme and has been discussed in terms of a competition between OH^- and inhibitory anion for a common site on the metal.[7c] Overall the situation is best visualized in terms of the process depicted in eq. 6. At low pH, **9** exists as a protonated form not coordinated to Co^{2+}. Addition of OH^- deprotonates ImH^+ and allows full coordination by pH 7 since all added H^+ is accounted for by that point. However, this complex must be octahedral since no blue colors are evident. Further addition of OH^- causes a deprotonation of Co^{2+}-OH_2 (or displacement of H_2O by OH^-) with concomitant reduction in the coordination numbers. That the consumption of OH^- is not tied to deprotonation of the imidazole pyrrole NH, is evidenced by the fact that the **9**:Zn^{2+} complexes show no large bathochromic shifts of the imidazole absorbances in the basic complex as would be expected if pyrrole NH ionization is occurring.[24] In this interpretation, anions coordinate with **9**:Co^{2+} to form a 4(5) coordinate species which, at higher [OH^-] give rise to the same basic complex. Importantly, since the only anion common to all titrations is OH^-, the basic form of the complex must have an associated OH^-.

6.
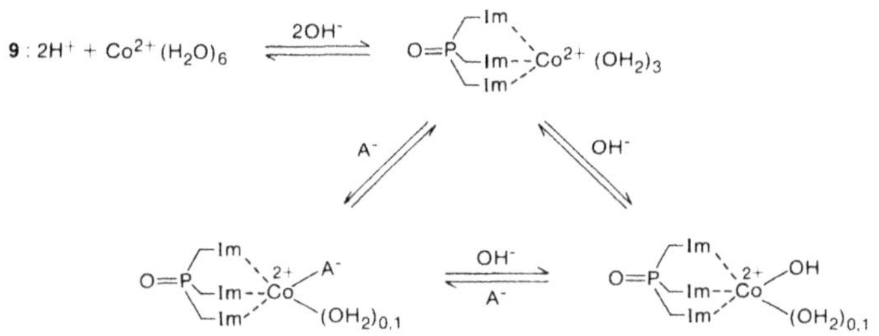

Figure 11 Spectrophotometric pH dependent titration curves of **9**:Co(II) in the presence of various anions showing the general tendency to increasing pK_a in the presence of coordinating anions (80% ethanol/H_2O, redrawn from ref. 21).

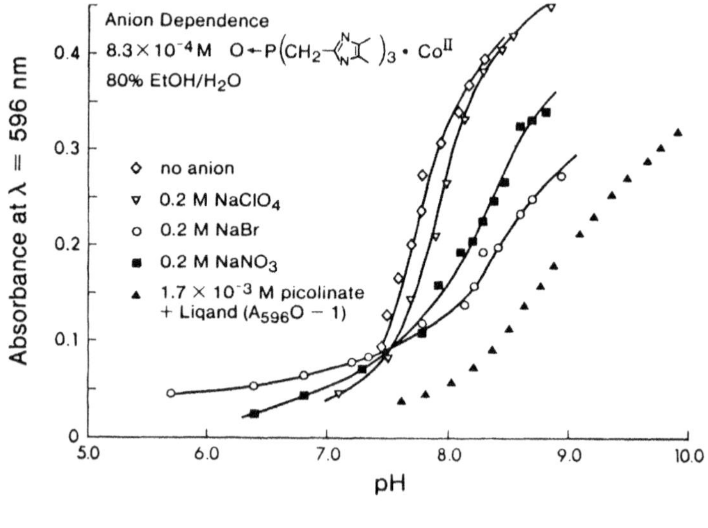

3. Catalytic Models

It is well at the outset of this section to consider the uncatalysed processes for hydration of CO_2 and the microscopic reverse reaction, dehydration of bicarbonate. At pH values larger than the pK_a of carbonic acid (3.76[25a]) eq. 1 can be restated as eq. 1a below, where[25a,b] k_{hyd} and k_{dehyd} at 25°C have currently accepted values of $0.037 s^{-1}$,

1a. 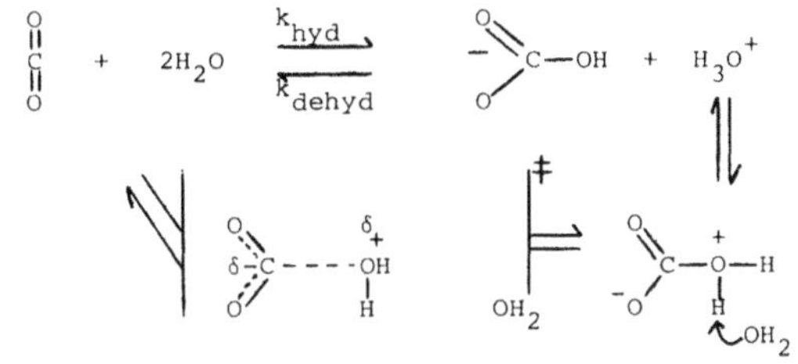

and $4.1 \times 10^4\ M^{-1}s^{-1}$[25b] respectively. In D_2O, the corresponding values are ≃ $0.02\ s^{-1}$ and $7.3 \times 10^4\ M^{-1}s^{-1}$ respectively.[25b] The solvent isotope effects on the hydration (k_{H_2O}/k_{D_2O} ≃ 2) and dehydration ($k_{H_3O^+}/k_{D_3O^+} = 0.56$) processes have been discussed at length by Pocker and Bjorkquist[25b] and are consistent with a reversible pathway proceeding via a zwitterionic form of carbonic acid. Unfortunately these data are not sufficient to distinguish the specific acid pathways from general acid pathways. These two processes differ in the timing of the proton transfer steps with the latter process apparently having protonation, (deprotonation) concerted with C-O bond formation (cleavage). Very recently some interesting $k(^{12}C)/k(^{13}C)$ kinetic isotope studies were performed on the dehydration of $H^{12}CO_3^-$ and $H^{13}CO_3^-$ by Paneth and O'Leary.[25c] Without going into detail about the methodology, their data is best interpreted in terms of eq. 1a in which dehydration involves stepwise protonation to form an intermediate (presumably the zwitterion) having similar rates of formation and decomposition.

Above pH 8, direct nucleophilic attack of hydroxide on CO_2 becomes the dominant process for formation of bicarbonate, the values of k_{OH^-} and k_{OD^-} being 8500 and 13500 $M^{-1}s^{-1}$ respectively.[25b] Although the only known physiological purpose of CA is to facilitate the interconversion of HCO_3^- and CO_2,[1] it will at reduced rates catalyse other hydration and hydrolysis reactions. The unifying feature of all reactions catalysed by CA is that they involve nucleophilic addition of OH^- to an Y-X=M unit, generally a C=O bond or an analogue thereof. For each of these, the kinetic behavior can be formally analysed in terms of an elementary Michaelis-Menten scheme as in eq. 7. Given in Table II are some k_{cat} and K_M terms for certain of the enzyme catalysed processes under the indicated conditions.

$$7.\quad E + S \underset{k_{-1}}{\overset{k_1}{\rightleftharpoons}} E:S \xrightarrow{k_{cat}} E + P$$

$$7a.\quad \frac{V_0}{[E_0]} = \frac{k_{cat}[S]}{(K_M + [S])}$$

Table II k_{cat} and K_M values for various hydration and hydrolysis processes catalysed by bovine carbonic anyhdrase.[a]

Substrate	pH	$k_{cat}(s^{-1})$	$K_M(M)$	k_{cat}/K_M $(M^{-1}s^{-1})$
carbon dioxide	6.75	4.16×10^5	1.55×10^{-2}	2.68×10^7
acetaldehyde	7.20	8.00×10^2	6.5×10^{-1}	1.23×10^3
2-pyridine carbox-aldehyde	7.15	7.68×10^1	1.4×10^{-2}	5.48×10^3
4-pyridine carbox-aldehyde	7.18	1.91×10^2	1.1×10^{-2}	5.5×10^2
_p_-nitrophenylacetate	7.58	6.6×10^0	1.2×10^{-2}	5.5×10^2

[a] As reported in ref. 1f.

The most difficult aspect of modelling the catalytic sequences of CA and indeed any enzyme mediated process is duplicating the preequilibrium binding of the substrate to the enzyme. This stems from the fact that the models are far smaller than the enzymes and do not have the structural complementarity for the substrate that the enzyme cavity has. Hence the models are generally studied as biomolecular reagents or catalysts for the reaction in question. (The distinction between a reagent role or catalytic role will become obvious later when it is seen that for some of the reactions, the models do not turn over). In order to compare the models and the enzyme as bimolecular catalysts the k_{cat}/K_M ratio for the latter is used. From the data given in Table II for bovine CA, it is clear that the enzyme is highly evolved for the hydration of CO_2, the k_{cat}/K_M ratio being $2.7 \times 10^7 M^{-1}s^{-1}$. By way of comparison, rate constant for OH^- attack on CO_2 is $\sim 8500\ M^{-1}s^{-1}$[25] while that for H_2O attack is $\sim 0.037\ s^{-1}$.[26]

The generally accepted process for CA catalysed hydration of CO_2 involves delivery of a Zn^{2+}-coordinated hydroxide as is illustrated in Fig. 12. The exact role of the Zn^{2+} is still debated but possibly it serves a dual function of providing a metal-based nucleophile as well as a Lewis acid to stabilize the incipient (-) charge on the developing bicarbonate.[2a] The fact that monovalent anions inhibit the enzyme activity and are known to be coordinated to Zn^{2+} suggests that they interfere with the access of OH^- and/or CO_2 to the metal surface. The major initial questions centered upon whether the Zn^{2+}-OH_2 in the active site of CA could have as low a pK_a as 7 or less and if so, does the Zn^{2+}-OH^- retain sufficient nucleophilicity to attack CO_2.

Figure 12 A stylized representation of the zinc hydroxide mechanism for CA catalysed hydration of CO_2 (adapted from ref. 2a).

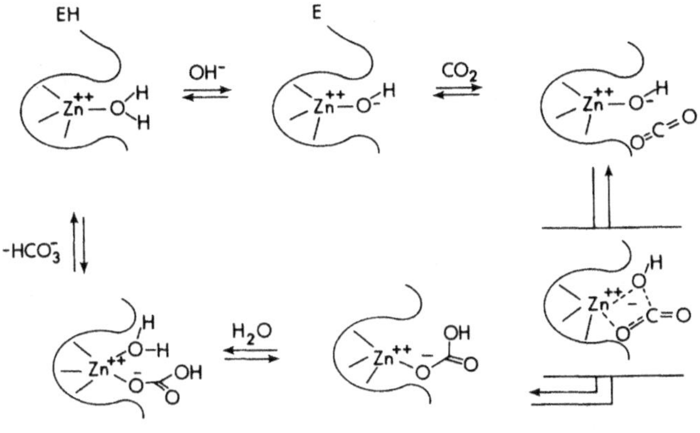

The answer to the first question rests upon two features embodied by the enzyme, namely a low coordinate Zn^{2+}, as well as a hydrophobic environment. Determination of the pK_a values for $Zn^{2+}(H_2O)_x$ ionization is complicated by the complexity of the equilibria involved and the formation of hydroxide precipitates at pH 7 or lower.[11,12] Nevertheless, the value of ~ 9[12] or ~ 10[27] is currently accepted. ^1H-NMR studies of histamine complexes of Zn^{2+}, indicate that Zn^{2+}-OH_2 begins to ionize at ~ pH 8.[14] Wooley has reported [27a] that 5-coordinate Zn^{2+}-OH_2 complex of **16a** at 25°C and 0°C has pK_a values of 8.69 and 9.12 respectively while for the desmethyl analogue **16b** the pK_a values at 25°C and 0°C are 8.13 and 8.53 respectively. The stronger acidity of these Zn^{2+}-complexes relative to $Zn^{2+}(H_2O)_6$ was reasoned to result from the lower coordination numbers of the former[27a,28] which increases the net positive charge on the metal and resulting polarization of the bound H_2O. Accepting this, it would appear that a further reduction in

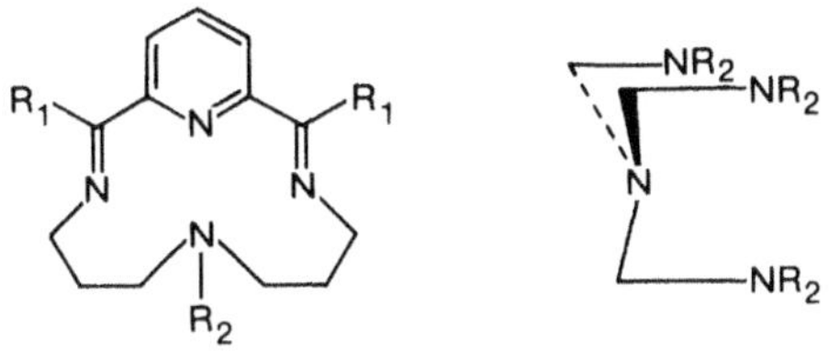

16 a R_1=CH_3, R_2=H

b R_1=H=R_2

17 a R=H

b R=CH_3

coordination number as is found for the active site metal in CA, would produce a further lowering of the pK_a. This appears to be the case for the Zn^{2+}-complex of the tridentate ligand **9** for which quantitative titration data show the complete formation of **9**: Zn^{2+}-OH^- by pH 7 in a solvent of 80% ethanol/H_2O.[21]

A second feature of the active site region which could lead to a reduction in pK_a is the hydrophobicity. In part, this may be a significant influence for the low pK_a values observed for the Zn^{2+}-OH_2 and Co^{2+}-OH_2 complexes of **9**.[21] Direct evidence for this has been provided by Coates, Gentle and Lincoln[29] who studied the pK_a values of M^{2+}-complexes of 2,2',2"-triaminotriethylamine (tren, **17a**) and 2,2',2"-tri(N,N-dimethylamino)triethyl amine (Me_6 tren, **17b**). Divalent Co, Cu and Zn all form stable 1:1 5-coordinate complexes with both **17a** and **17b** and the pK_a values corresponding to ionization of the bound H_2O are given in Table III. Molecular models suggest that in the Me_6 tren complexes, the M^{2+}-OH_2 is sterically crowded by the hydrophobic CH_3 groups while in the corresponding tren complexes, the OH_2 is close to the hydrophilic NH_2 groups. Inspection of the data in Table III indicate that for each metal, the pK_a in the Me_6 tren complex is reduced from that of the corresponding tren complex by 0.8-1.4 units. It would

Table III Apparent titrimetric pK_a value for acid dissociation of H_2O coordinated to tren and Me_6 tren complexes of divalent M^{2+} at 25°C in 1M $NaClO_4$.[a]

Complex	pK_a	ΔH°(kcal/mol)	ΔS°(cal/°K/mol)
$[CotrenH_2O]^{2+}$	10.22±0.01	0.22±0.65	-46.06±2.20
$[CutrenH_2O]^{2+}$	9.37±0.01	-3.25±0.45	-53.77±1.50
$[ZntrenH_2O]^{2+}$	10.26±0.02	2.84±0.45	-37.40±1.50
$[CoMe_6trenH_2O]^{2+}$	8.80±0.04	8.10±0.45	-13.03±1.50
$[CuMe_6trenH_2O]^{2+}$	8.52±0.01	6.88±0.45	-15.89±1.50
$[ZnMe_6trenH_2O]^{2+}$	9.00±0.01	11.76±0.45	- 1.72±1.50

[a] Ref. 29

appear that this is strongly influenced by the entropies of ionization which suggest that in the Me_6tren complexes, less reordering of the structure of the adjacent solvent molecules accompanies ionization. This probably stems from the fact that the hydrophobic face of $[MMe_6trenH_2O]^{2+}$ leads to a situation generally considered to favor an ordered ice-like structure for the adjacent solvent water. An ice-like structure of H_2O in the active site of CA has been observed by x-ray crystallography.[2] It seems likely that solvent H_2O adjacent to $[MtrenH_2O]^{2+}$ will be less ordered and the decrease in entropy greater upon protolysis than will be the case with $[MMe_6trenH_2O]^{2+}$.

The second major question concerning the viability of the Zn^{2+}-OH^- involvement was whether the hydroxy group retained sufficient nucleophilicity to attack CO_2.[27,30] One might have envisioned that a

Brønsted relationship between pK_a of the M^{2+}-OH_2 and nucleophilicity would be followed. In that event, reduction of the pK_a of H_2O from 15.7 in free water to ~ 7 as in the active site of CA should produce a significant drop in nucleophilicity. Martin has provided a significant analysis of that Brønsted relationship,[31] and the pK_a and second order rate constants for attack of various M-OH^- on CO_2 are presented in Table IV. The surprising result is that the Brønsted β-value is 0.15[36]-0.2[31]. The second order rate constants are relatively insensitive to pK_a. The same sort of insensitivity toward nucleophilic attack on propionic anhydride (β = 0.25[37]) and acetaldehyde (β = 0.35[36]) is observed and appears to be a general trend for nucleophilic attack of M-OH^- on electrophilic centers. This is graphically represented in Fig. 13.[36] The obvious advantage of the low β for the enzyme is that it allows production of large local concentrations of OH^- in the active site, without concomitant loss of nucleophilicity.

Figure 13 Graphic representation of the Brønsted relationship between pK_a of the conjugate acids of various M-OH^- and the log of the second order rate constants for attack on carbonyl species.[36] (Identity of species corresponding to #'s can be found in ref 36.)

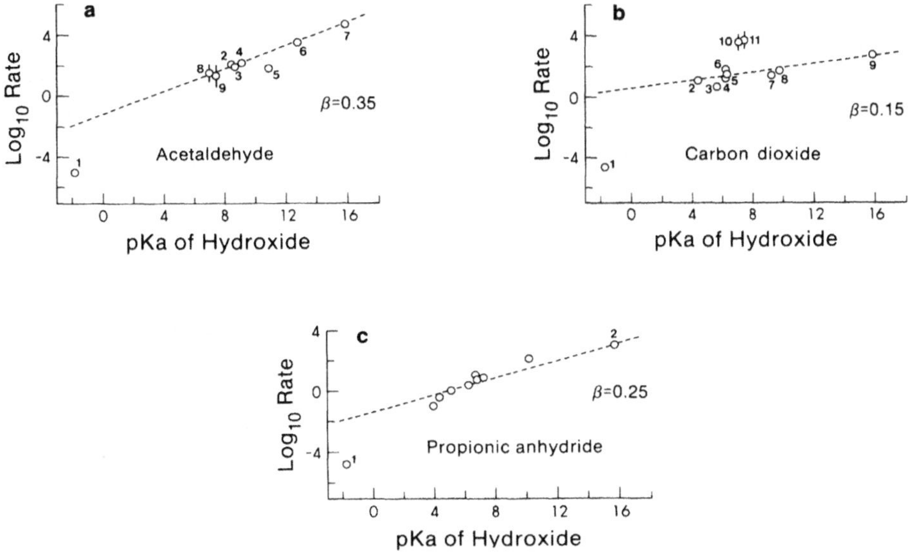

Table IV Second order rate constants and pK_a values of conjugate acids for hydroxide nucleophiles with CO_2 at 25°C.[a]

Nucleophile	pK_a conjugate acid	k_2 ($M^{-1}s^{-1}$)
OH^-	15.7	8500[b]
glycylglycine $Cu\overset{+}{O}H$	9.4	580[c]
$(NH_3)_5RhOH^{2+}$	6.8	470[d]
$(NH_3)_5IrOH^{2+}$	6.7	590[d]
$(NH_3)_5CoOH^{2+}$	6.4	220[e]; 310[f]
$(NH_3)_5CrOH^{2+}$	5.3	> 11[f]

[a] as reported in ref 31; [b] ref 25; [c] originally obained as 56 $M^{-1}s^{-1}$ at 0°C in ref 32 but corrected to 25°C; [d] ref 33; [e] ref 34a; [f] ref 34b; [g] ref 35.

Although aquopentamine Co(III) complexes presented above show well defined pK_a values[33-39] for OH_2 ionization and good nucleophilic activity toward CO_2, they are inner sphere exchange inert and cannot be considered as true turnover catalysts since the Co(III)-OCO_2^- bond is not cleaved to carbonate or bicarbonate.[40] This is shown in a variety of studies probing the mechanism of aquation of carbonato complexes of Co(III) which generally proceeds as in eq. 8 (* representing ^{18}O).[34,41]

8.

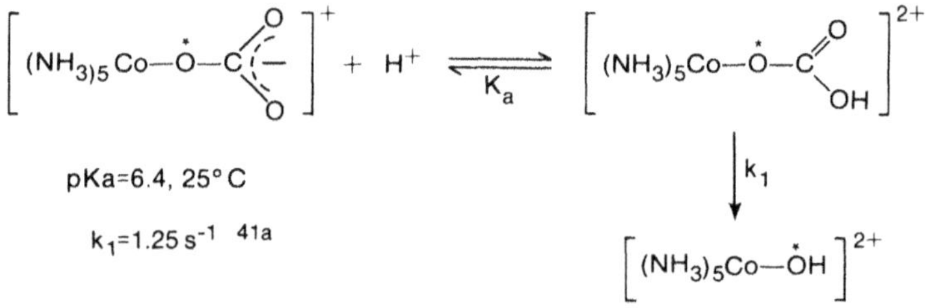

The real question of interest is whether Zn^{2+} complexes which undergo rapid Zn^{2+}---X association-dissociation reactions will facilitate the interconversion of HCO_3^- and CO_2. Pocker and Bjorkquist[42] and later Tabushi and Kuroda[43] have reported that Zn^{2+} in imidazole buffers show only a small catalytic effect in CO_2 hydration ($k_{Zn^{2+}} \sim 8\ M^{-1}s^{-1}$, 25°C, pH = 7[43]). The miniscule activity probably is a result of several deleterious factors such as the multiple forms of the Zn^{2+} with respect to numbers and types of ligands, and the fact that the concentration of any Zn^{2+}-OH^- forms at pH 7 is very low. Given that a reasonable value for the pK_a of $(im)_y Zn(OH_2)_x$ is ~ 9-10, at pH 7, only ~ 1% of the Zn^{2+} is associated with $Zn^{2+}OH^-$. Hence the k_{cat}^{max} is probably on the order of ~ 1000 $M^{-1}s^{-1}$, a value which fits nicely with the data in Table IV.

The multiple equilibria problem can be eliminated by providing polydentate ligands which reduce the M^{2+}-coordination. (A number of these were presented previously in the discussion of the spectroscopic models for CA.) An anticipated benefit of this approach is that the so-called "chelate effect" will allow the metal to be bound strongly enough by the ligand to maintain solubility at pH values where $M^{2+}(OH^-)_x$ precipitation would normally be a problem. One of the first such systems investigated[45] was the Cu^{2+} complex of glycylglycinate (**18**) for which a pK_a of 9.7 (0°C) is known for dissociation of the Cu^{2+}-OH_2.[46] The role of the Cu^{2+}-OH^- in this example was said to be either as a nucleophile or general base (shown in eq. 9) with stabilization of the resultant HCO_3^- in the transition state.[45] The $k_2^{CO_2}$ value for **18** was dependent upon the basic form and attained a value of ~ 50 $M^{-1}s^{-1}$ at

9. 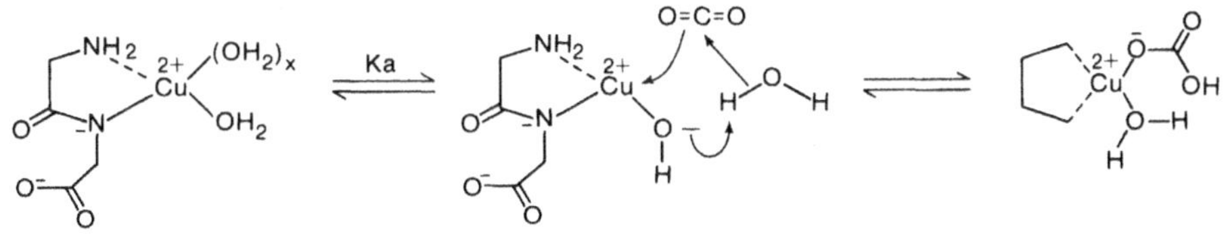

18

0°C, and ~ 580 $M^{-1}s^{-1}$ at 25°C.[31] Although the author favored the general base role, that pathway in the enzyme seems unlikely. (Experiments of Silverman and Tu indicated that dehydration of ^{18}O-labelled HCO_3^- yields CO_2 and 1/3 of the ^{18}O left on the metal, probably as Zn-^{18}OH which is capable of being transferred to $^{13}CO_2$. If the general base mechanism were operative, dehydration would directly yield $H_2{}^{18}O$ which is anticipated to be infinitely diluted by bulk solvent such that $H^{12}C^{18}OO_2^-$ cannot be formed.[47] Also the equilibrium rate of interconversion of HCO_3^- and CO_2 catalysed by the enzyme exhibits a solvent deuterium isotope effect of close to unity.[47c,48])

Woolley[27a,b] has reported catalysis of the hydration of acetaldehyde by the Zn^{2+} complex of **16a** (Zn^{2+}CR). The catalysed process follows a sigmoidal pH vs k_2^{obsd} profile with a point of inflection at pH 9.07 (0°C). This value is close enough to the thermodynamic pK_a of 9.17 (0°C) for ionization of the CR Zn^{2+}-OH_2, that it lends confidence to the interpretation that the Zn^{2+}-OH^- is indeed the active form. CRZn^{2+}-OH^- also appears to catalyse the hydration of CO_2 with a k_2^{cat} value of ~ 25 $M^{-1}s^{-1}$ being reported at 0°C.[27a]

In an effort to approximate the active site metal binding cavity of CA, Tabushi and co-workers[43,49] studied the Zn^{2+}-complexes of bis(histamino)-β-cyclodextrin ($(His)_2$CD-Zn^{2+} **19**). The M^{2+}-association constants for both Zn^{2+} and Co^{2+} are quite low, being (4.5±2) x 10^2 M^{-1} and (2.4±0.6) x 10^2 M^{-1} respectively. In the absence of additional imidazole (supplied as the buffer), **19** showed virtually no activity in

CO_2 hydration. However with imidazole present as the buffer, k_{cat}^{max} was observed to be 3100±800 $M^{-1}s^{-1}$ at pH 7.5 and 3500±800 $M^{-1}s^{-1}$ at pH 7.8.[43] Hence, Tabushi concluded that the catalytically viable form of the complex required a Zn^{2+}-associated imidazole (**19** x = im). Although the exact catalytic pathway remains elusive, the somewhat hydrophobic cyclodextrin cavity appears to be necessary since Zn^{2+} N-methyl histamine complexes do not catalyse the reaction. While it was not explicitly addressed by Tabushi's analysis, it seems quite possible that the cyclodextrin cavity may form a binding site[50] for the buffer imidazole which, when in place, provides a third Zn^{2+} binding group approximating the situation in CA. Whether $(His)_2CD-Zn^{2+}$ is a true catalyst is open to interpretation since the complex does not catalyse the dehydration of bicarbonate. This was rationalized as arising from a disruption of the integrity of the complex on binding bicarbonate, perhaps by displacement of imidazole.

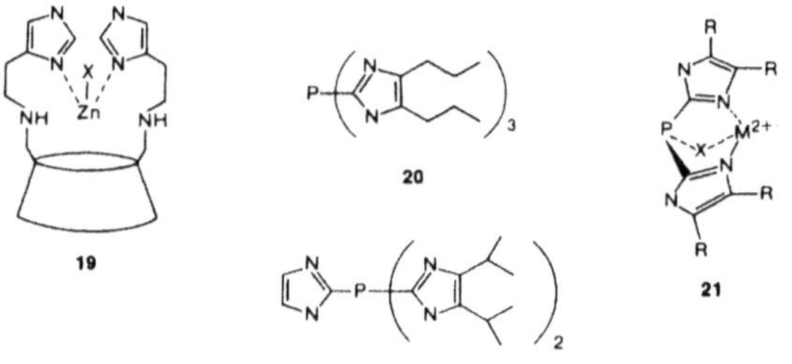

19 20 21 22

The first real demonstration of catalysis the interconversion of HCO_3^- and CO_2 by a ligand with three imidazole groups approximating the M^{2+}-binding of CA appears to be made by Brown and coworkers who studied tris-imidazolyl phosphines **6a,b,c**[18] and **20.**[51] As previously discussed, physical and spectroscopic studies of the Zn^{2+} and Co^{2+} complexes showed symmetrical tridentate coordination. The evaluation of the catalytic viability was studied in terms of the pseudo first order rate constant (k_{obsd}) for catalysed attainment of equilibrium of eq. 10 where $k_{obsd} = (k_1 + k_{-1})$.[18b] According to the analysis, catalytic activity is judged by a large increase in the value of k_{obsd} at a given pH over the value obtained in the absence of catalyst. Several important observations were made. Firstly, in the presence of Zn^{2+} or ligand alone, no enhancement of k_{obsd} was evident at any pH. Secondly **6a** and **6b** in the presence of equimolar Zn^{2+} exist in solution as depicted by **21**, but neither is catalytically active. However **6c** (**21** R = $CH(CH_3)_2$) has substantial activity. The latter has k_{cat} value on the order of 900±60 $M^{-1}s^{-1}$ from pH 6.4-6.8 (80% $EtOH/H_2O$, HEPES, 25°C, μ = 0.2 $MNaClO_4$.[18b]) Apparently there is a correlation between the ability of the complex to enforce 4 or 5 coordination about the metal, and the catalytic viability. Neither of **6a** or **6b** as their Co^{2+} complexes give evidence for distorted 4/5 coordinate geometries but **6c** does. Also the x-ray crystal structure of **6c**: $Zn^{2+}-Cl^-$ showed a strict 4-coordinate environment and that the isopropyl groups would severely retard access of a bulky ligand to the fourth coordination site.

$$10.\quad H^+ + HCO_3^- \underset{k_{-1}}{\overset{k_1}{\rightleftharpoons}} H_2O + CO_2$$

Subsequent evolution of the catalytic models[51] rested upon the latter point. It seemed unlikely that even with low β values known for attack of M^{x+}-OH^- on CO_2,[27b,31] that the second order rate constant for attack of CA upon CO_2 ($k_{cat}^{CO_2}/K_M \sim 10^8\ M^{-1}s^{-1}$) should exceed so vastly that of OH^- ($k_2^{OH} = 8500\ M^{-1}s^{-1}$)[25] unless a second, vital role for the active site Zn^{2+} were present. From Fig. 12, the dual role involves both a CO_2 polarization and Zn^{2+}-OH^- delivery passing through a 5-coordinate transition state. For the model **6c**: Zn^{2+}, such a transition state would be unlikely unless one of the Zn^{2+}-imidazole units either debound or reoriented the isopropyl groups away from the cavity. In order to address this question, ligands **20** and **22** were synthesized, each of which when bound to Zn^{2+} as in **21** would allow a greater accessibility of reactants to the exposed Zn^{2+} surface. That the accessibility factor is important is evidenced by the fact that the catalytic activity of both **22**:Zn^{2+} and **20**:Zn^{2+} exceed that of **6c**:Zn^{2+} their values being 1420 $M^{-1}s^{-1}$ (pH 6.55, HEPES, 25°C, μ = 0.2 $MNaClO_4$, 80% EtOH/H_2O) and 2700 $M^{-1}s^{-1}$ (pH 6.45, HEPES, 25°C, μ = 0.2 $MNaClO_4$, 80% EtOH/H_2O) respectively.[51]

It has long been known that the presence of monovalent anions inhibits the catalytic viability of CA.[1] These apparently bind to the active site metal and inhibit access of the reagents. Interestingly, at low concentration (5×10^{-4} - 2×10^{-3}), monovalent anions such as Cl^-, Br^-, I^- and NO_3- inhibit the ability of **6c**:Zn^{2+}, **20**:Zn^{2+} and **22**:Zn^{2+} to facilitate the interconversion of HCO_3^- and CO_2. In fact, Cl^- is such a good inhibitor, all activity is lost when [**6c**:Zn^{2+}] = [Cl^-]. In these cases, the inhibitory effects of the anions also clearly stems from their association with the metal which again blocks access of HCO_3^-, H_2O and CO_2 to the metal.

4. Future Directions and Considerations

From the above account one can conclude that the state of the art of modelling the active site and activity of CA is more advanced in the areas of structural models than catalytic models. As far as the structural and spectroscopic models are concerned, fairly reasonable approximations of the tridentate binding by trisimidazolyl and other tris-N-containing heterocycles such as pyridyl or pyrazolyl are available.

These have provided evidence that a Zn^{2+} or Co^{2+}-coordinated H_2O can have a pK_a value of 8 or lower provided that the coordination of the metal is reduced to 4. In addition, accompanying the ionization of the coordinated Co^{2+}-OH_2 are spectral changes that in some ways mimic those occuring with CoCA. It must be expected that the behavior of the models with respect to the above will be somewhat different in fine detail from that of the enzyme since the model is necessarily imperfect in terms of exact geometry of coordination and environment. It should be kept in mind that the enzyme utilizes 3 histidine imidazole units to bind the Zn^{2+} so that one might consider that information germane to the enzyme might be gathered more safely from studies of tris-imidazolyl complexes. Understandably, model studies are often dictated not so much by what one would like to make, but by what is conveniently synthesized. One obvious deficiency of structural models vis à vis the enzyme is that for the latter, the binding site is apparently "pre-formed" and that the active site of geometry of the apo-enzyme is similar to that of the holo enzyme. On the other hand, the geometry of the smaller ligands in part is determined by the binding of the metal. The most obvious disparity between models and enzyme becomes evident when one considers the pH-dependence of the Co^{2+} derivatives. For the enzyme, changes from one low coordinate environment to a different low coordinate species (4 or 5) accompanies the change in pH. However, reductions in pH for the small complexes completely destroy the low coordinate environment since the ligands, when protonated debind from the metal and allow access of several H_2O ligands.

Far less developed are models for the catalytic events occurring during enzyme-mediated interconversion of CO_2 and HCO_3^-. Apparently the preliminary feelings that the significant reduction in pK_a of the Zn^{2+}-OH_2 unit would result in a commensurate decrease in nucleophilicity of the bound OH have been somewhat resolved. Nevertheless, no M^{2+}-OH^- model identified to date exhibits a nucleophilicity toward CO_2 which exceeds free OH, although in the enzyme the active site Zn^{2+}-OH^- must be roughly 10^4-fold more reactive than OH^-. It seems likely that if a Lewis acid role for the Zn^{2+} could be invoked at the time of delivery of the Zn^{2+}-OH^- to CO_2, some significant enhancement in rate might be achieved. Thus future models must be designed which allow access of both CO_2 and a Zn^{2+}-coordinated (H)OH at the same time while maintaining a low coordinate environment about the metal if the above notions are to be tested. The second major deficiency of the models concerns their affinity for M^{2+} at high pH. This is a consequence of the competition of OH^- or ligand for M^{2+}, the former winning out at high pH. Precipitation is invariably a problem for the models and this prevents study of the catalytic reactions over a wide pH range. At low pH, ligand protonation and debinding occurs. Thus the M^{2+} binding capabilities for the ligands will have to be increased considerably in the future in order to study catalysis.

5. References

1. For pertinent reviews of the enzymes structure and function:
   a. Maren, T.H. Physiol. Rev. 1967, **47**, 595.
   b. Lindskog, S. Adv. Inorg. Biochem. 1982, **4**, 115.
   c. Lindskog, S.; Henderson, L.E.; Kannan, K.K.; Liljas, A.; Nyman, P.O.; Strandberg, B. in "The Enzymes", Vol. 5, Boyer, P.D., Ed., Academic Press: New York, 1971, 587.
   d. Lindskog, S.; Ibrahim, S.A.; Jonsson, B.-H.; Simonsson, J. in

"The Coordination Chemistry of Metalloenzymes", Bertini, I.; Drago, R.S.; Luchinat, C., Eds., D. Reidel Co., Dordrecht, Holland, 1983, 49.
e. Silverman, D.; Vincent, S.H. CRC Crit. Rev. Biochem. 1983, **14**, 207.
f. Pocker, Y.; Sarkanen, S. Advan. Enzymol. 1978, **47**, 149.
g. Bundy, H.F. Comp. Biochem. Phsyiol. B 1977, **57**, 1.
h. Prince, R.H. Adv. Inorg. Chem. Radiochem. 1979, **22**, 349.
i. "Biophysics and Physiology of Carbon Dioxide", Bauer, C.; Gros, G.; Bartels, H., Eds., Springer-Verlag, Berling: Heidelberg, 1980.
j. Carter, M.; Jeffrey, S. Biochem. Soc. Trans. 1985, **13**, 531.

2. a. Kannan, K.K.; Petef, M.; Cid-Dresdner, H.; Lovgren, S. FEBS Lett. 1977, **73**, 115.
b. Notstrand, B.; Vaara, I.; Kannan, K.K. in "The Isozymes", Vol. 1, Markert, C.I., Ed., Academic Press: New York, 1975, 575.
c. Kannan, K.K.; Notstrand, B.; Fridborg, K.; Lovgren, S.; Ohlssen, A.; Petef, M. Proc. Natl. Acad. Sci. U.S.A. 1975, **72**, 51.
d. Kannan, K.K. in "Biophysics and Physiology of Carbon Dioxide", Bauer, C.; Gros, G.; Bartels, H., Eds., Springer-Verlag, Berlin: Heidelberg, 1980, 84.
e. Vendani, A.; Dunitz, J.D. J. Am. Chem. Soc. 1985, **107**, 7653.
3. a. Pocker, Y.; Meany, J.E. J. Am. Chem. Soc. 1965, **87**, 1809.
b. idem. J. Phys. Chem. 1970, **74**, 1486.
c. Pocker, Y.; Meany, J.E.; Davis, B.C. Biochemistry 1974, **13**, 1411.
4. a. Tashian, R.E.; Plato, C.C.; Shows, T.B. Science (Washington, D.C.) 1963, **140**, 53.
b. Pocker, Y.; Bjorkquist, L.C.; Bjorkquist, D.W. Biochemistry 1977, **16**, 3967.
c. Malmstrom, B.G.; Nyman, P.O.; Strandberg, B.; Tilander, B. in "Stucture and Activity of Enzymes", FEBS Symp. 1964, No. 1, 121.
d. Lo, K.W.; Kaiser, E.T. J. Am. Chem. Soc. 1969, **91**, 4912.
5. a. Lindskog, S.; Malmstrom, B.G. J. Biol. Chem. 1962, **237**, 1129.
b. Hunt, J.B.; Rhee, M.-J.; Storm, C.B. Anal. Biochem. 1977, **79**, 614.
6. a. Lindskog, S. J. Biol. Chem. 1963, **238**, 945.
b. Lindskog, S.; Nyman, P.O. Biochim. Biophys. Acta 1964, **85**, 462.
7. a. Bertini, I.; Canti, G.; Luchinat, C.; Scozzafava, A. J. Am. Chem. Soc. 1978, **100**, 4873.
b. Bertini, I.; Luchinat, C.; Scozzafava, A. Inorg. Chim. Acta 1980, **46**, 85.
c. Jacob, G.S.; Brown, R.D.; Koenig, S.H. Biochemistry 1980, **19**, 3754.
d. Bertini, I.; Luchinat, C. Accts. Chem. Res. 1983, **16**, 272.
e. Simonsson, I.; Lindskog, S. Eur. J. Biochem. 1981, **123**, 29.
f. Bertini, I.; Luchinat, C.; Scozzafava, A. Struct. Bonding (Berlin) 1982, **48**, 45.
8. Brown, R.S.; Huguet, J.; Curtis, N.J. in "Metal Ions in Biological Systems", Vol. 15, Sigel, H., Ed., Marcel Dekker Inc., New York 1985, 55.
9. a. Lindskog, S. Struct. Bonding (Berlin) 1970, **8**, 153.

b. Ciampolini, M. Struct. Bonding (Berlin) 1969, **6**, 52.
c. Coleman, J.E. Prog. Bioorg. Chem. 1971, **1**, 159.
10. Morris, P.J.; Martin, R.B. J. Am. Chem. Soc. 1970, **92**, 1543.
11. Appleton, D.W.; Sarkar, B. Proc. Natl. Acad. Sci. U.S.A. 1974, **71**, 1686.
12. Martin, R.B. Proc. Natl. Acad. Sci. U.S.A. 1974, **71**, 4346.
13. Sovago, I.; Kiss, T.; Gergely, A. J. Chem. Soc. Dalton 1978, 964.
14. Tang, C.C.; Davalian, D.; Huang, P.; Breslow, R. J. Am. Chem. Soc. 1978, **100**, 3918.
15. Brown, R.S.; Huguet, J. Can. J. Chem. 1980, **58**, 889.
16. Bogess, R.K.; Boberg, S.J. J. Inorg. Nucl. Chem. 1980, **42**, 21.
17. a. McWhinnie, W.R.; Kulasingam, G.C.; Draper, J.C. J. Chem. Soc. A 1966, 1199.
b. McWhinnie, W.R.; Lancaster, J.C. Inorg. Chim. Acta 1975, **5**, 515.
18. a. Huguet, J.; Brown, R.S. J. Am. Chem. Soc. 1980, **102**, 7571.
b. Brown, R.S.; Curtis, N.J.; Huguet, J. J. Am. Chem. Soc. 1981, **103**, 6953.
c. Curtis, N.J.; Brown, R.S. J. Org. Chem. 1980, **45**, 4038.
19. a. Mani, F. Inorg. Nucl. Chem. Lett. 1981, **17**, 45.
b. Mani, F.; Scapacci, G. Inorg. Chim. Acta 1980, **38**, 151.
c. Bertini, I.; Canti, G.; Luchinat, C.; Mani, F. Inorg. Chim. Acta 1980, **46**, L91.
d. Benelli, C.; Bertini, I.; Di Vaira, M.; Mani, F. Inorg. Chem. 1984, **23**, 1422.
20. a. Thompson, L.K.; Ramaswamy, B.S.; Seymour, E.A. Can. J. Chem. 1977, **55**, 878.
b. Thompson, L.K.; Ball, R.G.; Trotter, J. Can. J. Chem. 1980, **58**, 1566.
c. Addison, A.W.; Hendriks, H.M.J.; Reedijk, J.; Thompson, L.K. Inorg. Chem. 1981, **20**, 103.
21. Brown, R.S.; Salmon, D.; Curtis, N.J.; Kusuma, S. J. Am. Chem. Soc. 1982, **104**, 3188.
22. Breslow, R.; Hunt, J.T.; Smiley, R.; Tarnowski, T. J. Am. Chem. Soc. 1983, **105**, 5337.
23. Read, R.J.; James, M.N.G. J. Am. Chem. Soc. 1981, **103**, 6947.
24. a. Boggess, R.K.; Martin, R.B. Inorg. Chem. 1974, **13**, 1525.
b. Sundberg, R.J.; Martin, R.B. Chem. Rev. 1974, **74**, 471.
c. Pyrrole N-H ionizations of imidazoles coordinated to divalent metal ions have pK_a values in excess of 10.[24]
25. a. Sirs, J.A. Trans. Faraday Soc. 1957, **54**, 201.
b. Pocker, Y.; Bjorkquist, D.W. J. Am. Chem. Soc., 1977, **99**, 6537.
c. Paneth, P.; O'Leary, M.H. J. Am. Chem. Soc., 1985, **107**, 7381.
26. a. Ho, C.; Sturtevant, J.M. J. Biol. Chem. 1963, **238**, 3499.
b. Gibbons, B.H.; Edsall, J.T. J. Biol. Chem. 1963, **238**, 3502.
c. Kernshan, H.C. Biochim. Biophys. Acta 1965, **96**, 304.
d. Khalifah, R.G. J. Biol. Chem. 1971, **246**, 2561.
27. a. Woolley, P. Nature (London) 1975, **258**, 677.
b. Woolley, P. J. Chem. Soc. Perkin II 1977, 318.
c. Sillen, L.G.; Martell, A.E. in "Stability Constants of Metal Ion Complexes", Chemical Society Special Publication No. **17**, 1964 and supplement No. 1, 1971.

28. Prince, R.H.; Woolley, P.R. Nature Phys. Sci. 1972, **240**, 117.
29. Coates, J.H.; Gentle, G.J.; Lincoln, S.F. Nature (London) 1974, **249**, 773.
30. a. Wang, J.H. Proc. Natl. Acad. Sci. U.S.A. 1970, **66**, 874.
b. Pesando, J.M. Biochemistry 1975, **14**, 681.
31. Martin, R.B. J. Inorg. Nucl. Chem. 1976, **38**, 511.
32. Breslow, E. in "The Biochemistry of Copper", Academic Press: New York, 1966, 149.
33. Palmer, D.A.; Harris, G.M. Inorg. Chem. 1974, **13**, 965.
34. Chaffee, E.; Dasgupta, T.P.; Harris, J.M. J. Am. Chem. Soc. 1973, **95**, 4169.
35. Early, J.E.; Alexander, W. J. Am. Chem. Soc. 1970, **92**, 2294.
36. Woolley, P. J. Chem. Soc. Perkin II 1977, 318.
37. Buckingham, D.A.; Engelhardt, L.M. J. Am. Chem. Soc. 1975, **97**, 5915.
38. Harrowfield, J.; Mac, B.; Sargeson, A.M. J. Am. Chem. Soc. 1976, **98**, 7282.
39. Mahapatra, S.K.; Prakash, K.S.; Prasana, U.L. J. Inorg. Biochem. 1984, **21**, 287.
40. a. Hunt, J.B.; Rutenberg, A.C.; Taube, H. J. Am. Chem. Soc. 1983, **74**, 268.
b. Bunton, C.A.; Llewellyn, D.R. J. Chem. Soc. 1953, 1692.
41. a. Dasgupta, T.P.; Harris, G.M. J. Am. Chem. Soc. 1968, **90**, 6360.
b. idem Inorg. Chem. 1978, **17**, 2313.
c. idem ibid. 1978, **17**, 3304.
42. Pocker, Y.; Bjorkquist, D.W. Biochemistry 1977, **16**, 5698.
43. Tabushi, I.; Kuroda, Y. J. Am. Chem. Soc. 1984, **106**, 4580.
44. Cotton, F.A.; Wilkinson, G. "Advanced Inorganic Chemistry", 4th Ed., Wiley and Sons, New York, 1980, 71.
45. Breslow, E. in "The Biochemistry of Copper", Peisach, J.; Aisen, P.; Blumberg, W.E., Eds., Academic Press: New York, 1966, 149.
46. Koltun, W.L.; Fried, M.; Gurd, F.R.N. J. Am. Chem. Soc. 1960, **82**, 233.
47. a. Silverman, D.N.; Tu, C.K. J. Am. Chem. Soc. 1975, **97**, 2263.
b. Tu, C.K.; Silverman, D.N. ibid. 1975, **97**, 5935.
c. Silverman, D.N.; Tu, C.K.; Lindskog, S.; Wyns, G.C. ibid 1979, **101**, 6734.
48. Simonsson, I.; Jonsson, B.-H.; Lindskog, S. Eur. J. Biochem. 1979, **93**, 409.
49. Tabushi, I.; Kuroda, Y.; Mochizuki, A. J. Am. Chem. Soc. 1980, **102**, 1152.
50. Dugas, H.; Penney, C. "Bioorganic Chemistry", Springer-Verlag, New York, 1981, pg. 290.
51. Slebocka-Tilk, H.; Cocho, J.L.; Frakman, Z.; Brown, R.S. J. Am. Chem. Soc. 1984, **106**, 2421.

C_3 PHOTOSYNTHESIS IN ALGAE: THE IMPORTANCE OF INORGANIC CARBON CONCENTRATING MECHANISMS

D.T. Canvin, A.G. Miller and G.S. Espie
Department of Biology
Queen's University
Kingston, Ontario K7L 3N6
Canada

ABSTRACT. At low concentrations of inorganic carbon, unicellular algae and cyanobacteria require a 'CO_2-concentrating' mechanism to sustain high rates of photosynthesis. Air-grown cells of Synechococcus possess well-developed transport systems for CO_2 and HCO_3^-. HCO_3^- transport requires Na^+ and is inhibited by Li^+ whereas CO_2 transport is not affected by either metal. Intracellular CO_2 concentrations can be over 1000-fold higher than extracellular CO_2 concentrations. High-CO_2 grown Synechococcus has a well developed CO_2 transport system that is dependent on Na^+ and inhibited by Li^+. A HCO_3^- transport system appears to be poorly developed. Standing culture cells of Synechococcus possess CO_2 transport and Na^+ dependent HCO_3^- transport systems similar to those in air-grown cells, but also develop a Na^+-independent HCO_3^- transport system. The multiple transport systems may be indicative of the essentiality of these transport systems to the survival of the organism.

In terrestrial C_3 plants, the rate of photosynthesis under ambient conditions is primarily determined by the affinity of ribulose-bisphosphate carboxylase/oxygenase (Rubisco) for carbon dioxide and oxygen (6). Carbon dioxide is always limiting and because oxygen competes with carbon dioxide at the catalytic site of the enzyme, maximum photosynthetic rates cannot be achieved (17). While the inorganic carbon supply (or CO_2 concentration) can be experimentally altered for such plants, the plants normally experience a relatively constant external inorganic carbon and oxygen concentration. In contrast, plants such as algae, which grow in an aquatic environment, experience wide variations in both inorganic carbon (CO_2 + HCO_3^- + CO_3^{-2} = DIC) and oxygen concentrations. In addition, depending on the pH of the water, little of the inorganic carbon may be present as CO_2 and the CO_2/O_2 ratio would be such that photosynthesis would frequently be prohibited in these algae (17).

However, algal cells survive and grow over a wide range of extracellular inorganic carbon concentrations and flourish at

M. Aresta and G. Forti (eds.), Carbon Dioxide as a Source of Carbon, 199–212.

concentrations of CO_2 where direct fixation of CO_2 by Rubisco would be inconsequential (7). In addition, little, if any, effect of the oxygen concentration on photosynthesis of algal cells can be observed (10).

This atypical behaviour from C_3 plant characteristics is not due to any differences in the pattern of carbon fixation nor to a Rubisco with increased affinity for CO_2. In fact, the Rubisco of algae has an affinity for CO_2 that is less than that of the higher plant enzyme (1). Rather, these characteristics are due to the development in algae of inorganic carbon transport systems which actively transport inorganic carbon and result in accumulations of inorganic carbon inside the cell several orders of magnitude above the external concentration (2, 3, 8, 12, 14). It is the affinity of the transport system for inorganic carbon that determines, in large part, the relationship between the rate of photosynthesis and the external inorganic carbon concentration in these organisms. But the presence and properties of the transport system are not constant; rather, there seems to be a number of transport systems that are specified by the environmental conditions under which the algae grow.

In chemostat cultures of cyanobacteria (blue-green algae), where carbon is the limiting element for growth, one of two equilibrium conditions is eventually established (14). In one, the extracellular DIC concentration experienced during growth is less than 80 μM, and in the other it is 1 - 4 mM. The apparent photosynthetic affinity for DIC ($K_{½}^{DIC}$) of the cells from the low DIC chemostat is 2.5 μM whereas the $K_{½}^{DIC}$ of the cells from the high DIC chemostat is 1500 μM (Fig. 1). Cells grown in batch culture with apparent affinities for inorganic carbon between these two values can be produced when inorganic carbon concentrations are experimentally controlled at intermediate levels (11).

The maximum photosynthetic rates of Synechococcus from low DIC and high DIC cultures are different (Fig. 1) and this difference can be equated with changes in the Rubisco content of the cells (21, 22). But changes in the Rubisco content cannot change the affinity of the enzyme for CO_2. The K_m (CO_2) of Rubisco from a marine Synechococcus was about 250 μM while the K_m (O_2) was about 1 mM (1). Calculations, using these kinetic constants shows that actual photosynthetic rates could be achieved in these cells only if there were several fold concentrations of inorganic carbon over that present in the external medium (Fig. 2). For the low DIC cells, concentration of a 1000-fold is required, but even for the high DIC cells concentration of 10-fold or greater is required.

To obtain and sustain differences between external and internal concentrations, active transport of inorganic carbon must occur. Early work focused on the transport of the bicarbonate ion and several authors (3, 8, 12, 13, 14) clearly showed the transport of this species and accumulations of inorganic carbon in the cells at the level necessary to sustain photosynthesis. But further investigation, primarily with Synechococcus, revealed that more than one transport system was distinguishable (3, 16). These systems could be differentiated as to whether they transported CO_2 or HCO_3^- and whether or not Na^+ was required for the transport process.

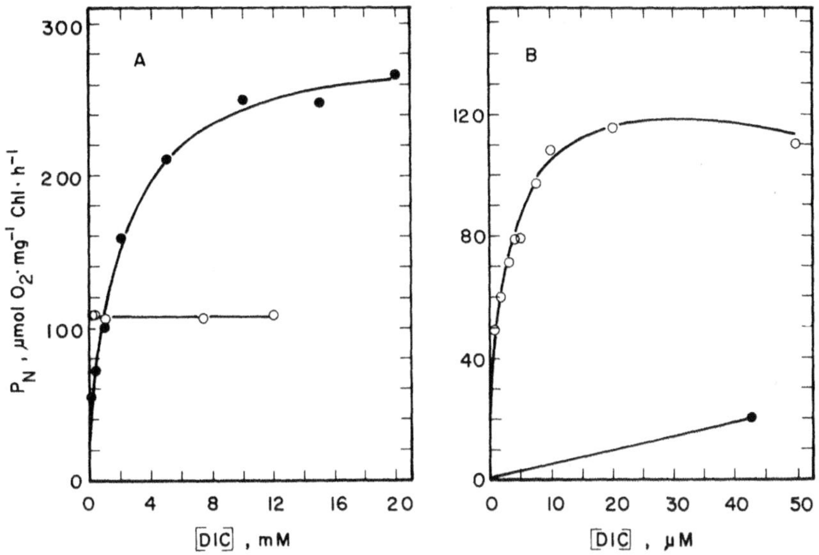

Figure 1. The response of photosynthesis to dissolved inorganic carbon [DIC] concentration with cells of <u>Synechococcus leopoliensis</u> harvested from carbon limited (Culture vessel DIC was 4.4 μM) and carbon sufficient (Culture vessel DIC was 1700 μM) chemostats. Photosynthetic rate (P_N) was determined with washed cells in an oxygen electrode in the presence of Na^+. The data for [DIC] below 60 μM has been replotted in (B). The $K_{1/2}^{DIC}$ for the carbon limited chemostat cells was 2.5 μM while the $K_{1/2}^{DIC}$ for the carbon sufficient cells was 1500 μM. (Reproduced from 14).

<u>Synechococcus</u> cells, similar in characteristics to the low DIC cells mentioned earlier, can be grown in cultures bubbled with air (air-grown cells). In such cells, photosynthesis at alkaline pH shows a marked dependency on the presence of Na (15). Such dependency on Na^+ can also be shown at pH 8.0, but if carbonic anhydrase (CA) is added to the medium, high rates of photosynthesis proceed in the absence of Na^+ (Fig. 3). The $K_{1/2}$ for Na^+ at pH 8.0 was 3 - 5 mM and the effect of Na^+ on photosynthesis could be specifically inhibited by 20 mM Li^+. The addition of carbonic anhydrase to the medium maintains the equilibrium between CO_2 and HCO_3^-. The rate of CO_2 production and the amount of CO_2 present under these conditions is apparently sufficient to maintain high photosynthetic rates. If ethoxyzolamide (EA, Fig. 3) were added to inhibit carbonic anhydrase, photosynthesis was greatly reduced. However, the addition of 10 mM NaCl restored high rates of photosynthesis. This stimulation was entirely inhibited by the addition of 20 mM LiCl. But the photosynthetic capacity of the cells was not impaired as the addition of saturating levels of inorganic carbon again produced high rates of photosynthesis.

The accumulation of dissolved inorganic carbon (DIC) in the Synechococcus cells was determined (Fig. 4A) using the silicone oil centrifugation technique (12). In the presence of Na^+ or CA

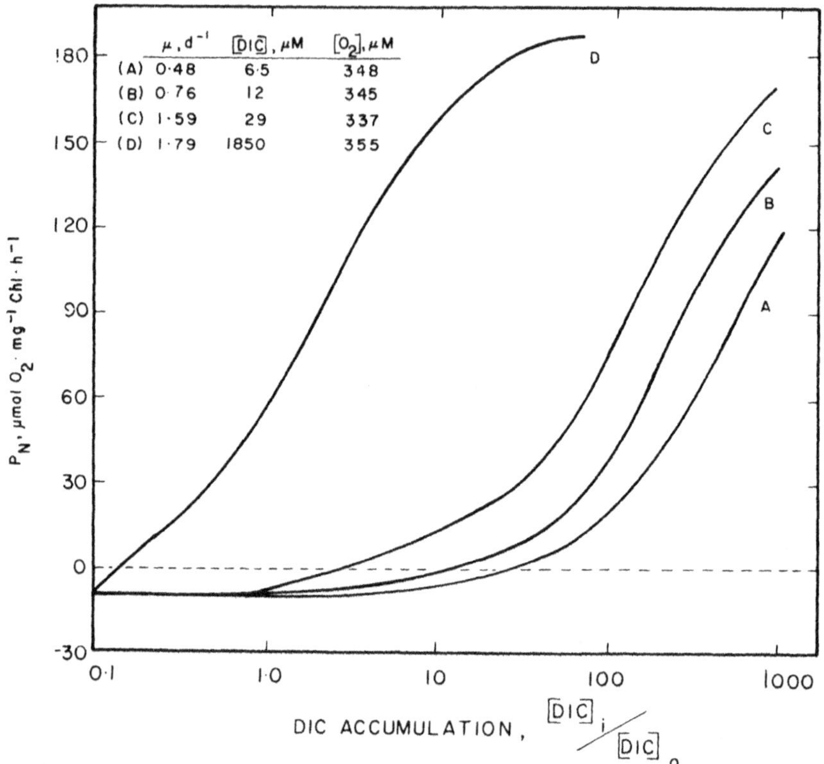

Figure 2. The requirement for 'CO_2-concentration' for photosynthesis and growth in Synechococcus. With the measured kinetic properties of Rubisco (1) the net rate of CO_2 fixation can be calculated for various $[O_2]/[CO_2]$ ratios by the method of Farquhar et al. (6). From (1) $V_c^{max} = 194\ \mu mol\ mg^{-1}$ Rubisco protein h^{-1}; $V_o^{max} = 22\ \mu mol\ mg^{-1}$ protein h^{-1}; $K_c = 250\ \mu M$; $K_o = 1010\ \mu M$. From (6)

$$V_c = \frac{V_c^{max} \cdot [CO_2]}{K_c \left[1 + \frac{O_2}{K_o}\right] + [CO_2]}$$

which assumes RuBP to be saturating and Rubisco fully activated. The net rate of photosynthesis (6) is given as

$$P_N = V_c - 0.5\ V_o$$

and

$$V_o = V_c \cdot \left[\frac{V_o^{max}}{V_c^{max}} \cdot \frac{[O_2]}{[CO_2]} \cdot \frac{K_c}{K_o} \right]$$

A value of 1 mg Rubisco protein mg^{-1} Chl has been assumed for these calculations. (Reproduced from 14).

concentrations of DIC within the cells was over a 1000-fold higher than external concentrations. The rate of fixation of inorganic carbon into acid-stable (photosynthetic) products was proportional to the size of the internal 'pools' (Fig. 4B).

These observations are consistent with 'air-grown' Synechococcus cells possessing both a CO_2 and HCO_3^- transport system, each with high capacity. CO_2 transport did not require Na^+ and was not inhibited by Li^+ whereas HCO_3^- transport did require Na^+ and was inhibited by Li^+. Active CO_2 transport in cyanobacteria has also been indicated or demonstrated in other cyanobacteria (3, 20).

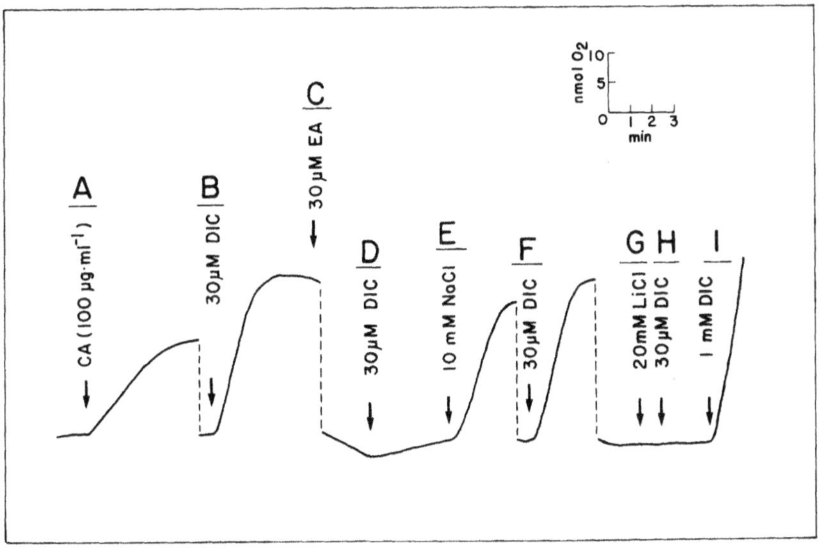

Figure 3. The effect of carbonic anhydrase (CA) and Na^+ on photosynthetic O_2 evolution in air-grown Synechococcus leopoliensis at pH 8.0. Cells were harvested, washed and suspended in Na^+-free medium, and additions were made as indicated to a reaction mixture in an oxygen electrode assembly. EA is ethoxyzolamide, an inhibitor of carbonic anhydrase. Dotted lines indicate adjustment of recorder pen. (Reproduced from 16).

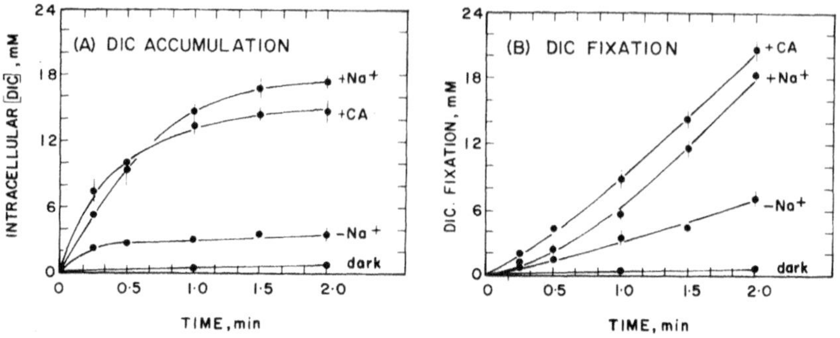

Figure 4. Effect of Na^+ and carbonic anhydrase (CA) on DIC transport, accumulation and fixation in air-grown Synechococcus at pH 8.0. The initial external [DIC] was 75 μM and fixation and accumulation were measured using the silicone oil centrifugation technique (12). (A) DIC accumulation refers to the amount of unmetabolized, acid labile DIC within the cells and was measured as in (15). Concentrations (mM) were calculated on the basis of the sorbitol impermeable space (15) which was 95 μl mg^{-1} Chl. (b) DIC fixation represents acid-stable photosynthetic products. The total amount of DIC transported at any time point is the sum of 'accumulation' and 'fixation'. After 2 min. the cells incubated in CA had an internal concentration of CO_2 of 1100 μM (assuming internal pH 7.5) compared to 0.86 μM in the medium. The calculated accumulation ratio for CO_2 would be 1279. (Reproduced from 16).

Synechococcus cells which are grown with bubbling of 5% CO_2 (high-CO_2 cells) through the culture are similar in photosynthetic characteristics to the high DIC cells from chemostats (Fig. 1). Such cells must also concentrate inorganic carbon to sustain high photosynthetic rates (Fig. 2).

Such cells also showed a requirement for Na^+ for maximum rates of photosynthesis (Fig. 5), but the concentration of Na^+ that was required was only one-hundredth of that required in air-grown cells. Lithium also inhibited this Na^+ stimulation.

One method to determine the form of inorganic carbon that is transported is to measure the accumulation of inorganic carbon (DIC) and photosynthesis in the cells, using the silicone oil centrifugation technique, when the carbon is supplied either as CO_2 or HCO_3^-. At pH 8.0 and 30° the hydration of CO_2 is a relatively slow process with a half-time of 12 sec. When CO_2 is supplied to the cells, CO_2 will be the dominant species in the initial time period, and observed uptake will likely be due to the use of this form. If carbonic anhydrase (CA) is simultaneously supplied with the CO_2, the CO_2 will rapidly be in equilibrium with HCO_3^- and the latter will be the dominant form in the reaction medium. The results of such an experiment are shown in Fig. 6.

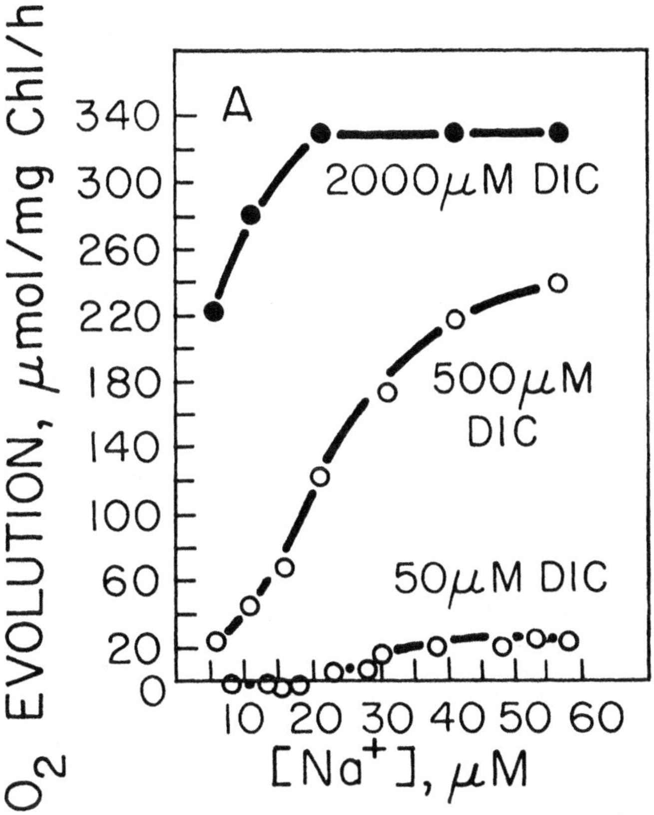

Figure 5. The relationship between [Na^+] and photosynthetic rate at pH 8.0 and several concentrations of DIC in Synechococcus grown on 5% CO_2. (Results of A.G. Miller).

All of the reaction mixtures contain Na^+ but it has already been shown that only low rates of photosynthesis occur in the absence of Na^+ (Fig. 5) and it can also be shown that, in these conditions, there is little accumulation of inorganic carbon in the cells. In the presence of Na^+, the intracellular inorganic carbon concentration was much higher when CO_2 was supplied than when HCO_3^- was supplied (Fig. 6A). Photosynthetic fixation was also greater with CO_2 than with HCO_3^- (Fig. 6B). The effect of pH on the $K_{\frac{1}{2}}^{DIC}$ for photosynthesis was also consistent with CO_2 use by this organism, as over a pH range of 6 to 9 the $K_{1/2}^{CO_2}$ remained relatively constant, whereas that for HCO_3^- changed 600-fold. In addition, the rate of photosynthesis in the high-CO_2 Synechococcus cells was always about equal to the rate of CO_2 generation from HCO_3^- in the reaction mixture.

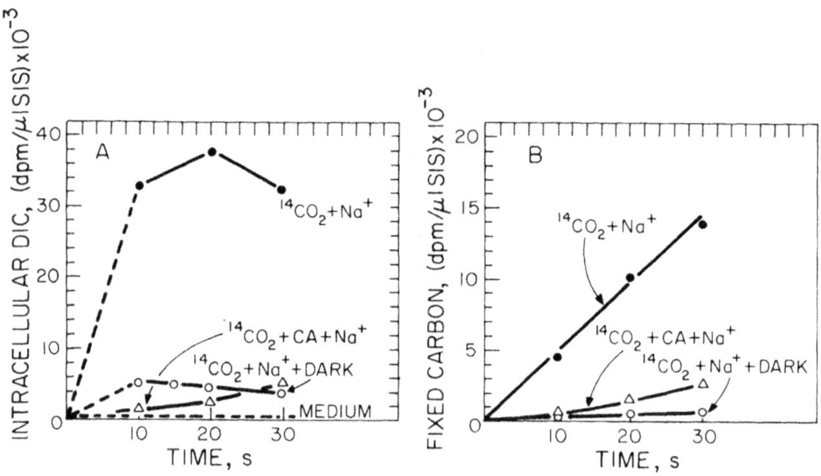

Figure 6. DIC accumulation and fixation by 5% CO_2-grown cells of Synechococcus under isotopic disequilibrium conditions (5). Washed cells were incubated at pH 8.0 with 100 μM Na^+ in the light in an oxygen electrode. Samples of 100 μl were layered over silicon fluid and covered with mineral oil and kept in the light. $DI^{14}C$ was injected into the algal suspension as $^{14}CO_2$ or $^{14}CO_2$ + 100 μg/ml CA to give a final concentration of 15 μM. The initial [DIC] of the buffer was 31 μM. Results are given in dpm because of the unequal specific activities of CO_2 and HCO_3^- in the medium. The dashed line in (A) represents the initial dpm ^{14}C/μl medium. If it is assumed that only CO_2 was taken up (CO_2 + Na^+) then intracellular [DIC] would have been 4.1 mM (257 μM CO_2 assuming intracellular pH 7.5) and the fixation rate 25.9 μmol $C.mg^{-1}.Chl.h^{-1}$. (Results of A.G. Miller).

These three lines of evidence support the view that high CO_2 cells of Synechococcus transport and utilize predominantly CO_2 as the form of inorganic carbon, and use sparingly the HCO_3^- ion. High CO_2 cells would appear to have a well developed Na^+ dependent CO_2-transport system, but have only a poorly developed HCO_3^--transport system.

Occasionally, with air-grown cells of Synechococcus, high rates of photosynthesis were obtained with cells suspended in Na^+ free medium and only a modest stimulation of photosynthesis was observed

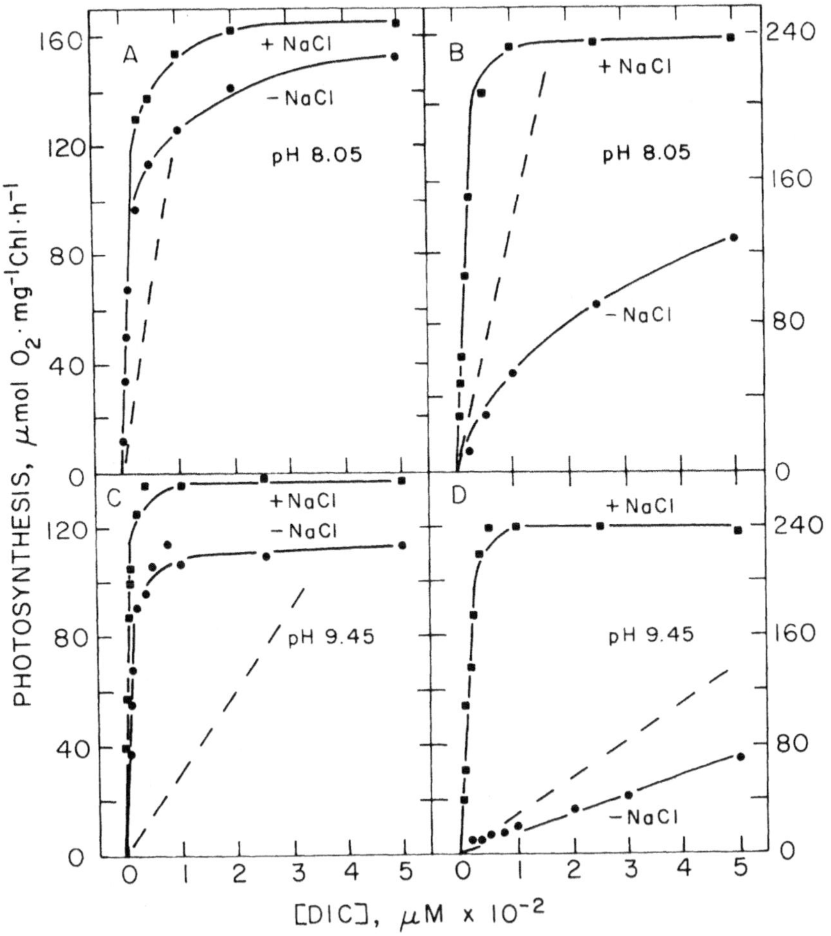

Figure 7. The response of photosynthesis to extracellular [DIC] (as $KHCO_3$) for standing culture (A and C) and air-grown cells (B and D) of Synechococcus in the presence and absence of 20 mM NaCl and pH 8.05 or 9.45 at 30°C. The maximum rate of photosynthesis that could be maintained by the CO_2 supply rate at the various [DIC] is shown (---). (Results of G.S. Espie).

on the addition of Na^+. After some exploratory investigation, we discovered that cells of Synechococcus grown in standing culture (from which the air-grown cells were derived) without any gas bubbling developed a system of inorganic carbon transport that was not

Na^+ dependent. Using isotopic disequilibrium experiments (similar to those described in Fig. 6, where CO_2 or HCO_3^- are supplied during steady-state photosynthesis) (5), it could be shown that the cells possessed an active CO_2-transport system which did not depend on the presence of Na^+. But it was also apparent, especially at pH 9.45, that CO_2 generation from HCO_3^- could not sustain the observed rates of photosynthesis, and hence HCO_3^- transport was also required (Fig. 7). In contrast to the air-grown cells, the $K_{1/2}^{DIC}$ of the standing culture cells was not markedly influenced by Na^+, but was of the same order of magnitude (6 to 40 μM at pH 8.05) as that observed in the air-grown cells in the presence of Na^+. Sodium additions to standing culture cells did increase the maximum rate of photosynthesis (Fig. 7) and the Na^+ concentration required for this effect was similar [$K_{1/2}(Na^+)$ = 4.8 mM] to that required for Na^+-dependent HCO_3^- transport in air-

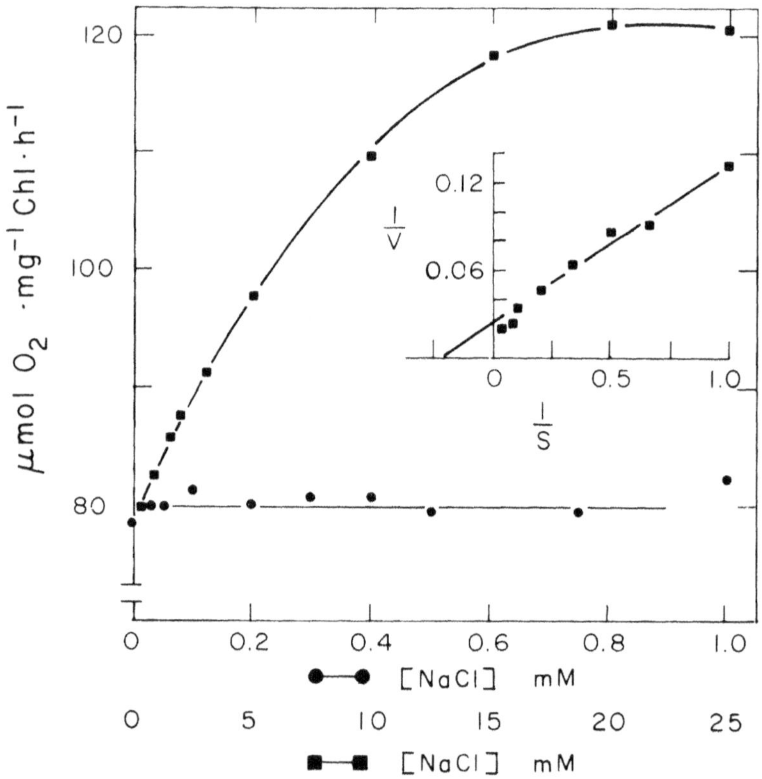

Figure 8. The effect of NaCl on photosynthesis of standing culture cells of Synechococcus in the presence of 20 μM DIC at pH 8.05 and 30°. Inset shows a double-reciprocal plot of the Na^+ dependent photosynthesis after subtraction of the Na^+ independent photosynthesis. (Results of G.S. Espie).

grown cells (Fig. 8). The Na dependent portion of photosynthesis was inhibited by Li^+ but Li^+ had little effect on the Na^+ independent HCO_3^- transport. Standing culture cells, if bubbled with air for two days, did not retain the Na^+-independent transport system for HCO_3^- but became identical to air-grown cells. Air-grown cells converted to the characteristics of standing culture cells only after a 5 to 7 day period of standing.

Active inorganic carbon transport and the accumulation of inorganic carbon in the cells is essential for the maintenance of photosynthesis and growth in cyanobacteria and eukaryotic algae (2), especially under conditions of low inorganic carbon supply. Concentrations of inorganic carbon in the cells can be 1000-times or more higher than the external concentrations. These high concentrations not only keep ribulose bis-phosphate carboxylase/oxygenase supplied with CO_2 at near saturation levels but also, apparently, prevent the oxygenase reaction, so that the cells no longer experience the stress due to atmospheric oxygen levels. In the cells, the equilibrium between CO_2 and HCO_3^- is apparently maintained through carbonic anhydrase, even though it is difficult to detect or isolate the enzyme from cyanobacteria (4). In order for the accumulation of inorganic carbon to be effective, the leakage of CO_2 from the cells must be prevented or decreased. This is apparently accomplished through the low conductance of the cell envelopes to the passive diffusion of CO_2 (4). Conductance values of about 10^{-5}cm sec^{-1} have been calculated (4) compared to conductance values of 10^{-1} to 10^{-2} cm sec^{-1} for leaves of terrestrial plants.

The molecular mechanism of CO_2 or HCO_3^- transport is largely unknown, although it is widely assumed that there are transporter proteins or 'pumps' in the cell envelope. A polypeptide has been isolated from the cytoplasmic membrane of a cyanobacterium that increases in parallel with the adaptation from high CO_2 conditions to low CO_2 conditions and it is thought that this polypeptide is involved in inorganic carbon transport (18), although no direct evidence on this latter point is yet available. If HCO_3^- is transported, either H^+ must also accompany it or OH^- must be extruded to maintain internal cell pH (15). The possible role that Na^+ plays in the transport process has been discussed (9, 19) but simultaneous transport of Na^+ with CO_2 in the high CO_2 grown cells could not be detected. This observation would support the view that, in this case, Na^+-binding to a 'CO_2' carrier protein modifies the transport kinetics of the carrier without concomitant Na^+ transport, but it is not known if such is the case for the other Na^+-dependent inorganic carbon transport systems.

The presence of multiple transport systems may be indicative of the importance of the inorganic carbon transport process to the survival of the organism. In Synechococcus, at least four distinguishable transport systems can be identified. Carbon dioxide can be transported by Na^+-independent or dependent systems, and HCO_3^- can also be transported by Na^+-dependent or independent systems. Some of these systems occur in the same cells (e.g., Na^+-independent CO_2 transport and Na^+-dependent HCO_3^- transport in air-grown cells) but others apparently only occur in different cell types which are produced under different environmental conditions. Whether each of the systems are separate systems or modifications of one or two systems, is not known at the present time.

LITERATURE CITED

1. Andrews, TJ, and KM Abel (1981) Kinetics and subunit interactions of ribulose bisphosphate carboxylase-oxygenase from the cyanobacterium, Synechococcus sp.. J. Biol. Chem. 256:8445-8451.

2. Badger, MR, A Kaplan and JA Berry (1978) A mechanism for concentrating CO_2 in chlamydomonas reinhardtii and Anabaena variabilis and its role in photosynthetic CO_2 fixation. Carnegie Inst. Yearbook 77:251-261.

3. Badger, MR, TJ Andrews (1982) Photosynthesis and inorganic carbon usage by the marine cyanobacterium Synechococcus sp. Plant Physiol. 70:517-523.

4. Badger, MR, M. Bassett and HN Comins (1985) A model for HCO_3^- accumulation and photosynthesis in the cyanobacterium Synechococcus sp. Plant Physiol. 77:465-471.

5. Espie, GS and B Colman (1986) Inorganic carbon uptake during photosynthesis 1. A theoretical analysis using the isotope disequilibrium technique. Plant Physiol. 80:863-869.

6. Farquhar, GD, S von Caemmerer and JA Berry (1980) A biochemical model of photosynthetic CO_2 assimilation in leaves of C_3 species. Planta 149:78-90.

7. Jensen, RG and JT Bahr (1977) Ribulose 1.5-bisphosphate carboxylase-oxygenase. Ann. Rev. Plant Physiol. 28:379-400.

8. Kaplan, A, MR Badger, JA Berry (1980) Photosynthesis and the intracellular inorganic carbon pool in the blue-green alga Anabaena variabilis: Response to external CO concentration. Planta 149:219-226.

9. Kaplan, A, M. Volokita, D. Zenvirth and L Reinhold (1984) An essential role for sodium in the bicarbonate transporting system of the cyanobacterium Anabaena variabilis FEBS Lett 176:166-168.

10. Lloyd, ND, DT Canvin and DA Culver (1977) Photosynthesis and photorespiration in algae. Plant Physiol. 59:936-940.

11. Mayo, WP, TG Williams, DG Birch and DH Turpin (1986) Photosynthetic adaptation by Synechococcus leopoliensis in response to exogenous dissolved inorganic carbon. Plant Physiol. 80:1038-1040.

12. Miller, AG and B Colman (1980) Active transport and accumulation of bicarbonate by a unicellular cyanobacterium. J. Bacteriol. 143:1253-1259.

13. Miller, AG, B Colman (1980) Evidence for HCO_3^--transport by the blue-green alga (cyanobacterium) Coccochloris peniocystis. Plant Physiol. 65:397-402.

14. Miller, AG, DH Turpin and DT Canvin (1984) Growth and photosynthesis of the cyanobacterium Synechococcus leopoliensis in HCO_3^--limited chemostats. Plant Physiol. 75:1064-1070.

15. Miller, AG, DH Turpin and DT Canvin (1984) Na^+ requirement for growth, photosynthesis and pH regulation in the alkalotolerant cyanobacterium Synechococcus leopoliensis. J. Bacteriol. 159:100-106.

16. Miller, AG and DT Canvin (1985) Distinction between HCO_3^- and CO_2-dependent photosynthesis in the cyanobacterium Synechococcus leopoliensis based on the selective response of HCO_3^- transport to Na^+. FEBS Lett 187:29-32.

17. Ogren, WL (1984) Photorespiration: Pathways, regulation and modification. Ann. Rev. Plant Physiol. 35:415-442.

18. Omata, T and T Ogawa (1986) Biosynthesis of a 42-kD polypeptide in the cytoplasmic membrane of the cyanobacterium Anacystis nidulans Strain R2 during adaptation to low CO_2 concentration. Plant Physiol. 880:525-530.

19. Reinhold, L, M Volokita, D. Zenvirth and A Kaplan (1984) Is HCO_3^- transport in Anabaena a Na^+ symport? Plant Physiol. 76:1090-1092.

20. Volokita, M, D Zenvirth, A Kaplan and L Reinhold (1984) Nature of the inorganic carbon species actively taken up by the cyanobacterium Anabaena variabilis. Plant Physiol. 176:599-602.

21. Yokota, A and DT Canvin (1985) Ribulose bisophosphate carboxylase/oxygenase content determined with [^{14}C] carboxypentitol bisphosphate in plants and algae. Plant Physiol. 77:735-739.

22. Yokota, A and DT Canvin (1986) Changes of ribulose bisphosphate concentration and photosynthetic activity during adaptation of high-CO_2 grown cells to low-CO_2 conditions in Chlorella pyrenoidosa. Plant Physiol. 80:341-345.

PAST AND PRESENT: LONG TERM CHANGES IN ATMOSPHERIC CO_2 CONCENTRATION AND PLANT STRATEGIES FOR THE OPTIMISATION OF PHOTOSYNTHESIS

M.N. Sivak
Research Institute for Photosynthesis,
University of Sheffield,
Sheffield S10 2TN,
U.K.

ABSTRACT Carbon dioxide concentrations diminished and oxygen concentrations increased over milenia of plant photosynthesis on earth. CO_2 is a substrate of plant photosynthesis and its present concentration is very low and limiting for most plants in natural conditions. Some plants evolved additional metabolic pathways (e.g. C_4 metabolism) which optimise photosynthesis in these unfavourable conditions. During this century carbon dioxide concentration has been steadily increasing as a result of man's activities. These changes are likely to affect climate in the long term as a direct result of the greenhouse effect and through complex and poorly understood effects on global photosynthesis.
Photorespiratory, C_4 and Crassulacean acid metabolism, and CO_2 accumulation mechanisms in some algae are briefly described. The response of different aspects of C_3 photosynthesis to increased CO_2 concentration are considered in the context of present knowledge of factors and processes which limit photosynthesis and of the role of regulatory mechanisms. The influence that the expected increases in atmospheric CO_2 concentration and of genetic manipulation of plants may have on the present balance between the partial processes of photosynthesis, is discussed. Bases of expression of photosynthetic rates used in plant physiology are described and a new one, *efficacy*, is proposed. It is argued that adequate understanding of the long term effects of changes in CO_2 concentration and of other factors on photosynthesis and growth may require improved techniques and the establishment of standards for the measurement of photosynthesis.

1. INTRODUCTION

It seems now clear that atmospheric carbon dioxide concentration has been increasing steadily during this century as a result of human activities, mainly as a result of the combustion of carbon-based fuels. Although the full extent of the changes occurring in the earth's atmosphere is still a matter for investigation, it is certainly far reaching. For example, the future changes in the concentrations of CO_2 and other gases that strongly absorb in the infrared part of the spectrum (the "green-house" gases) are expected to bring about a global warming, and research in this field indicates that this effect will become unambiguously clear by the end of the 1980s. The expected changes in CO_2 concentration, a substrate in photosynthesis, should have a very complex effect on plants and they attract

M. Aresta and G. Forti (eds.), Carbon Dioxide as a Source of Carbon, 213–236.

increasing attention from scientists and research agencies. The predicted changes in the composition of earth's atmosphere are, however, relatively small when compared with those that plants had to face during evolution. The main factor in these changes was photosynthesis. In photosynthesis, light energy is used to reduce CO_2 into CH_2O (this being a simple representation of carbohydrate) and O_2 is formed. Over milenia, photosynthesis has progressively increased O_2 concentration and decreased CO_2 concentration in the atmosphere. The following is a brief description of the different strategies that allow plants to photosinthesize and grow in an atmosphere that has very little CO_2 and high $[O_2]$.

2. PHOTOSYNTHETIC CARBON FIXATION AND PHOTORESPIRATION

2.1. Carboxylation

Carboxylation is an energy requiring reaction. There is also very little CO_2 (about 0.034 %) in the atmosphere. The "ideal" carboxylation reaction would involve, in these circumstances, an "energy rich" substrate which, when combined with CO_2 in the presence of a suitable catalyst, would conveniently form a new carboxylated compound in a reaction accompanied by a large decrease in free energy. This would ensure an extremely favourable equilibrium position. The "ideal" catalyst would have an extremely high affinity for CO_2, such that the reaction would continue at a relatively constant rate in extremely small concentrations of carbon dioxide.

When the carboxylation of 1,5 ribulose bisphosphate (RuBP) was first described it was soon obvious that the first criterion had been met. The decrease in free energy was some 8 Kcal. The reaction was virtually irreversible. On the other hand, the affinity for CO_2 was ridiculously low with some 6% CO_2 required for half saturation. This enigma was resolved by Lorimer *et al.* (1976) who found that the purified enzyme requires activation by CO_2, Mg^{++} and high pH. At about the same time it became clear that chloroplast stroma, released into suitable media, also displayed carboxylase activity with believable characteristics. The whole picture was rounded out once it was understood (Bowes and Ogren, 1972) that the carboxylase is also an extremely active oxygenase.

The carboxylation of RuBP by RuBP carboxylase/oxygenase (Rubisco) is a unique reaction in a number of respects, one of them the large number of chemical steps involved in transforming substrates to products. The overall reaction can be dissected into five steps (Lorimer et al. 1986): enolization, carboxylation (which forms the six-carbon intermediate, 2'-carboxy-3-keto-D-arabinitol 1,5-bisphosphate), hydration (of the six carbon intermediate to the gem-diol form), deprotonation and carbon-carbon cleavage. The mechanisms by which Rubisco catalyses carboxylation and oxygenation of RuBP are currently the object of intensive research. Both activities depend on the binding of a CO_2 molecule followed by a Mg^{++} and then by RuBP. The requirements of the enzyme to act as an oxygenase are the same as for the carboxylase activity, and CO_2 and O_2 are competitive inhibitors each for the reaction of the other.

2.2 Photorespiration

Like animals, plants respire, oxidising CH_2O, and other metabolites, to CO_2

and H_2O. In the light, C_3 plants also lose newly fixed CO_2 to the atmosphere. The source of this CO_2 is photorespiration. Carbon dioxide fixation in the Benson-Calvin cycle (or reductive pentose phosphate pathway, RPPP) is catalysed by Rubisco. In the presence of excess CO_2, the normal carboxylation occurs, but because Rubisco also has oxygenase activity, i.e., O_2 can compete with CO_2, in the presence of air concentrations of CO_2 and O_2, oxygenation occurs which leads to the formation of phosphoglycollate (Figure 1).

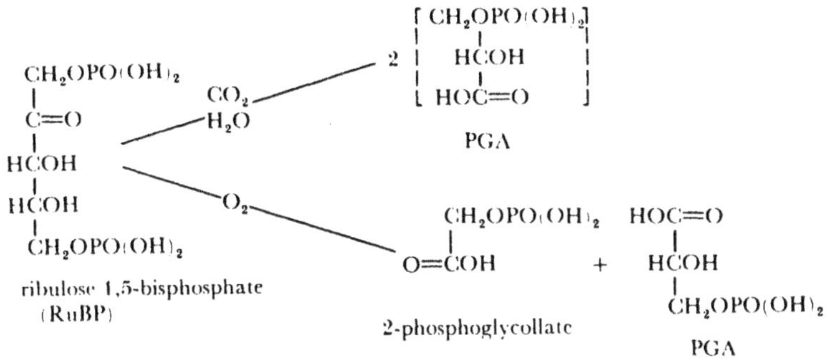

Figure 1. Carboxylation and oxygenation: the two reactions catalysed by Rubisco.

The phosphoglycollate so formed gives rise to other C2 compounds such as glycollate, glyoxylate and glycine and one or more of these may be broken down to release CO_2. Since a significant fraction, which varies with the relative amounts of CO_2 and O_2, of the normal product of CO_2 assimilation is lost in the oxygenase reaction (C3 + C2 rather than 2 x C3) regeneration of the CO_2-acceptor and accumulation of CH_2O both proceed at slower rates (some serine carbon may re-enter the RPPP as phosphoglycerate but CO_2 has already been lost in serine formation). Peroxisomes contain some key enzymes of the glycolate pathway, those needed to convert glycollate to glycine and serine to hydroxypyruvate to glycerate. The mitochondria contain the enzymes which catalyse decarboxylation of glycine to serine + CO_2. Although the loss of CO_2 associated with photorespiration occurs in this organelle, the synthesis of glycollate and its subsequent metabolism depends on partial reactions of the pathway that occur in chloroplasts, peroxisomes, and mitochondria and therefore, on substantial metabolite transport between them. In the pathway, glycollate is synthesised in the chloroplasts, transported to peroxisomes, and there converted to glycine. Glycine is then transported to mitochondria and decarboxylated to serine + CO_2 + NH_3 through glycine synthase - serine hydroxymethyltransferase. The serine is further metabolised to glycerate in the peroxisomes and the glycerate is metabolized to triose phosphate in the chloroplasts. Metabolite exchange of glycolate, glycine, serine and glycerate is proposed to occur in the peroxisomes and metabolite exchange of glycine, serine and CO_2 is proposed to occur in the mitochondria.

The massive concentration of Rubisco (the most abundant protein in the world) in the chloroplast stroma could be regarded as an evolutionary response to decreasing CO_2 (and increasing O_2) in the atmosphere. The photorespiratory pathway can be regarded as a classic example of making a virtue out of a necessity. Some at least of the "lost" carbon finds its way back into the cycle and some useful by-products are formed.

The above is certainly true for C_3 plants. But other plants have solved the problem of decreasing CO_2 concentration in the atmosphere in other ways.

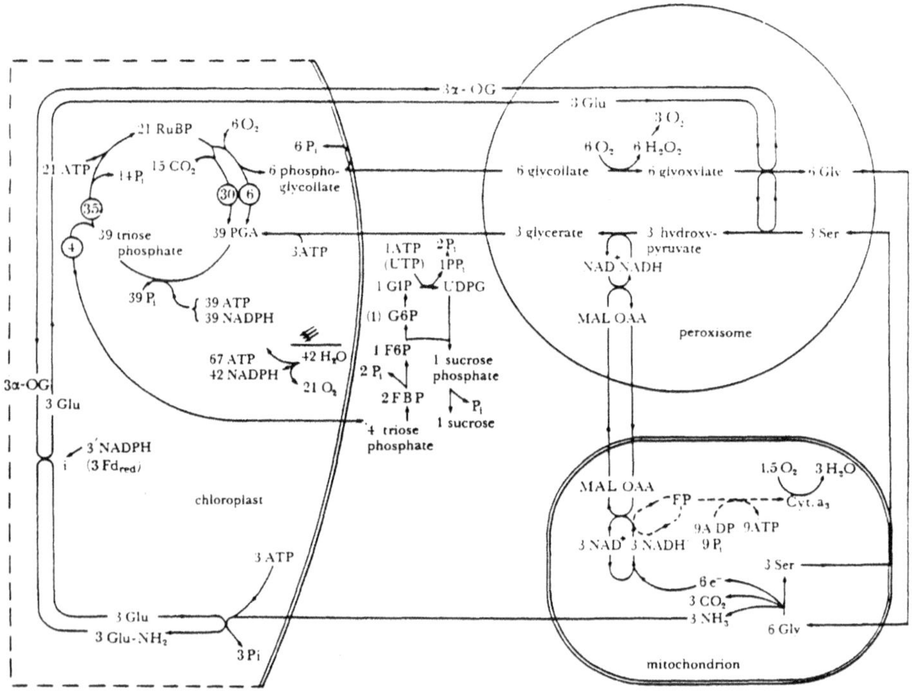

Figure 2. Possible shuttle of metabolites between peroxisomes, mitochondria and chloroplasts during metabolism in the photorespiratory pathway (from Keys, 1986).

C_4 METABOLISM

2.1. An additional sequence

The immediate end-product of the RPPP is a three-carbon (C-3) compound. All green plants operate this sequence of carbon assimilation. Most temperate species depend upon it entirely and are, therefore, called C_3 plants. Others (often tropical or semi-tropical grasses such as maize and sugar cane) operate an additional sequence, which yields a four-carbon compound as the first product of CO_2 fixation, and are therefore called C_4-plants. In order to provide a high, saturating CO_2 concentration to the RPPP, these plants have evolved a metabolic CO_2 pump which is driven by ATP. It functions to trap atmospheric CO_2 in an outer layer of photosynthetic cells and shuttle it into a inner layer of the leaf where it is assimilated in the RPPP. Many of the worst weeds of the world are C_4 species, whilst most crops are C_3. The most economically important C_4 crops are sorghum, maize and sugar-cane, millets (e.g. *Penisetum typhoides*, *Panicum miliaceum*), and a number of pasture grasses.

The process of carbon assimilation, first elucidated by Kortschak, Hatch and Slack, is based partly upon modified biochemistry and partly on modified anatomy. The typical C_4-plant such as sugar cane has Kranz-type anatomy (Fig. 2). Kranz is German for "wreath" and is used to describe leaf anatomy in which the vascular bundle (leaf "vein") is surrounded by two concentric layers (or "wreaths") of photosynthetic tissue. The inner-layer is termed the bundle-sheath, the outer is the mesophyll.

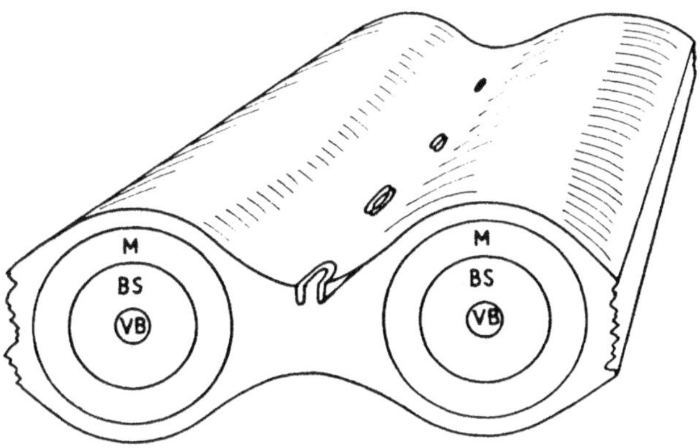

Figure 3. The Kranz-type anatomy. A diagrammatic representation of part of a sugar-cane leaf. The ridges in the leaf mark the position of the "veins" or vascular bundles (VB). The photosynthetic tissue is restricted to two concentric cylinders, the bundle sheath (BS) and the mesophyll (M) surrounding the vascular tissue. Carbon dioxide enters the leaf through stomata located in the "troughs" between the ridges and therefore passes into the photosynthetic tissue via the mesophyll (From Walker, 1983).

The C_4 dicarboxylic acid pathway of photosynthetic carbon assimilation (the C_4 pathway) is a complex biochemical and physiological elaboration of the common photosynthetic carbon reduction cycle (RPPP, C_3 pathway). The C_4 pathway has been defined (Hatch and Osmond, 1976) as the complete reaction sequence in which CO_2 is transferred via the C-4 carboxyl of C_4 acids to the reactions of the RPPP and there reduced to the level of carbohydrate. The distinctive biochemical features of this process are the carboxylation and associated reactions leading to the synthesis of C_4 acids, and those concerned with the subsequent decarboxylation of these C_4 acids to supply CO_2 for the RPPP. Unlike the C_3 cycle, in which carboxylation and carbon reduction *are* restricted to the chloroplast, the C_4 pathway involves the operation of reactions in the cytosol, mitochondria and chloroplasts, and the transport of intermediates between intracellular compartments. In this sense, it may be compared with another elaboration of the C_3 cycle, the photorespiratory pathway. However, a distinctive feature of the C_4 pathway is the *mandatory* exchange of photosynthetic intermediates between adjacent cells.

The lack of aparent photorespiration in C_4 plants is mostly attributed to CO_2 enrichment in the inner (bundle-sheath) compartment as a consequence of the decarboxylation process ($C4 \rightarrow C3 + CO_2$). If C_4 photosynthesis is seen as a modification of the C_3 pathway which alleviates photorespiratory loss, it may be asked why evolution has not favoured the development of plants which rely entirely on the carboxylation of PEP (since this is not subject to photorespiratory loss). But the major carboxylation mechanism in photosynthesis might be expected not only to regenerate its CO_2 acceptor, but also to work autocatalytically, as a breeder reaction, producing more potential substrate than it utilizes. The carboxylation of phosphoenolpyruvate (PEP) in C_4 photosynthesis has associated with it a mechanism for regenerating PEP but only in the same quantities that are consumed. Thus in C_4 species, as in C_3, autocatalysis is achieved only by the Benson-Calvin cycle. For this reason, PEP carboxylation can only function as an adjunct to the main cycle and not as an alternative to it.

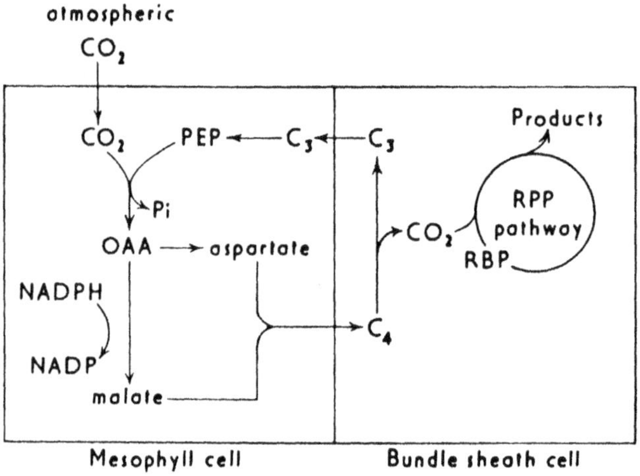

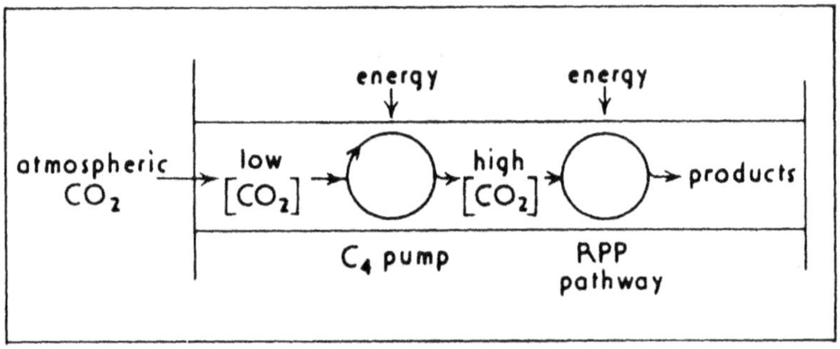

Figure 4. C_4 photosynthesis. Top: Carbon dioxide, entering the leaf via the stomatal apertures, passes into the mesophyll where it is fixed into C4 acids. These pass into the bundle sheath where they are broken down. This releases CO_2 which is fixed into the Benson-Calvin cycle. The C3 residue is returned to the mesophyll where it is reconverted into the original C3 acceptor (using additional ATP). Bottom: The C_4 cycle as a CO_2 pump (from Edwards and Walker, 1983).

2.1. Carboxylation in C_4 plants

Like RuBP, phosphoenolpyruvate (PEP) is a 'high energy' molecule with a free energy of hydrolysis of about 13 kcal. In the presence of ADP and PEP transphosphorylase, half of this energy is conserved in the formation of ATP. Alternatively, in the presence of PEP carboxylase the available energy can be use to drive a carboxylation.

$$PEP + HCO_3^- + H_2O \rightarrow \text{oxaloacetate} + P_i$$

This reaction is strongly exergonic ($\Delta G^{o\prime}$= -6 to -8 Kcal/mole) and is essentially irreversible. Carboxylation of PEP is catalised by PEP carboxylase. The enzyme requires divalent metal ions and they seem to bind PEP and Mg^{++} separately. There is PEP carboxylase in C_3 plants, but the amounts are relatively small, and the enzymatic forms are probably different.The characteristics of PEP carboxylase from C_4 plants are: 1) it uses HCO_3^- as the substrate (not free CO_2), 2) it appears to have a Km for HCO_3^- around 0.2 - 0.4 mM, 3) its V_{max} with saturating PEP and HCO_3^- is 800-1200 µmol. mg^{-1} Chl. h^{-1} at 30°C and, 4) unlike Rubisco, is not inhibited by O_2. Thus, the advantage of PEP carboxylase over Rubisco in efficiency of carboxylation may lie in its lack of oxygenase activity and possibly higher V_{max}. Certainly, the V_{max} of PEP carboxylase at saturating PEP and HCO_3^- is well above the rates of leaf photosynthesis. What is the concentration of HCO_3^- *in vivo*? Several factors influence the availability of HCO_3^- at the site of carboxylation by the PEP carboxylase. These factors include the [CO_2] in the intracellular air spaces inside the leaf, the leaf temperature, the pH of the cytosol and the possible role of carbonic anhydrase. Both PEP carboxylase and carbonic anhydrase are localized in the cytosol. Once CO_2 enters the cytosol of the cell, it rapidly hydrates to HCO_3^- + H^+. Both CO_2 and HCO_3^- diffuse throughout the cytosol along concentration gradients. As the potential gradient for HCO_3^- is much larger than for CO_2, the former is readily available for PEP carboxylase. If the catalytic capacity of carbonic anhydrase greatly exceeds that of PEP carboxylase, the final concentration of HCO_3^- in the cytosol is then dependent on the intercellular [CO_2] in the air space around the mesophyll cells, the temperature and the pH of the cytosol. For example, at a concentration of 250 µl/l CO_2 in the leaf air space, and at a leaf temperature of 30 °C, the solubility of free CO_2 would be 7.4 µM. At a pH of 8.0 in the cytosol, this CO_2 would be in equilibrium with 330 µM HCO_3^-, a value close to the Km for PEP carboxylase. This would allow rates of about 500 µmol CO_2 fixed mg^{-1} Chl. h^{-1} at saturating PEP (Edwards and Walker, 1983).

2.3. Three C_4 subgroups.

The C_4 cycle can be divided into two phases: the carboxylation phase, which occurs in the mesophyll cells, and the decarboxylation phase which occurs in the bundle sheath cells. CO_2 fixation through PEP carboylase is common to the carboxylation phase in all C_4 species. C_4 species can be divided into three subgroups on the basis of differences in the mode of C_4 acid decarboxylation. NADP-malic enzyme type, NADP-malic enzyme type and PEP carboxykinase type. The major metabolites of the C_4 cycle transported between cells are malate pyruvate in NADP-ME species, aspartate alanine in NAD-ME species, and aspartate and PEP in PEP-CK species.

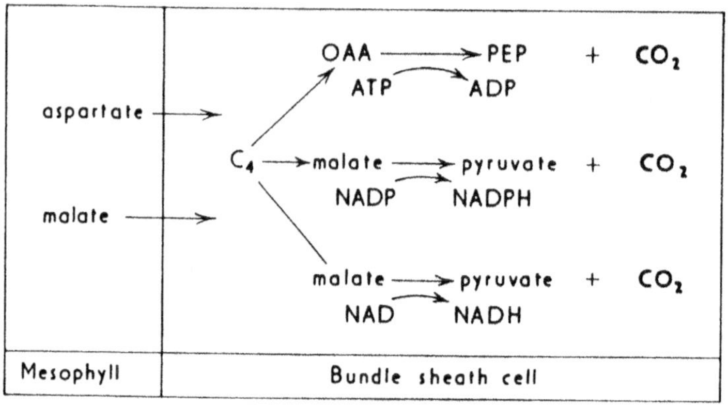

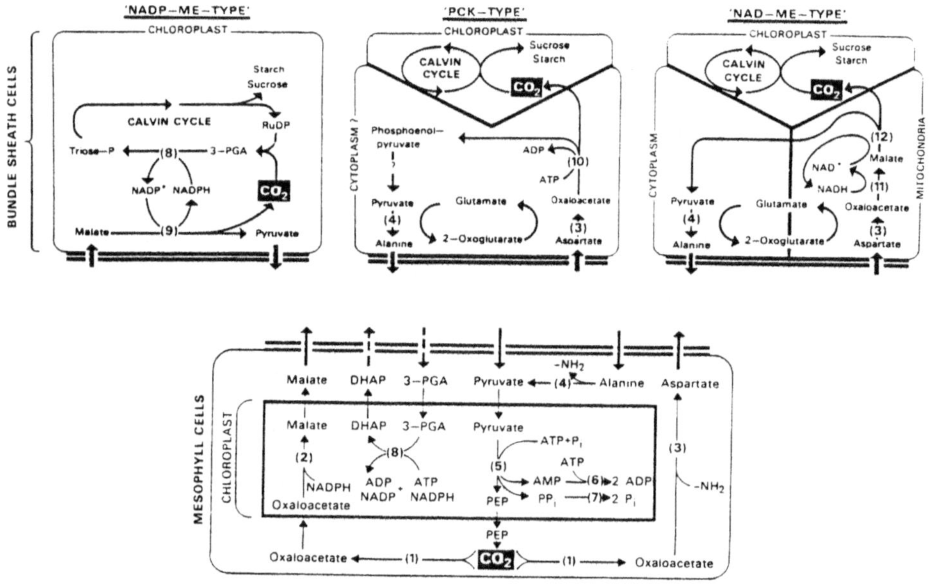

Figure 5. Diversity in decarboxylating mechanisms systems in C_4 photosynthesis. There are different types of "CO_2 pumps" in different C_4 species. The fixation of atmospheric CO_2 into C_4 acids in mesophyll cells, is followed by transport of mlate and aspartate to bundle sheath cells where CO_2 is released by decarboxyaltion. Some species use PEP carboxykinase, some NADP-malic enzyme and others NAD-malic enzyme as the dcarboxylating enzyme. These variations in the C_4 cycle are linked to differences in photochemistry, leaf anatomy and taxonomy (From Hatch and Osmond, 1976).

Photochemical differences in the mesophyll and bundle sheath chloroplasts in the three subgroups are consistent with differences in energetics of the C_4 cycle. In NADP-ME species, malate transport enables the transfer of reductive power from mesophyll to bundle sheath cells. The bundle sheath cells of these species are deficient in photosystem II, and may shuttle up to half of the PGA (product of Rubisco actvity in the bundle sheath) to mesophyll cells for reduction. Thus, in these species the mesophyll cells are the primary site of light-dependent generation of reductive power (NADPH). In PEP-CK and NAD-ME species, only ATP is required for aspartate synthesis in the mesophyll cells. The bundle sheath chloroplasts are the primary site for generating NADPH for reduction of PGA. The energy requirements per CO_2 fixed are similar for all three subgroups. The net additional energy beyond that required for the RPPP is the ATP needed to drive the C_4 cycle.

2.4. The C_4 syndrome

The characteristics which typify C_4 plants are collectively referred to as the C_4-syndrome. These are summarised in Table 1 where they are compared to those exhibited by C_3 and CAM plants. Perhaps the oddest of these is the fact that C_3 plants discriminate against the heavy isotope of carbon (^{13}C) more effectively than C_4 plants. Most of the CO_2 in the atmosphere is $^{12}CO_2$ but there is some $^{13}CO_2$ and this is not as readily fixed as its lighter isotope. This discrimination is particularly marked in the carboxylation of RuBP. In consequence, a forensic scientist could distinguish, if necessary, between sugar from cane (C_4) and sugar from beet (C_3). Thus only a slight discrimination will occur in the outer (mesophyll) compartment of sugar cane and then all of the CO_2 subsequently released in the bundle sheath will be ultimately re-fixed despite the fact that the Rubisco in this inner compartment would prefer $^{12}CO_2$. In sugar beet, on the other hand (where all of the CO_2 available for fixation is in free equilibrium with the atmosphere), the $^{13}CO_2$ rejected by Rubisco will escape and the sucrose formed will contain more ^{12}C.

All C_4 plants lose much less water per CO_2 fixed than C_3 plants and are therefore at an advantage in semi-arid environments. This may follow from the fact that the diffusion gradient for CO_2 is steeper in C_4 plants. PEP carboxylase can provide a very efficient "sink" for CO_2 and this enables C_4 species to photosynthesise at fast rates with narrower stomatal apertures than C_3 species and hence with diminished water loss. Similarly, C_4 species often display faster rates of photosynthesis than C_3 and if water loss and stomatal resistance is unchanged this would, in itself, improve the ratio of CO_2 fixed to water loss.

Up to a point, plants photosynthesise better as the temperature increase. A broad optimum is soon reached, however, as the advantage derived from faster carbon assimilation is offset by various deleterious factors. Photorespiration is important amongst these and therefore C_4 plants tend to have higher temperature optima. C_3 species photosaturate at about 1/5 full sunlight (i.e. photosynthesis does not increase much as the light intensity is raised above this level). Conversely, photosynthesis by C_4 species (or by C_3 species in high CO_2 or low O_2) increases with increasing light intensity up to, and beyond, full sunlight. Again, the difference is principally attributed to photorespiration, and again it can be seen that C_4 species are likely to be at an advantage under these particular conditions.

C3	C4	CAM
Typically temperate species such as spinach, wheat, potato. Usually moderately productive.	Typically semi-tropical species such as sugar cane, maize, amaranth. Often highly productive.	Typically succulent species such as cacti, stonecrop. Usually very poorly productive.
Usually lack Kranz-type anatomy.	Kranz-type anatomy is an essential feature.	Lack Kranz-type anatomy.
Initial CO_2 acceptor ribulose bisphosphate (5-carbon sugar).	Initial CO_2 acceptor phosphoenolpyruvate (3-carbon acid).	CO_2 acceptor is is phosphoenolpyruvate in dark, ribulose bisphosphate in light.
First formed product, phosphoglycerate (3-carbon acid).	First formed product oxaloacetate (4-carbon acid).	As for C4.
Only one main CO_2 fixation process.	Two main fixation processes separated in space.	Two main fixation processes separated in time.
C3 plants photorespire.	C4 plants do not photorespire.	Photorespiration sometimes repressed.
Mostly exhibit moderate rates of photosynthesis.	Often exhibit rapid rates of photosynthesis.	Usually have slow rates of photosynthesis.
Do well in temperate climates.	Do well in high light, high temperatures and in semi-arid environments.	Do well in very arid environments.
Lose water readily.	Lose water less readily.	Conserve water effectively.
Photosaturate at about 1/5 full sunlight.	Do not readily photosaturate.	May not photosaturate
Have relatively high CO_2 compensation points.	Have relatively low CO_2-compensation points.	Have high affinity for CO_2 by night.
Open stomata by day.	As for C3.	Open stomata by night.
Discriminate against $^{13}CO_2$	Discriminate against $^{13}CO_2$ less readily than C3.	Discrimination usually like C4.
Usually only have one type of chloroplast.	May have two distinct types of chloroplast.	As for C3.

Table 1: Some characteristics of C3, C4 and CAM plants

3. CAM PLANTS

Crassulacean acid metabolism (CAM) is a pathway which is utilised mostly by succulent species growing in arid environments. The history of studies on plants with CAM is longer than either that of photorespiration or C_4 photosynthesis. The earliest observations showed dark acidification, and a dark respiratory quotient indicative of both CO_2 and O_2 uptake in the dark. However, it took many years of research until the link was made between dark fixation of CO_2 and dark acidification.

Dark acidification is known to be accompanied by starch degradation, glycolitic formation of PEP and carboxylation through PEP carboxylase utilizing atmospheric CO_2. The oxaloacetate formed is reduced to malate and stored in the vacuole. In the light, malate is decarboxylated through the "malic enzyme" or converted to oxaloacetate and decarboxylated through PEP carboxykinase. The CO_2 is then donated to the RPPP. Some CAM plants can switch to a C_3 like photosynthesis under certain environmental conditions. In the "CAM mode", the stomata are open by night and closed by day, while the reverse is true in the C_3 mode. The isotope composition ($^{13}C/^{12}C$) of a plant is useful to distinguish between these modes of CO_2 assimilation. PEP carboxylase shows little discrimination against fixing $^{13}CO_2$ whilst discrimination is more pronounced with Rubisco (see section).

Ecologically, CAM is best described as an adaptation to water stress and water conservation. CAM plants have a higher water-use efficiency than C_3 or C_4 plants but this is accompanied by a lower productivity. Water is conserved during night fixation because the water vapour concentration gradient between the leaf and the atmosphere is low. The relatively high stomatal resistant and low mesophyll resistance (the latter due to CO_2 fixation through PEP carboxylase) provide favourable conditions for minimizing the transpiration/ photosynthesis ratio.

In C_4 photosynthesis, the first and second carboxylation occur in outer and inner compartments. In CAM the reactions are separated in time instead of in space (Fig. 6). CAM plants normally close their stomatal apertures during the day preventing water loss. This would severely curtail photosynthesis by limiting CO_2 if fixation by PEP carboxylase did not occur during the night when the stomata are open. Malate formed as a result of this carboxylation is stored in large vacuoles and decarboxylated during the next light period so that the Benson-Calvin cycle can then operate behind "closed doors". High CO_2 concentrations, trapped by closed stomata, supress photorespiration. In this way CAM plants can survive in very arid environments.

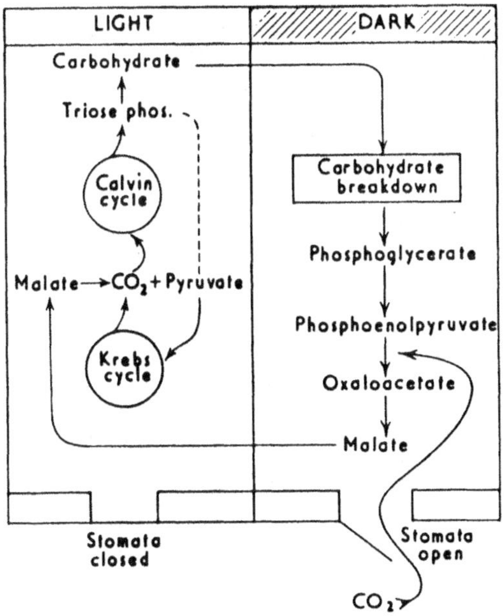

Figure 6. In CAM the stomata open by night, CO_2 enters and is fixed into malate (C1 + C3 → C4). The next day, behind closed stomata (which limit water loss) the CO_2 is released (C4 → C3 + C1). This sequence is similar to that in C_4 plants except that the CO_2-fixation and the CO_2-releasing processes are separated in time rather than in space. (From Walker, 1983)

4. CO_2 ACCUMULATING MECHANISMS IN ALGAE

CO_2 accumulating mechanisms different to C_4 metabolism have been of great interest in the last few years to plant physiologists, biochemists and molecular biologists. These processes substantially decrease photorespiration under natural conditions. There is a group of cyanobacteria (blue-green algae), which possess the ability to concentrate CO_2 via an energy requiring pump (Miller and Colman, 1980). These photosynthetic microorganisms can change their relative affinities for external inorganic carbon (CO_2 and CO_3^-) depending on the level in the external medium, and there is some active accumulation mechanism which allows CO_2 to be concentrated (accumulation in excess of 10^3 that of the external medium has been demonstrated) inside the cell. Recently, Omata and Ogawa (1986) demonstrated that induction of the CO_2-concentrating pump in a cyanobacterium is associated with the appearance of a 42 kDa polypeptide in the cytoplasmic membrane, which can be accumulated in great amounts, comprising up to 24 % of the total protein in this membrane.

The information available suggests that this mechanism is dependent on light energy and that the membrane permeability to CO_2 is high, implying that the energy costs of a CO_2 accumulating mechanism are likely to be high (Anderson and Badger, 1985). As for the C_4 pathway, this is not in itself a major disadvantage when light intensity is high and "spare" energy is available and this CO_2 accumulating mechanism permits the organism to maintain high growth rates in environments where CO_2 is limiting.

5. WHAT LIMITS PHOTOSYNTHESIS

5.1. The three substrates of photosynthesis

In C_3 plants and in air, the characteristics of Rubisco seem to be paramount and herein lies the current importance attached to the possibility of genetic manipulation of these characteristics. In higher concentrations of CO_2, or at low temperatures, or in otherwise stressed leaves, there is almost certainly co-limitation and therefore a real prospect that any engineered improvement in the characteristics of the carboxylase might simply replace one limitation by another. Tampering with reactions subsequent to the oxygenation of RuBP is now generally thought to be counter-productive because it ensures that once carbon is lost as glycollate it will stay lost rather than be scavenged or recycled (see section 2).

Synthesis of triose phosphate by the chloroplast requires three substrates: light, CO_2 and orthophosphate (P_i) (for reviews see Edwards and Walker, 1983 and Walker and Sivak, 1986). In the response of the rate of carbon assimilation to the concentration of CO_2, the kinetic properties of the RuBP carboxylase/oxygenase constitute the main limitation at low CO_2 concentrations, while at higher concentrations of CO_2 the limitation is shifted towards the reactions leading to the regeneration of the substrate, RuBP, driven by electron transport. In these circumstances, light and/or P_i can become limiting.

5.2. Limitation vs. regulation

Before 1970, Rubisco appeared to be inadequate for its task. Once the requirement for activation had been established there was no doubt that, at least in some species, there was more than enough RUBISCO to account for *maximal* photosynthesis (Lilley & Walker 1975). Inevitably there will always be doubts about the adequacy of extraction from many species and the degree of activation but there is growing evidence that, in many circumstances, Rubisco is not fully activated *in vivo* (see e.g. Perchorowicz and Jensen, 1983). Viewed teleologically, there is also the fact that since RuBP carboxylase is a major resource investment (which may constitute 50% or more of leaf protein in C_3 species) it is surprising that a plant can afford more than it appears to need. In other words, it does not make sense that, having made this investment, the same plant may not utilise all of its carboxylating capacity. Within certain limits, a plant may be capable of adjusting the composition of its photosynthetic machinery to the growth conditions which it encounters during its development (for a review see Björkman, 1981). However, environmental conditions also change markedly during the day, sometimes in a matter of seconds. Short term under-utilisation of parts of this photosynthetic machinery may therefore be inevitable, the price the plant pays for its immobility and total dependence on a changing environment.

When light is plentiful but conditions do not otherwise favour carbon assimilation, mechanisms exist which can dissipate light energy (e.g. pseudocyclic electron transport), although the phenomenon of photoinhibition demonstrates that their capacity may be insufficient in extreme conditions. Conversely, the activity of enzymes can be adjusted to the prevailing light intensity. The multiplicity of regulatory mechanisms known to affect the activity of all the enzymes involved in photosynthesis (with many new factors found every year), is probably part of this seemingly wasteful adjustment of every part of the machinery to the prevailing rates of the others. A single factor (light intensity, metabolites, pH) or a combination of these can decrease the activity of an enzyme substantially. However, a low activity of a particular enzyme is not necessarily proof that the photosynthetic process is limited by it. It can equally well be interpreted as a regulatory and useful adjustment to the activity of the rest of the photosynthetic apparatus. Thus, regulation is not synonymous with limitation and, although it is easy to accept that certain waste of enzymic capacity occurs in the short term, it seems unlikely that photosynthetic rates could be limited during long periods by the activation state of a single enzyme, especially when the enzymic protein is already there and the nitrogen and energy needed to form it has already been invested. Seen in this context, the reported low activation status of the carboxylase (Perchorowicz & Jensen 1983) could well be an example of short term under-utilisation, with the carboxylase only partly activated when the regeneration of the substrate is limiting. In an absolute sense, photosynthesis is ultimately limited by what Arnon called assimilatory power (i.e. light-generated ATP and NADPH). "Non-cyclic photophosphorylation" and associated NADP reduction, if freed of all external constraints, will proceed at the same rates as whole-leaf photosynthesis whereas *maximal* carboxylation capacity exceeds this rate by a factor of 2 to 3 (Lilley & Walker 1975). This implies that the ultimate ceiling is imposed by the intrinsic rate of electron transport. Photosynthetic phosphorylation is, however, constrained by the availability of NADP as well as by the availability of ADP and P_i. When other constraints are removed, P_i feeding can bring about substantial increases in the rate of photosynthesis (see below and Walker & Sivak 1985*b*; Sivak & Walker 1986).

In certain circumstances (c.f. Fig. 7), e.g. in saturating light and saturating CO_2, major oscillations in rate can be observed if photosynthesis is interrupted or perturbed by a short period of darkness or a change in the gas-phase surrounding the leaf (Walker *et al.* 1983; Sivak & Walker 1985*a*, and see refs. therein). The peaks in CO_2 uptake or O_2 evolution then observed can exceed quasi steady-state photosynthesis by 50% or more (figures 7 and 8, and see Walker *et al.* 1983; Sivak & Walker 1985*a,b*). These peaks are of interest because they (and not the quasi steady-state values) must reflect the total electron transport and carboxylation capacity of the system. In the short-term, these oscillations dampen to lower values implying that photosynthesis is regulated *down* to lower values.

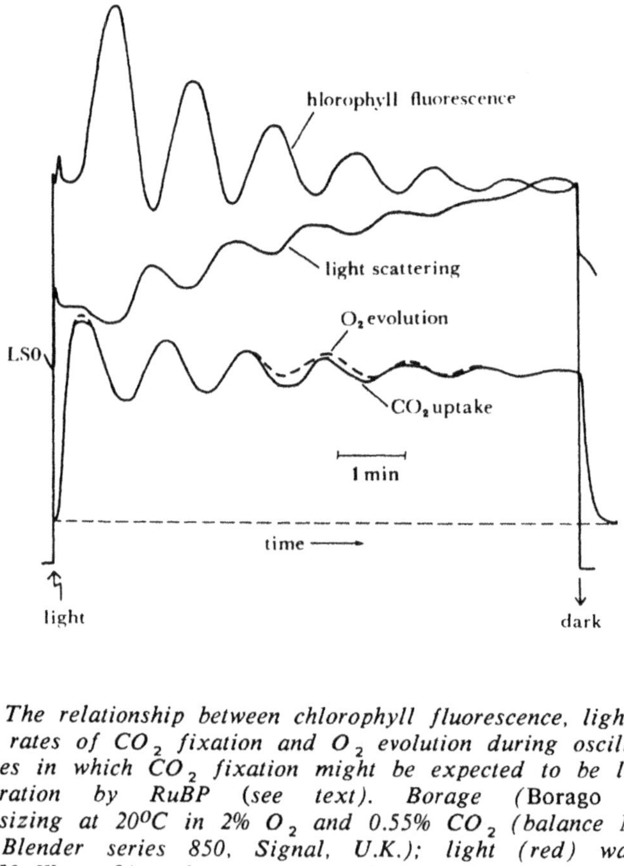

*Figure 7. The relationship between chlorophyll fluorescence, light-scattering (Heber 1969), and rates of CO_2 fixation and O_2 evolution during oscillatory behaviour in circumstances in which CO_2 fixation might be expected to be limited by the rate of regeneration by RuBP (see text). Borage (*Borago officinalis*) leaf, photosynthesizing at 20°C in 2% O_2 and 0.55% CO_2 (balance N_2, obtained using a Signal Blender series 850, Signal, U.K.); light (red) was near saturation (approx. 150 W. m^{-2}) and gas was supplied at a flow of 75 $ml.min^{-1}$. Carbon dioxide was measured by infra red gas analysis in a specially modified apparatus (ADC (Hoddesdon, U.K.). Oxygen was measured using a Draeger electrode (Draeger Transoxode, Draegerwerk, AG Luebeck, Germany) but similar results were obtained using a Clark type electrode plus extensive, stable pre-amplification, stable suppression of a large fraction of the signal is required with either electrode. See also Sivak* et al. *1985*a, *Sivak & Walker 1985*a,b; *Walker & Sivak 1985*b.

5.3. Photosynthesis and phosphate

It is well established (for a review see Edwards and Walker, 1983) that the major transport of elaborated carbon from the chloroplast to the cytosol involves a stoichiometric exchange of triose phosphate and orthophosphate (P_i). In order to optimise photosynthesis by isolated chloroplasts, [P_i] must be maintained between very narrow limits. Too little causes P_i- limitation of triose phosphate synthesis, too much limits photosynthesis by draining off too much triose phosphate, thereby limiting regeneration of the CO_2 acceptor. The response of isolated chloroplasts to external P_i is clear and unequivocal. It had been thought unlikely, however, that P_i supply would ever limit photosynthesis in a healthy plant grown in complete nutrient medium. The leaves of such plants normally contain large reserves of P_i in their vacuoles. Also, experimental evidence indicated that P_i could not limit photophosphorylation (the process by which P_i enters organic combination) *in vivo*: the affinity of the thylakoid ATPase and the concentration of [P_i] in the chloroplast were too high. These considerations greatly influenced the thinking of plant scientists, strongly affecting interpretation of experimental data. Only recently (Sivak and Walker 1986, Walker and Sivak 1986) the evidence became too strong to be ingnored and it was recognised that P_i *can*, in some circumstances, limit photosynthesis *in vivo* (c.f. Figs. 8 and 9). This opened the way for re-examination of previous data, some of them relevant to agricultural problems.

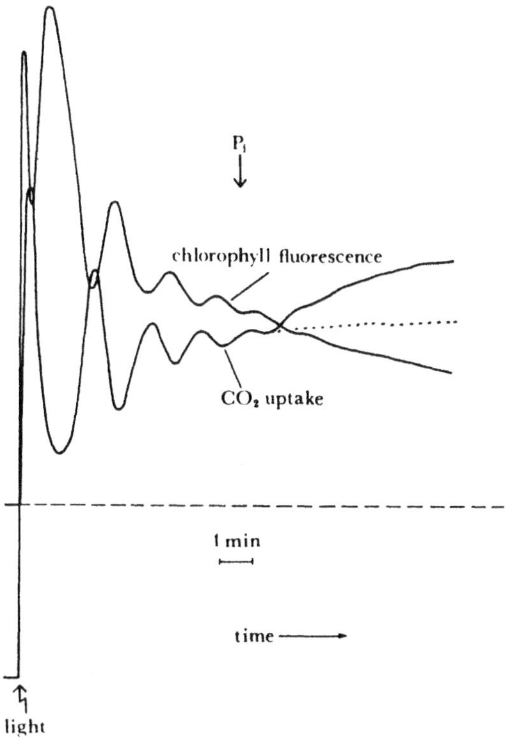

Figure 8. Stimulation of quasi *steady-state photosynthesis by P_i feeding (via the petiole) to a young barley leaf in saturating light and saturating CO_2 (0.55 CO_2, 20% O_2, balance N_2, at a temperature of 20°C). Under these conditions re-illumination induces marked oscillations in CO_2 uptake (note that rate at the first trough is about 15 % of the rate at the peak) and broadly reciprocal but phase-shifted oscillations in fluorescence (figure 5 and see Walker* et al., *1983; Sivak & Walker, 1985). Following P_i feeding the amplitude and frequency of the oscillations is decreased (not shown but see Walker & Sivak 1985*b *and Sivak & Walker 1986). CO_2 uptake measured by infrared gas analysis, as in figure 5, at 20°C. Gas flow 100 ml/min, illumination red light, 180 $W.m^{-2}$.*

Although there are many unkowns in the complex regulation of photosynthesis, it is now clear that it is a whole-cell affair, in the sense that other cellular compartments, besides the chloroplast, contribute to a large extent to this regulation. For example, the vacuole contains much of the cellular P_i. Photosynthesis depends on a control of cytosolic P_i concentration, and it is therefore self-evident that efficient regulation depends on the precise and regulated movement of P_i from the vacuole, the largest reservoir of P_i in the cell. Photorespiratory metabolism, a process which also recycles P_i, depends on the interaction of chloroplast, mitochondria, glyoxysomes and cytosol. Sucrose synthesis is a cytosolic event. If export of sucrose from the source to the sinks in also considered, it becomes clear that regulation of photosynthesis is a whole plant affair.

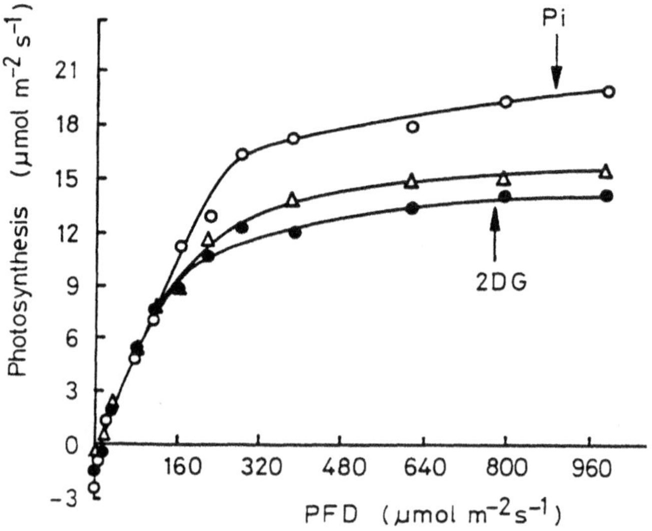

Figure 9. Stimulation of photosynthesis by P_i feeding and inhibition by 2-deoxyglucose (2-DG) feeding, shown as a function of light intensity (PFD: photon flux density). No changes in rate were observed at the lower intensities (i.e. the quantum yield, which can be determined from the initial slope, was unchanged). This experiment was carried out at 20°C on a 10 cm² disc cut from a spinach leaf. The leaf disc was fed (between measurements) for 20 min; first with 20 mM 2-deoxyglucose and then with 20mM P_i at pH 5.8, through its mid-rib. No increase in rate was seen in otherwise similar experiments in which discs were provided with water rather than P_i or 2-DG. [Oxygen measured with a Hansatech leaf disc electrode (Delieu & Walker 1983) at 20°C in 5% CO_2. Illumination with red light (Sivak et al., *1985*a,b *& Walker, 1985)]. See also Walker & Osmond 1986; Sivak & Walker 1986).*

We have been concerned for many years with the study of induction phenomena in photosynthesis. Induction *per se* refers to the slow increase in photosynthesis that follows re-illumination but the use of the term in a broader sense could include the response of the photosynthetic apparatus to other changes in the environment: temperature, light intensity, [CO_2], [O_2]. The complexity of these induction phenomena, however, is so great that many researchers opted for ignoring them, looking only at steady-state photosynthesis. But this (expedient) solution involved the "elimination" of a large amount of information. It probably contributed to the ignorance of the limitation of photosynthesis by inadequate supply of P_i. We believe that P_i supply is involved in several enigmatic adaptation phenomena (e.g. oscillatory behaviour and slow induction, described already in the 1950's by Rabinowich). This may be related to the ability of some plants to grow better at relatively low temperatures. For example, the slow photosynthetic induction which may be observed in some circumstances (Sivak and Walker 1986 and Sivak, unpublished) and which, eventually, leads to the steady-state rate, is much slower at low than at high temperatures (e.g. 5 hours at 5°C and 1 h at 20 °C). Also, it can be shortened to a few minutes if P_i is fed to the leaf via the petiole (Fig. 8).

6. TOWARDS IMPROVED PLANTS

6.1. Genetic intervention

It is perhaps unlikely that any attempt to introduce C_4 characteristics into C_3 crops will ever be successful. From the point of view of the breeder or the genetic engineer, the simultaneous modification of many proteins, or a major modification of leaf anatomy, is a very complicated matter. More feasible, but not less attractive, is the possibility of introducing to higher plants a CO_2 accumulating mechanism of the sort present in many algae (likely to consist of no more than a few proteins). Re-targetting of a single protein into any membrane in the cell is now a real possibility. Plant physiologists, biochemists and molecular biologists will have to collaborate in the tasks of choosing the most desirable characteristics, and of evaluating the feasibility of their introduction and the effects of this genetic engineering on the overall performance of the plant. Evaluation of the success of genetic intervention is not as easy as it may seem to the non specialist. This difficulty is also reflected in the lack of agreement between plant scientists about how best to express photosynthetic rates and whether a plant that displays high photosynthetic rates is synonymous with a "good" plant.

In the past, different criteria have been of importance (Evans, 1981) and in the future, others might be more relevant but the criterion of crop yield which is currently most used is, simply, the amount of plant material harvested per unit area of land per crop. Increase in the yield of crop plants could come in many ways such as better adaptation to environmental conditions, greater resistance to pests and diseases, improved agronomic practices, etc. All of these factors might result in an increase in the overall rate of dry matter production by the whole plant i.e. in an increase in growth. An additional improvement in yield could also derive from an increase in the proportion of the dry weight which accumulates in the organs harvested by man, i.e. in the harvest index.

Gifford and Evans (1981) believe that the relative growth rate of young plants does not appear to have increased in the course of domestication (references are supplied for wheat, maize, tomato and cow-pea). Thus, in the absence of genetically based increases in photosynthesis and growth, past improvements in yield potential have largely derived from increases in the harvest index.

Although some further increases in harvest index are still possible, there is obviously a limit. Austin *et al* (1980) suggest that, for wheat, this limit is about 62%. The best contemporary wheats have harvest indices of about 50% (Austin *et al*, 1980). If this is so, increasing attention will have to be given to attempts to increase photosynthesis and growth rates. One approach would be to breed and to select empirically for growth itself. A more challenging and intellectually satisfying approach, which offers the possibility of faster advance, would be to identify the characteristics that cause a larger overall rate of dry matter production. The present need for a better understanding of crop growth and yield, in biochemical terms, is now more urgent than ever before because, in addition to genetic manipulation by plant breeding, new possibilities are being opened up by recent developments in molecular biology and recombinant DNA technology. The potential use of biotechnology includes plant improvement by transference of specific DNAs but, if this is to become a practical proposition, it will be necessary to identify the DNA to be transfered. Day (1981) says that "*molecular blueprints for constructing the crops we would like to grow do not exist. We know neither how to draw them nor even to specify our major requirements in terms of replacing or adding specific DNA, and this is precisely because plant sciences have been the Cinderella for past decades*".

6.2. Photosynthesis and growth: a paradox

We need, then, to identify the genes of agricultural importance but first we must describe, in anatomical and biochemical terms, those features that we desire. This must precede any attempt to identify the "productivity genes" that we wish to introduce.

From a practical point of view, any survey of large numbers of plants would almost certainly require short-term photosynthetic measurements in the laboratory rather than field trials. This would be cheaper, quicker and simpler. Here, however, we are immediately faced with a paradox. Since Van Helmont's experiments in the seventeenth century it has been clear that green plants only derive a small part of their mass, other than water, from the soil. They derive their dry mass almost entirely from the atmosphere by photosynthetic carbon assimilation. If we take two plants which respire at equal rates and one of these assimilates carbon faster than the other it will accumulate dry matter more quickly; it will grow faster. The crop physiologist, on the other hand, is quick to remind us that an absolute correlation between growth and photosynthesis is tenuous at best and that, at worst, it may be non-existent.

It is essential, if much effort is to be invested in trying to improve plant productivity via photosynthesis, to ask not only if improvement is possible but also if "improved" photosynthesis would necessarily translate into increased productivity. Above all, we need to know the relationship between growth and other (short-term) measurements of photosynthesis and to understand the nature of the apparent contradictions referred to above. Here, we shall suggest that the contradictions may lie in the ambiguity of expression.

When we say that a larger overall rate of photosynthesis should lead to faster growth, we are talking about the performance of the whole plant. Apparent paradoxes may arise if we slide from that use into equating "photosynthesis" with "photosynthetic rate per unit leaf area".

Leaf area, as a basis of expression of photosynthetic rate, is widely used in plant science. It is convenient and sensible, if only in as much as there is usually a clear correlation between area and the interception of light. Conversely,the use of "photosynthesis per unit leaf area" as a synonym for "photosynthesis" is too dangerous to be overlooked, *particularly if it likely to be applied to breeding programmes or to decisions about research.* Let us look at wheat, for example. Primitive forms of wheat, with narrow leaves, show photosynthetic rates (per unit area of flag leaves) higher than the modern hexaploid varieties, which have broad flag leaves. Austin *et al* (1982) give the following values (as *means*) for groups of hexaploids, tetraploids and diploid varieties of wheat (Table 2).

Table 2: Leaf areas and maximum rates of photosynthesis per unit leaf area of flag leaves of Triticum *and* Aegilops *species as means for the different genotypes (from Tables 1 and 2 in Austin* et al, *1982)*

Genotype	Leaf area (cm^2)	Net photosynthetic rate per unit leaf area (mg CO_2 dm^{-1} h^{-1})
6X	18.4	28
4X	14.6	32
2X	8.3	38

Table 3: calculated from the data in Table 1, taken from Austin et al, *1982.*

Genotype	Photosynthesis *per leaf* (mg CO_2 h^{-1})
6X	5.15
4X	4.67
2X	3.15

On the basis of rate per unit area, it is evident that the best wheat in Table 2 has to be amongst the primitive diploids. If, on the other hand, we use the total mean flag leaf area to calculate the total photosynthesis *per leaf* in mg CO_2 h^{-1} we get the following figures (Table 3). In short, the order of excellence is reversed and we arrive at the unsurprising (but not the only possible) conclusion that a larger leaf is likely to accomplish more photosynthesis than a smaller leaf.

It now becomes immediately apparent why no correlation between "photosynthesis per unit area" and growth can be expected. Such a basis of expression is *only valid for leaves.* It can not be extrapolated to plants or to crops.

6.3. Better bases for the expression of photosynthetic rates are needed.

If we accept that leaf area, however convenient, is nevertheless a potentially misleading basis of expression for photosynthetic rates, we are still left with a need for a satisfactory alternative. Why is there a need? A plant biochemist, who usually works with fairly constant plant material (for instance, spinach chloroplasts or pea chloroplasts but not, usually, with both simultaneously) has the possibility of changing one experimental factor at a time and analysing the effect of that change on the system under consideration. A plant physiologist, or a crop physiologist, faces a different sort of problem: he has to compare plants that are different, either because they are genetically different or because they were grown in disimilar environments. If he wished to study, for example, the effect of anatomical differences on the photosynthetic performance of these plants, he would have to bear in mind the fact that these plants will also differ in many characteristics other than those which are immediately obvious (or those that interest him).

The use of leaf area as a basis of expression implies, in a way, that a given leaf area covers a given quantity of photosynthetic machinery and that a leaf displaying higher rates (expressed on an area basis) is making a better use of that machinery. This, of course, is wrong. Plants have different leaf and cell structures and they solve the problem of adaptation to different environments in very different ways. One adaptation, for example, is the adjustment of leaf thickness according to the light intensity received during growth. Clearly, it could not be said that a thin leaf grown under low light is a "bad" leaf when compared to a much thicker leaf grown under high light. Probably, the same is true for wheat. Diploids may have more photosynthetic machinery under a given area of leaf than the hexaploids. The distribution of photosynthetic machinery within the leaf would then be different and this could account for, or result in, the hexaploid having broader and "thinner" leaves and the diploid having smaller and "thicker" leaves. The use of the words "thicker" and "thinner" between quote marks is absolutely necessary. Volume might be a good measure of quantity of photosynthetic machinery in certain cases (c.f. Charles-Edwards, 1979), but there is no reason to assume that the wheats under consideration are so similar that the same leaf volume implies the same quantity of photosynthetic machinery. Accordingly, leaf volume as a basis of expression of photosynthetic rates could be as misleading as leaf area.

6.4. A "fair" basis

Let us analyse the need for a basis of expression, not only for photosynthetic rates but for any process we want to measure. Underlying the use of such a basis, there is the need to standarise, that is,to make comparisons on a "fair" basis. We want to compare the behaviour of *equal quantities*. When we wish to invest our money we compare the interest that different banks offer for the same quantity of capital and time of investment.

A "perfect" basis of expression would ensure a comparison of identical quantities when comparing yields, a basis which would allow us to distinguish clearly between capital and interest, quality and quantity.

Leaf area does not fit this requirement. Any anatomical study will show how dissimilar leaf structures can be and quantitative anatomy puts numbers to these differences (see, e.g. Jellings and Leech, 1984). This is one more reason why paradoxes are inevitable when "photosynthesis per unit leaf area" is used as though it were synonymous with "photosynthesis".

Other possible units, such as volume of mesophyll, cell, chloroplasts, etc., can be quantified using microscopy and other related techniques (c.f. Jellings and Leech, 1984). Do they provide us with a means of standarising photosynthetic rates, allowing us to isolate the *quantity* component of the comparison, in order to be "fair"? We do not believe so. These anatomical parameters have more to do with *distribution* of the photosynthetic machinery than with its quantity. *(It is likely that distribution of the photosynthetic machinery affects performance, but that is another matter).*

Components of the photosynthetic machinery are also used as a basis of expression of photosynthetic rate, the most popular being chlorophyll. Other components have been proposed, the usual approach being to find a photosynthetic factor (per unit area) that would show a close correlation with the photosynthetic rate (of the same area of tissue). It is apparent that this kind of investigation has involved a search for a limiting intrinsic factor of photosynthesis. However, the photosynthetic machinery is extremely complex and, theoretically, any component could be limiting.

Even so, there is a growing conviction (von Caemmerer and Farquhar, 1981, Björkman, 1981) that, in general, at low concentrations of CO_2, the kinetics of CO_2 assimilation *in vivo* approximate to the RuBP saturation kinetics of RuBP carboxylase-oxygenase, whereas at relatively high partial pressures of CO_2 (in some circumstances, air concentrations of CO_2 i.e. external 340 μl/l, is thought to be near the transition point) the most important limitation is the rate of regeneration of RuBP. This is governed, in turn, by the rate of electron transport. At relatively high partial pressures of CO_2 and *saturating* irradiances, the rate of regeneration of RuBP would be determined by the maximum capacity of the electron transport chain (Björkman, 1981). This means that the maximum photosynthetic activity would be determined by the maximum rate of photosynthetic electron transport displayed by the whole tissue.

Let us analyse this concept. The electron transport chain is a complex system. Several attempts have been made to find a close correlation between photosynthesis and biochemical components of the photosynthetic machinery, including correlations with one or more of the components of the light -collecting antennae pigments (e.g. chlorophyll) or the reaction centers themselves (Malkin *et al*, 1981, Falkowsky *et al.*, 1981).

If the maximum photosynthetic capacity were determined by the maximum rate of photosynthetic electron transport, a close correlation between maximum photosynthetic capacity (*in vivo*) and the components of the photochemical apparatus (if any) that are always in a stoichiometric relationship with that rate could be expected. That was probably the case in the plants studied by Malkin *et al.* (1981) who found a good quantitative correlation between the maximum photosynthetic capacity and the number of PSII photosynthetic units, but this might not be a general relationship (c.f. Björkman, 1981, Falkowsky *et al*, 1981, Mellis and Harvey, 1981). In a model proposed by Björkman (1981), adaptation by a leaf to different irradiances during growth includes a change in the number of PSI and PSII units serving each electron transport chain (experimental support for this model can be found in Mellis and Harvey, 1981). The number of chlorophyll molecules serving each photosystem varies with species (Malkin *et al.*, 1981) and it is not surprising that no quantitative correlation could be found between chlorophyll content and photosynthetic capacity. Since cytochrome *f* appears to be the electron carrier present in the smallest amount, it might be assumed that the number of electron transport chains equals the number of cytochrome *f* molecules, and that is what Björkman (1981) assumes for his model. If this is so, we might expect to find a quantitative relationship between photosynthesis and cytochrome *f* content. If, however, we are just looking for a quantification of photosynthetic machinery, there is no *practical* need to look for quantification of a *single* component of the system. We could just measure photosynthetic capacity, as the maximum photosynthetic rate obtained in saturating irradiance and saturating CO_2 concentration. This would permit us to quantify a system not by measuring of a single component but by measuring the overall activity displayed.

Why do we propose *maximum photosynthetic capacity* as a basis of expression of photosynthetic activity? If, as it seems, it reflects the number of electron chains present (*electron transport capacity*) that, in itself would be useful, as electron chain proteins comprise a large part of the proteins present in the photosynthetic tissue. Most of the remainder is carboxylase, and from the point of view of energy invested, these are the most important components to take into account.

How would this alternative basis of expression be used? It would imply separating photosynthetic *capacity* from photosynthetic *performance* and then relating the two. Here, *capacity* would be a measure of the amount of photosynthetic machinery present in a given amount of tissue. It could be measured by either of two methods: a) extrapolating to *saturating* irradiance and *saturating* CO_2 concentration, the rates obtained by conventional IRGA (as in Malkin *et al.* 1981) or b) measuring photosynthetic O_2 evolution polarographically in saturating irradiance and saturating CO_2 concentration (Delieu and Walker, 1981). Photosynthetic performance would be measured, as at present, by conventional infra-red gas analysis and could, by taking into account such factors as temperature and irradiance, enable one species or variety to be compared with another with due regard to environmental factors.

"*Photosynthetic efficacy*" as it might be called, would then be an expression of the extent to which photosynthetic *capacity* was realised in given circumstances (i.e. performance/capacity):

i.e ***Efficacy = Rate in air / Rate in saturating*** CO_2

Other gas phases could be used provided that all comparisons were made on an agreed basis and that obvious pit-falls, such as non-saturating light and inhibition by excess CO_2, were avoided. Similarly it would be immaterial whether the rates were expressed in terms of leaf area, chlorophyll or whatever, since these terms would be lost in the calculation of the efficacy ratio. The search for a better basis of expression would, in effect, be ended by disregarding the basis of rate measurement and relying on an index of efficient use. Internal regulation would be taken into account, as well as capacity, and the effectiveness with which different varieties, species or ecotypes deploy their photosynthetic machinery could be determined.

CONCLUSIONS

The short and long term effects of increased CO_2 concentrations on plant metabolism are relevant to a broad range of issues; for example, in the optimisation of crop growth in greenhouses with enriched CO_2, and in the prediction of the global effects of *increasing* carbon dioxide concentrations in the atmosphere. Also, genetic manipulation of plants may soon allow the introduction of mechanisms that will enable plants to increase their internal CO_2 concentrations.

Much is known about the basic aspects of the photosynthetic process. However, the understanding of the more fine details, like the role of regulatory mechanisms in the optimisation of photosynthesis, is still inadequate. Predictions of the likely effects of changes in the environment, or of genetic and chemical manipulation will require the concerted effort of plant scientists, agronomists and other researchers in related fields. Recent developments in instrumentation and in methods for the simultaneous measurement of different aspects of photosynthesis *in vivo* will certainly play an important role in this endeavour.

FURTHER READING

Clayton, R.K. (1970) *Light and Living Matter.* Mc Graw Hill, New York.

Edwards, G.E. and Huber, S.c. (1981) the C_4 pathway. In *The Biochemistry of Plants. A comprehensive treatise* (Hatch, M.D. & Boardman, N.K.) pp. 237-81. Academic Press, New York.

Edwards, G. and Walker, D.A. (1983) *C3, C4: mechanisms, and cellular and environmental regulation, of photosynthesis.* Blackwell Scientific Publications, Oxford.

Hall, D.O. and Rao, K.K. (1972) *Photosynthesis,* Edward Arnold, London.

Osmond, C.B. (1978) Crassulacean Acid Metabolism: A curiosity in context. *Ann. Rev. Plant Physiol.* 29, 379-414.

Walker, D.A. (1979) *Energy, Plants and Man.* Packard Pub. Ltd., Chichester.

Specific citations

Austin, R.B., Bingham, J., Blackwell, R.D., Evans, L.T., Ford, M.A., Morgan, C.L. and Taylor, M. (1980) *J. Agric. Sci.* 96: 675-689.

Austin, R.B., Morgan, C.L., Ford, M.A. and Bhagwat, S.G. (1982) *Annals Bot.* 49: 177-189.

Björkman, O. (1981). In: *Lange, O.L., Nobel, P.S., Osmond, C.B. and Ziegler, H. (eds). Encyclopedia of Plant Physiology, New series, Vol. 12A (Physiological Plant Ecology I, Responses to Plant Environment), pp.* 57-107. Springer, Berlin.

Bowes, G. and Ogren, W.L. (1972) *J. Biol. Chem.* 247, 2171-2176.

Charles-Edwards, D.A. (1979). In: *Photosynthesis and Plant Development. R Marcelle,H. Clijters and M.van Poucke (eds.) Dr. W. Junk Publishers,* The Hague.

Day, P.R. (1981) Plant Biochemistry - a cinderella too long. *Trends Biochem. Sci.*, 6: I-II.

Delieu,T.J. and Walker, D.A. (1981). *New Phytol.* 89, 165-178.

Ellis, J. (1985) *Nature* **313**: 353-354.

Evans, L.T. (1981) In: *"Wheat science- Today and Tomorrow" L.T. Evans and W.S. Peacock (eds.) Cambridge Univ. Press.* Cambridge.

Falkowsky, P.G., Owens, T.G., Ley, A.C. and Manzerall, D.C. (1981) *Plant Physiol.* **68**: 969-973.

Gifford, R.M. and Evans, L.T. (1981) Photosynthesis, carbon partitioning and yield. In: *Ann. Rev. Plant Physiol.* **32**: 485- 509.

Hatch, M.D. and Osmond, C.B. (1976) In: *Encyclopedia of Plant Physiology (New Series)* Vol. III. Transport in Plants. (eds. C.R. Stocking and U. Heber) pp. 144-184. New York, Springer Verlag.

Jellings, A.J. and Leech, R.M. (1984). *New Phytol.* **96**: 371-382.

Keys, A. (1986) *Phil. Trans. R. Soc. Lond. B*, in the press.

Lilley, R.McC. and Walker, D.A. (1975) *Plant. Physiol.* **55**: 1078-1092.

Lorimer, G.H., Badger, M.R. and Andrews, T.J. (1976) *Biochemistry* 12, 18-23.

Lorimer, G.H., Andrews, T.J., Pierce, J. and J.V. Schloss (1986). *Phil. Trans. R. Soc. Lond. B.* In press.

Malkin, S. Armond, P.A., Mooney, H.A. and Fork, D. (1981). *Plant Physiol.* 67: 570-579.

Melis, A. and Harvey, (198) *Biochim. Biophys. Acta* **637**: 138-145.

Miller, A.G. and Colman, B. (1980) *J. Bacteriol.* **143**: 1253-1259.

Perchorowicz, J.T. and Jensen, R.G. (1983) *Plant Physiol.* **71**: 955-960.

Omata, T. and Ogawa, T. (1986) *Plant Physiol.* **80**: 525-530.

Sivak, M.N. and Walker, D.A. (1985*a*) In: *Plant Products and the New Technology. Ann. Proc. Phytochem. Soc. Eur.* **26**: 29-44.

Sivak, M.N. and Walker, D.A. (1985*b*) *Plant Cell Environ.* **8**: 439-448.

Sivak and Walker (1986) *New Phytol.* **102**: 499-512.

Von Caemmerer, S. and Farquhar, G.D. (1981). *Planta* **153**: 376-387.

Walker D.A., Sivak M.N., Prinsley, R.T. and Cheesbrough, J.K. (1983)

Walker D.A. and Sivak, M.N. (1985) *Physiol. Veg.* **23**: 829-841.

Walker, D.A. and Sivak, M.N. (1986) *Trends Biochem. Sci.* **11**: 176-179.

PHOTOSYNTHETIC BACTERIA

Reidun Sirevåg
Department of biology
Division of Molecular Cell Biology
University of Oslo
Box 1050, Blindern
N-0316 Oslo 3
Norway

ABSTRACT. Based on their ability to grow autotrophically in the light, an overview of photosynthetic bacteria is presented. The common features of these bacteria as well as their differences with regard to metabolism, structure and ecology is emphasized. Resent results regarding mechanisms of CO_2-fixation, structure and function of the photosynthetic apparatus and their exploitation by recombinant-DNA techniques and genetics are reviewed.

1. INTRODUCTION

The formation of organic material from CO_2 by living organisms can be expressed in general terms in the following way:

$$CO_2 \xrightarrow{4(H) + ATP + enzymes} (CH_2O) + H_2O,$$

where (CH_2O) is a rough expression for cell material. The actual composition of cell material is approximately $C_5H_8O_2N$.

The ability to live autotrophically with CO_2 as the sole source of carbon is widespread in nature. It is the common feature among higher plants and algae which are eukaryotes, and it is also found in several groups of prokaryotic organisms.

According to the equation above, all organisms need ATP and reducing power in addition to the necessary enzymes, in order to reduce carbon dioxide to organic material. Except for some interesting special cases, which will be discussed below, the various autotrophic organisms share a common pathway, the reductive pentose phosphate cycle, or Calvin cycle, for the reduction of carbon dioxide (Fig. 1).

With regard to generation of energy and reducing power, autotrophic organisms are versatile and use many different mechanisms. One large, diverse group, the photoautotrophs, obtain energy by the absorbtion of light, and use reduced inorganic compounds (e.g. H_2O or H_2S) or organic compounds as the ultimate source of reducing power. Other organisms, the chemoautotrophs, cannot use light energy any obtain both energy and reducing power by the oxidation of reduced inorganic compounds like H_2S, NH_4, Fe^{2+} or H_2. Table I lists the major groups of autotrophic organisms, and their sources of reducing power and energy. Although all these organisms convert carbon dioxide to organic material, their contribution to the global carbon cycle varies considerably. Higher

M. Aresta and G. Forti (eds.), Carbon Dioxide as a Source of Carbon, 237–261.

TABLE I. Some characteristic features of the major groups of autotrophic organisms.

	Source of energy	Source of reducing power	Relation to oxygen	Products formed
Photoautotrophs				
Eukaryotes				
Higher plants	light	H_2O	aerobic	O_2
Algae	light	H_2O	aerobic	O_2
Prokaryotes				
Cyanobacteria	light	H_2O	aerobic	O_2
Purple bacteria	light	H_2S, S^o, H_2, organic	anaerobic (mostly)	SO_4^{2-}
Green bacteria	light	H_2S, S^o, $S_2O_3^{2-}$	anaerobic (mostly)	SO_4^{2-}
Chemoautotrophs				
Prokaryotes				
Nitrifying bacteria		NH_4^+, NO_2^-	aerobic	NO_2^-, NO_3^-
Sulfur oxidizers		H_2S, S^o, $S_2O_3^{2-}$	aerobic	SO_4^{2-}
Iron bacteria		Fe^{++}	aerobic	Fe^{3+}
Hydrogen bacteria		H_2	aerobic	H_2O

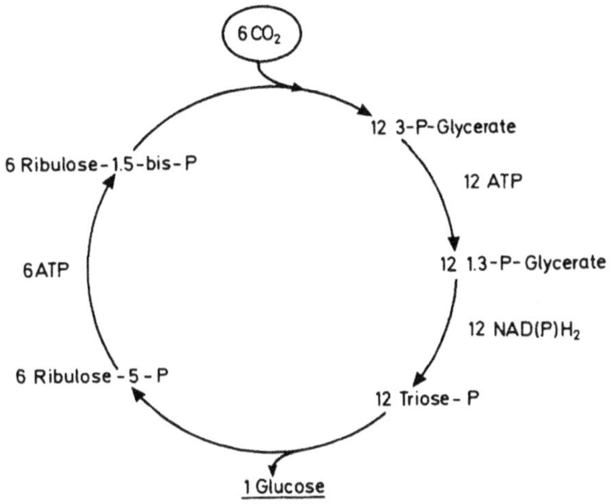

Figure 1. The reductive pentose phosphate cycle.

plants are responsible for most of the carbon dioxide reduction on land, where algae and bacteria play only minor roles. In the ocean however, algae and cyanobacteria (phytoplankton) are the major contributors. It has been estimated that 25% of the total net primary production in the world is due to phytoplankton, of which cyanobacteria comprise a major fraction.

Although the contribution of chemoautotrophic bacteria to the global carbon cycle is negligible, they are extremely important in the global cycles of nitrogen and sulfur.

2. PHOTOSYNTHETIC PROKARYOTES

2.1. General characteristics

The photosynthetic prokaryotes comprise a large and heterogeneous group, which have in common that they contain chlorophyll and are able to generate chemical energy from light.

Two major groups are recognized: the cyanobacteria or blue-green bacteria on the one hand, and the purple and green bacteria on the other. In addition to differences in their contents of photopigments and type of photosynthesis, the two groups also show major structural differences. The main properties of these organisms are summarized in Table II.

The cyanobacteria carry out the same type of photosynthesis as green plants: two light reactions are involved and H_2O is the electron donor for the reduction of carbon dioxide. As a result, water is split and molecular oxygen, is produced. This type of photosynthesis, which is termed <u>oxygenic</u>, is in contrast to the <u>anoxygenic</u> photosynthesis of

Table II. Characteristics and major properties of the different groups of photosynthetic bacteria.

	Cyanobacteria	Purple bacteria	Green bacteria
Type of photosynthesis	Oxygenic	Anoxygenic	Anoxygenic
Type of chlorophyll	Chlorophyll a	Bacteriochlorophyll a or b	Bacteriochlorophyll a and c, d or e
Accessory pigments	Carotenoids Phycobiliproteins	Carotenoids	Carotenoids
Location of photosynthetic apparatus	Thylacoids and phycobilisomes	Cell membrane	Chlorosomes and cell membrane
Reductant for CO_2-assimilation	H_2O	H_2, H_2S, organic compounds	H_2, H_2S, $S_2O_3^{2-}$
Examples of genera	Synechococcus Anabaena Oscillatoria	Rhodospirillum Rhodopseudomonas Chromatium	Chlorobium Chloroflexus

purple and green bactera. Here only one light reaction is involved, and O_2 is never produced. Reduced substances in the environment provide reducing power for the reduction of carbon dioxide, and these can be either reduced inorganic compounds like H_2S or H_2, or an organic compound.

Many members of the purple and green bacteria grow well photoheterotrophically using light as energy source and organic compounds as source of carbon. Under such conditions, carbon dioxide is often only a minor source of cell carbon. In addition, some purple and green bacteria are able to grow chemoheterotrophically in the dark, performing respiration at the expense of oxygen.

The phototrophic bacteria differ with regard to the types and content of chlorophyll. The cyanobacteria contain chlorophyll *a*, which is identical to that of algae and higher plants. The purple and green bacteria contain a different type of chlorophyll, termed bacteriochlorophyll (Bchl). The basic structure of chlorophyll and bacteriochlorophyll is similar, the differences beeing due to minor modifications of the porphyrin ring and its substituents. The structures of the various chlorophylls are shown in Fig. 2.

Although the chemical modifications of the porphyrin ring appear to be minor, they lead to dramatic changes in the characteristic absorption spectra of the pigments, the biological consequence beeing that an organism containing a certain type of chlorophyll is better suited to utilize light of a particular wavelength than other organisms. This is easy to demonstrate in the case of purple bacteria, which contain Bchl *a* or Bchl *b* as their sole chlorophyll. These pigments share a seemingly minor chemical feature (saturation between C-3 and C-4 of the porphyrin ring), which has a profound effect on their spectral properties. The positions of all the major absorbtion peaks of Bchl *a* or *b* are moved relative to those of other chlorophylls more than 100 nm towards the red end of the spectrum. The absorbtion properties of purified chlorophylls are described in Table III. The absorption peaks of the chlorophylls *in vivo* are displaced towards the red as compared to the spectra in organic solvents.

In addition to chlorophyll, all phototrophic bacteria contain a variety of carotenoids which can function as accessory light harvesting pigments. The carotenoids absorb light between 400 and 550 nm and thus extend the wavelength range of usable light for the organism.

In addition to chlorophyll *a* and carotenoids all cyanobacteria contain characteristic pigments, the *phycobilins*. These water soluble proteins that contain covalently bound linear tetrapyrrols as chromophores, function as accessory pigments in photosynthesis. Within the cells, they are located in special structures, phycobilisomes. One class of phycobilins, the phycocyanins, are blue, absorbing at approximately 625 nm. Together with chlorophyll *a* these are responsible for the blue green colour of these bacteria. Some cyanobacteria produce phycoerythrin, a red phycobilin which absorbs at 575 nm. Cyanobacteria containing this pigment are red or brown in colour.

Fig. 3 shows the absorption spectra of various phototrophic bacteria measured *in vivo*. It can be seen that some organisms absorb light in regions where others absorb light poorly. Algae and cyano-

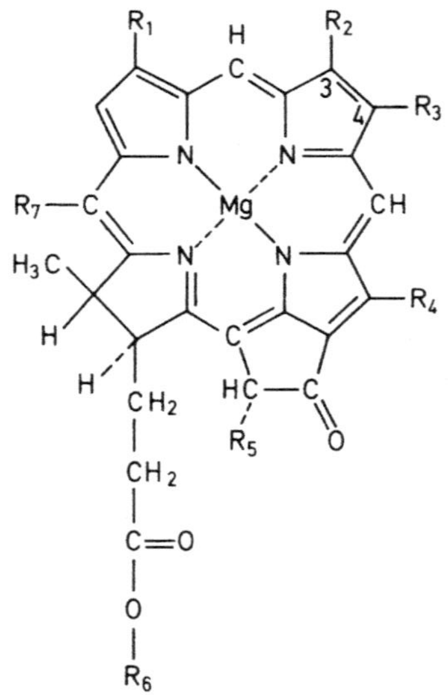

Figure 2. Chemical structure of various chlorophylls.

Substituent	chlorophyll a	bacteriochlorophyll a	b	c	d	e
R_1	$-CH{=}CH_2$	$-COCH_3$	$-COCH_3$	$-CHOHCH_3$	$-CHOHCH_3$	-CHOHCH3
R_2	$-CH_3$	$-CH_3$	$-CH_3$	$-CH_3$	$-CH_3$	-CHO
R_3	$-C_2H_5$	$-C_2H_5$	$-CHCH_3$	$-C_2H_5$	$-C_2H_5$	$-C_2H_5$
R_4	$-CH_3$	$-CH_3$	$-C_2H_5$	$-C_2H_5$	$-C_2H_5$	$-C_2H_5$
R_5	$-CO-OCH_3$	$-CO-OCH_3$	-H	-H	-H	-H
R_6	phytyl	phytyl	phytyl	farnesyl	farnesyl	farnesyl
R_7	-H	-H	-H	$-CH_3$	-H	$-CH_3$

In bacteriochlorophyll <u>a</u> and <u>b</u>, carbon atoms 3 and 4 are linked by a single bond.

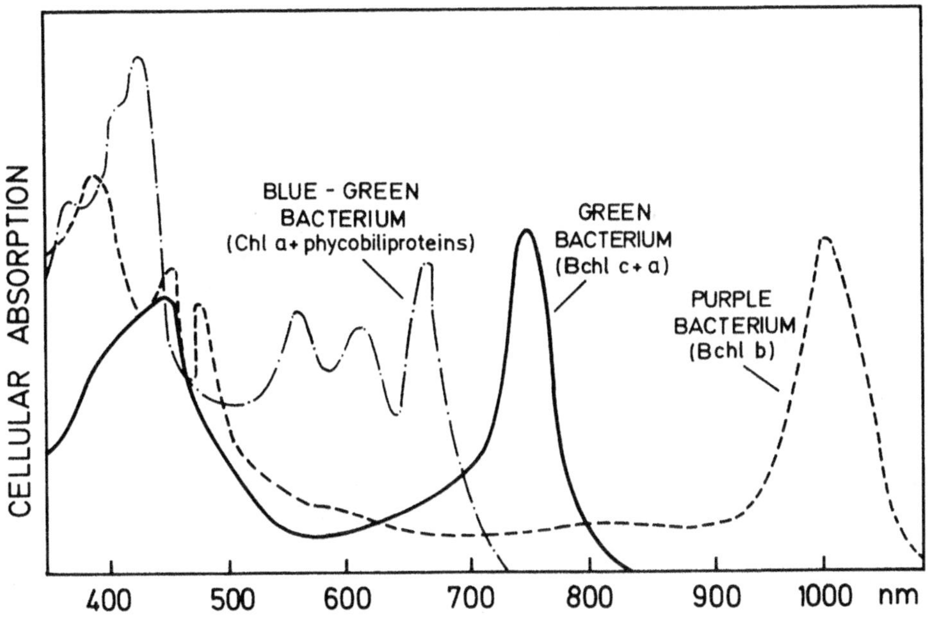

Figure 3. Absorption spectra of the main groups of phototrophic bacteria measured _in vivo_.

bacteria are aerobes and which grow in surface waters where they absorb much of the light in the blue and red region. Purple and green bacteria on the contrary are facultative or strict anaerobes and grow in deeper waters and in the surface of muds, where anaerobic conditions prevail. In muds these bacteria would underly the cyanobacteria and utilize light in the far red and infrared regions of the spetrum which the cyanobacteria and algae allow to pass. Thus, the different absorption spectra of the various groups of photosynthetic microorganisms have an ecological basis.

The cyanobacteria comprise a large and interesting heterogeneous group of organisms, which formerly were classified with the green algae. Thus, they have traditionally been termed blue-green algae. However, increased knowledge of their cellular structure shows clearly that they are prokaryotic organisms. Although a lot could be said about cyanobacteria, this article will be restricted mainly to the classical "phototrophic bacteria", the purple and green bacteria, which carry out anoxygenic photosynthesis. (It should be mentioned that under anerobic conditions in the presence of H_2S, some cyanobacteria can perform an anoxygenic photosynthesis, using only one photosystem. Thus, the distinction between the different types of photosynthetic bacteria is not as sharp as was once believed.)

TABLE III. Spectroscopic properties of purified chlorophylls in ether solution.

	Absorption maxima (nm) and relative peak heights						
Chlorophyll a			430(100)		615(13)	662(77)	
Bacteriochlorophyll c			428(200)		622(29)	660(63)	
Bacteriochlorophyll d			424(100)		608(17)	654(61)	
Bacteriochlorophyll e			458(100)		593(12)	647(32)	
Bacteriochlorophyll a	358(100)	391(68)		577(29)			773(126)
Bacteriochlorophyll b	368(100)	407(87)		582(30)			795(96)

TABLE VI. Carbon isotope ratios of photoautotrophically grown cells.

Organism	Carbon isotope ratios ^{13}C (0/00) ^{13}cells	^{13}medium	^{13}C	References
Chloroflexus aurantiacus	-22.5	-8.8	-13.7	Holo and Sirevåg (1986)
Chlorobium limicola	-28.7	-9.2	-19.5	Sirevåg et al. (1977)
Chromatium, strain D	-39.0	-9.2	-29.8	"
Chlamydomonas reinhardii	-49.4	-14.4	-35.0	"

3. ANOXYGENIC PHOTOTROPHIC BACTERIA

3.1. General considerations

The common features of the anoxygenic phototrophic bacteria are their anoxygenic photosynthesis and the presence of bacteriochlorophyll as one of several photopigments. With regard to internal structure and organization of the photosynthetic apparatus however, the purple and green bacteria differ considerably from each other. Furthermore, the species within each of the two groups vary in their carbon metabolism and in the nature of the electron donor in photosyntesis. On the basis of such properties, four families of phototrophic bacteria are distinguished: Rhodospirillaceae (purple non-sulfur bacteria), Chromatiaceae (purple sulfur bacteria), Chlorobiaceae (green sulfur bacteria) and Chloroflexaceae (green gliding bacteria). Table IV summarizes the main properties of these four families. The purple and green bacteria are very diverse, and for a variety of reasons they have been of interest to researchers from different fields. The anoxygenic phototrophic bacteria have proved extremely useful for studying fundamental aspects of photosynthesis, and they are of great evolutionary significance. Their diversity is challenging to biologists as well as to biochemists.

The phototrophic bacteria have been particularly useful for the study of the early photochemical events in photosynthesis. This is partly because they contain only one photosystem, and are thus "simpler" than green plants and cyanobacteria. In addition, many of these organisms are easy to grow, and mutants for genetic and other analysis are relatively easy to obtain. Thus, mutant strains of purple bacteria which lack carotenoids have been useful in the study of photosynthetic reactions by absorption spectroscopy, since carotenoids normally mask many of the spectral changes of cytochrome and other components.

Working with carotenoid-less mutants of purple bacteria, Clayton and coworkers originally formulated the concept of reaction centers and the photosynthetic unit in photosynthesis (Reed and Clayton, 1968). Using detergents they were able to isolate the reaction center as a defined unit from the purple nonsulfur bacterium Rhodopseudomonas sphaeroides, thus contributing enormously to the understanding of photosynthesis in general. More or less pure reaction centers have now been isolated from a variety of purple bacteria, and as will be shown later, the application of sophisticated analytical methods to the study of the photosynthetic apparatus in general and the reaction centers in particular, has given detailed structural information at the molecular level.

In photosynthetic organisms, light is absorbed by the light harvesting apparatus, which is assosiated with a photochemical reaction center in the photosynthetic membranes. When a light harvesting pigment molecule absorbs light, it is excited. The energy thus trapped is channelled to an adjacent pigment molecule and transfer continues randomly until the energy is either lost as fluorescence, or enters the reaction center. Only in the latter case is the energy useful for photosynthesis. Owing to strong interactions with proteins and other components, the chlorophyll of the reaction center is more reactive than

TABLE IV. Characteristics and major properties of the different groups of anoxygenic phototrophic bacteria.

	Purple bacteria		Green bacteria	
	Rhodospirillaceae (purple nonsulfur bacteria)	Chromatiaceae (purple sulfur bacteria)	Chlorobiaceae (green sulfur bacteria)	Chloroflexaceae (green gliding bacteria)
Location of photosynthetic apparatus	Cell membrane	Cell membrane	Chlorosomes	Chlorosomes
Bacteriochlorophylls	Bchl a or b	Bchl a or b	Bchl a and c, d or e	Bchl a and c or d
Electrondonor	H_2, organic substrates	H_2, H_2S, $S_2O_3^{2-}S^o$, organic substrates	H_2, H_2S, $S_2O_3^{2-}S^o$	H_2, H_2S, organic substrates
Accumulation of S^o	-	Intracellular	Extracellular	
CO_2-fixation	Calvin cycle	Calvin cycle	Reductive TCA-cycle	?
Aerobic dark growth	+	-	-	+

that in the light harvesting apparatus. Thus, when the reaction center chlorophyll is excited, an electron is transferred to a primary acceptor. This reduces a secondary acceptor, which in turn reacts with other electron carriers, one of which eventually transfers an electron back to the oxidized chlorophyll of the reaction center. This cyclic flow of electrons is associated with vectorial translocation of protons, across the photosynthetic membrane which is coupled to the synthesis of ATP. The products needed for the reduction of CO_2, reduced ferredoxin and $NAD(P)H_2$ are generated via a noncyclic flow of electrons or by reversed electron transport. The products of photosynthesis, ATP and reducing power can in addition be used to fix molecular nitrogen, a process which is widespread among purple and green bacteria (for review see Johansson et al., 1983).

3.2. The purple bacteria

3.2.1. Metabolism and structure. The purple sulfur bacteria and the purple nonsulfur bacteria differ in their relationship to sulfide. In nature, the electron donor in photosynthesis of purple sulfur bacteria is usually H_2S, which is partially oxidized and stored inside the cells as large globules of elemental sulfur until it is further oxidized to sulfate. The stored sulfur represents a reserve electron donor, which the cells carry and use for reduction of CO_2 when needed. Contrary to earlier beliefs, the purple nonsulfur bacteria are also able to use H_2S as an electron donor, provided it is present in low concentrations. However, these cells never store elemental sulfur endogenously, and are not able to oxidize it to sulfate.

Potentially, all purple bacteria are able to grow photoautotrophically wih CO_2 as the only carbon source, and in the species studied, the reductive pentose phosphate cycle, which is shown in Fig. 1. is the major mechanism for CO_2 reduction; ribulose-1.5-disphosphate carboxylase (RuBisCO) has been domonstrated in all cases. RuBisCO is a relatively large protein whose quaternary structure has been well characterized in a number of organisms (see Codd, 1984 for review). The enzyme from different sources varies with regard to molecular weight and subunit structure: it may contain subunits of two different types, a large (L) type of molecular weight 55 000 and a small (S) type of molecular weight 15 000. Generally RuBisCO from higher plants and algae comprises 8 subunits of each type and is designated L_8S_8, whereas more variation in the structure is found among anoxygenic phototrophic bacteria as can be seen in Table V.

The purple bacteria are metabolically versatile. Most species, especially among the purple nonsulfur bacteria, grow well photoheterotrophically on a variety of carbon sources, and simple organic acids like acetate and malate are easily metabolized by these organisms. Some species also utilize lower and higher fatty acids, methanol and ethanol. Most species of purple bacteria require one or more B-vitamins and many are able to grow aerobically in the dark on organic carbon sources.

In all purple bacteria, the photosynthetic pigments and the proteins associated with these are located in intracytoplasmic unit membrane systems, which are continuous with the cytoplasmic membrane and

TABLE V. Molecular properties of RuBisCO from some photosynthetic bacteria.

	Molecular weight	Subunits and quarternary structure	Reference
Purple nonsulfur bacteria			
Rhodospirillum rubrum	114.000	L_2S_0	Tabita and McFadden (1974)
Rhodopseudomonas sphaeroides	500.000		
Form I	360.000	L_6S_0	Gibson and Tabita (1977)
Form II	550.000	L_8S_8	Gibson and Tabita (1977)
Rhodopseudomonas capsulata			
Form I	360.000	L_6S_0	Gibson and Tabita (1977)
Form II	550.000	L_8S_8	Gibson and Tabita (1977)
Rhodomicrobium vannielii	430.000	L_6S_6	Taylor and Dow (1980)
Purple sulfur bacteria			
Chromatium D	520.000	L_8S_8	Tabaka and Akazawa (1975)
Ectothiorhodospira halophila	600.000	L_8S_8	Tabita and McFadden (1976)
Thiocapsa roseopersicina	500.000	L_8S_8	Purohit et al. (1979)
Cyanobacteria			
Anabaena cylindrica	500.000	L_8S_8	Takabe (1970), Okabe and Codd (1980)
Microcystis aeruginosa	518.000	L_8S_8	Stewart et al. (1977)
Oscillatoria limnetica	500.000	L_8S_8	Cook (1980)
Synechococcus PRIMPT/NI	530.000	L_8S_8	Andrews et al. (1981)

exhibit either tubular, vesicular or lamellar forms. Under certain conditions, the photosynthetic membranes may fill almost the entire volume of the cell, and as a consequence the photometabolism is not restricted to the periphery, but extends to its inner part. The fact that purple sulfur bacteria store sulfur inside their cells, is in agreement with the localization of their photosynthetic apparatus.

Many purple nonsulfur bacteria easily shift their metabolism from aerobic growth in the dark to anaerobic, phototrophic growth, and during this transition, the biochemical and structural changes that take place can be studied. Upon disruption of the cells by ultrasonic treatment or homogenization, the intracytoplamic membrane system breaks down into single vesicles, which contain the photosynthetic apparatus. Such chromatophores have been widely used for physical and chemical analysis of the photosynthetic apparatus (for review see Oelze, 1983).

3.2.2. The photosynthetic apparatus. The light harvesting complex of the purple bacteria is an integral part of the membrane. It consists of bacteriochlorophyll, carotenoids and several proteins, which have been isolated and characterized from a number of species (for review see Oelze, 1983). In some species only one type of light harvesting complex is found, which according to its absorbtion peak is designated B875. However, most species produce an additional complex, B850. Together these complexes contain more then 90% of the total bacteriochlorophyll in the organism.

The absorbed energy is transferred by inductive resonance from the light harvesting complexes to the reaction centers, which are also integral parts of the membrane (for review see Pierson and Olson, 1986). Each reaction center may be linked to many (20-40) light harvesting complexes in the purple bacteria.

The active chlorophyll of the reaction centers of the purple bacteria is referred to as P870. The reaction center is composed of 4 molecules of bacteriochlorophyll, 2 molecules of bacteriophaeophytin (bacteriochlorophyll without Mg), 3 polypeptides, a cytochrome, a non-heme iron and a quinone (Evans and Heathcote, 1983). Recently, well ordered three-dimensional crystals of the reaction center of *Rhodopseudomonas viridis* which contains Bchl *b* have been obtained (Michel, 1982). Using X-ray structure analysis in combination with available data from amino acid sequence analysis of the proteins, it has been possible to construct a model which shows the arrangement of the individual molecules and atoms in the reaction center (Deisenhofer *et al.*, 1985). According to this model, the arrangement of the pyrrol ring systems of the bacteriochlorophyll and bacteriophaeophytin molecules show local twofold symmetry. Near the axes of symmetry, two bacteriochlorophyll molecules are in close contact and form a dimer, the so-called "special pair", which function as the primary electron donor, where the charge separation occurs.

Integrated into the membrane close to the reaction center is a complex formed by cytochromes b and c, plus an iron-sulfur protein. Other components of the photosynthetic apparatus are ubiquinone and cytochrome C_2, the former being a mobile carrier within the membrane. Fig. 4 is a schematic drawing of the photosynthetic apparatus in the

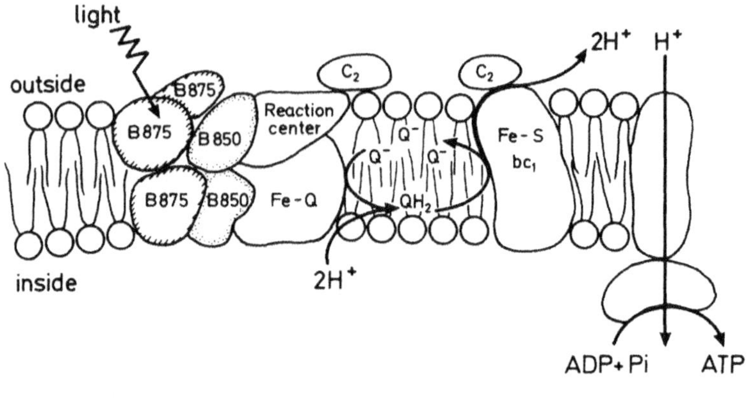

Figure 4. Localization of the photosynthetic apparatus in the membrane (Redrawn after Gottschalk, 1986).

membrane of purple bacteria. It shows the possible location of the compounds and how protons are translocated across the membrane as a consequence of electron transport. Fig. 5a shows what happens when light energy is transferred to the reaction center. One electron is translocated from the "special pair" via one of the other bacteriochlorophylls to bacteriophaeophytin and further to a ubiquinone-iron complex. At the same time P870 is reduced by cytochrome C_2. Initially, the photochemical event operates with high efficiency and the redox difference of the charge separation between P870 and the bacteriophaeophytin is about 1.05 V. However, a considerable part of this energy is lost in stabilizing the charge separation to the quinone with a final difference in the redox potential between P870 and secondary quinone of only 0.5 V (Evans and Heathcote, 1983). As a result of this low efficency, and since the redox potential of the acceptor is -0.15 V, NAD which has a redox potential of -0.32 V cannot be reduced directly by the acceptor. However, a proton gradient is produced across the membrane as a result of the initial charge separation, and the proton motive force thus created can be used both for the synthesis of ATP and for the reduction of NAD by reversed electron transport. In purple nonsulfur bacteria NADH is generated with either organic substrates or H_2 as hydrogen donors, whereas the natural source of reducing power for the purple sulfur bacteria is H_2S.

3.3. The green bacteria

3.3.1. Metabolism and structure. The common features of the green sulfur bacteria (i.e. the genus *Chlorobium*) and the green gliding bacteria (i.e. the genus *Chloroflexus*) are the types of bacteriochlorophyll they contain and the organization of their photosynthetic apparatus. However, in most other properties they differ from each other.

The green sulfur bacteria are strictly anaerobic, obligate

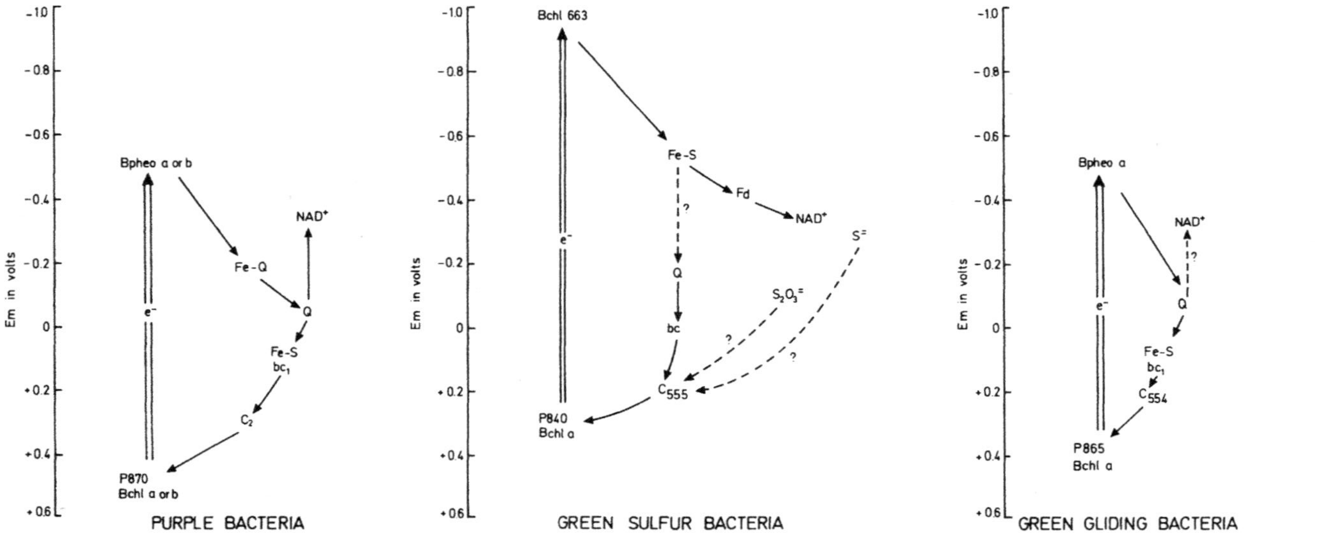

Figure 5. The electron transport in the reaction center and electron transport chain of a) purple bacteria, b) green sulfur bacteria and c) green gliding bacteria. The electron carriers are placed according to their midpoint potentials.

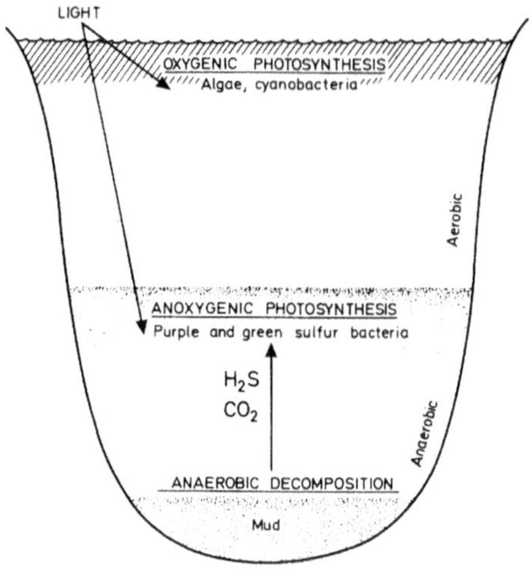

Figure 6. The structure of a meromictic lake, showing the distribution of photoautotrophic organisms.

autotrophs with a very limited capacity to assimilate organic compounds. In nature, members of this group are found underneath layers of green algae, cyanobacteria and purple bacteria in stagnant ponds and in meromictic lakes, with vertical gradients of light from above and H_2S from below, as is shown schematically in Fig. 6.

In contrast to other photosynthetic organisms, the green bacteria do not fix carbon dioxide *via* the reductive pentose phosphate cycle (Fig. 1). Instead, *Chlorobium* uses the reductive citric acid cycle proposed by Evans *et al*. (1966). Originally this cycle, which is shown in Fig. 7, was thought to be an additional mechanism to the reductive pentose phosphate cycle in *Chlorobium*, since the presence of RuBisCO had been reported (Smillie *et al*., 1962). However, several lines of evidence, including short-term labelling experiments (Sirevåg, 1974), use of washed cells and inhibitors (Sirevåg and Ormerod, 1970) and isotope fractionation studies (Table VI) (Sirevåg *et al*., 1977) indicate that it is the sole mechanism of CO_2 fixation in this organism. Because of inconsistency of data regarding the presence of RuBisCO (Tabita *et al*., 1974, Buchanan and Sirevåg, 1976), and citrate lyase, the key enzyme of the reductive citric acid cycle (Beuscher and Gottschalk, 1972) the presence of the latter mechanism has been questioned. However, recent evidence by Ivanovsky *et al*. (1980) and Fuchs *et al*. (1980) strongly supports its presence (for reviews see Ormerod and Sirevåg, 1983; Fuchs and Stupperich, 1985).

The green gliding bacterium, *Chloroflexus*, is thermophilic and was first isolated relatively recently (Pierson and Castenholz, 1974). Like the purple nonsulfur bacteria, it is metabolically versatile, growing anaerobically in the light on carbon dioxide or organic compounds and

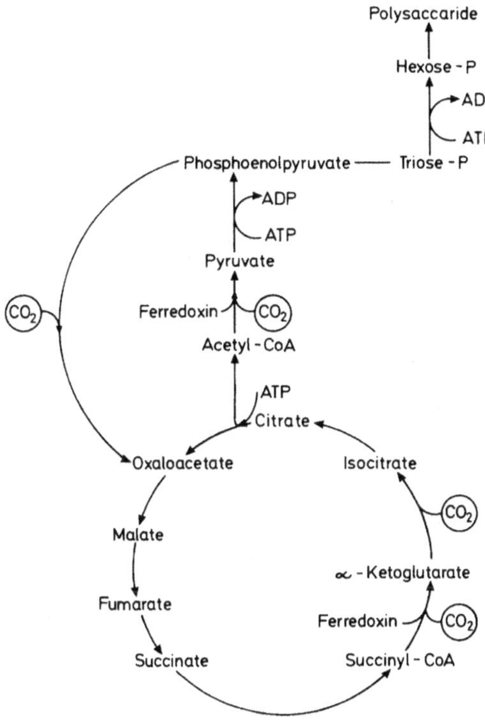

Figure 7. The reductive citric acid cycle.

aerobically in the dark (respiration). This metabolic versatility is not suprising, considering the natural habitat of the organism. In nature, Chloroflexus is abundant in hot springs where it grows in long filaments and forms thick mats in association with cyanobacteria, presumably utilizing organic compounds excreted by the latter.

In a very recent study, using cells grown photoautotrophically with H_2 as electron donor, RuBisCO was not detected in cell-free extracts of Chloroflexus. Short-term labelling experiments with $^{14}CO_2$, use of inhibitors and various ^{14}C-labelled substrates in addition to $^{13}C/^{12}C$ isotope fractionation studies (Table VI) shows that the reductive pentose phosphate cycle is not operating. Nor, does Chloroflexus have the enzymes necessary for a complete reductive citric acid cycle. Pyruvate synthase is present, and inhibitor studies indicate that CO_2 is fixed via this enzyme, which catalyzes the formation of pyruvate from acetyl-CoA and CO_2. How acetyl-CoA is formed from CO_2 is not clear, but the mechanism differs from those known in other bacteria (Holo and Sirevåg, 1986; Holo, 1986).

In green bacteria, the light harvesting complexes are located in specific, oblong structures, known as chlorosomes, which are easily detected on electron micrographs. Fractionation studies have shown that they contain some bacteriochlorophyll a, all the bacteriochlorophyll c, d or e in addition to carotenoids and proteins. The quantum of light harvesting bacteriochlorophyll in *Chlorobium* is large and enables it to grow at light intesities lower than those usable by any other known phototrophic organism.

The chlorosomes are firmly attached to the cytoplasmic membrane, where the reaction centers, which contain bacteriochlorophyll a, bacteriophaeophytin and carotenoids are located, together with the cytochromes of the electron transport chain.

On the basis of freeze-fracture studies of *Chlorobium limicola*, Staehelin *et al*. (1980) have described the sturctural organization of the pigments and proposed a model of the chlorosome and the cytoplasmic membrane. According to this model, which is shown in Fig. 8, the chlorosome is surrounded by a monolayer of galactolipid. Each chlorosome contains 10-30 rod elements consisting of 5-6 bacteriochlorophyll c-protein subunits arranged in a ring-like structure. Between each chlorosome and the cytoplasmic membrane is a crystalline baseplate, consisting of water-soluble bacteriochlorophyll *a*-protein complexes. At each chlorosome attachment site, the cytoplasmic membrane contains 20-45 intramembrane particles, which contain the reaction centers and water-insoluble bacteriochlorophyll a protein. In *Chloroflexus aurantiacus*, the organization of the light harvesting complex is basically similar to that of *Chlorobium*. However, the rod elements, which have been isolated and characterized at the molecular level are smaller and have a different structure (Wechsler *et al*., 1985).

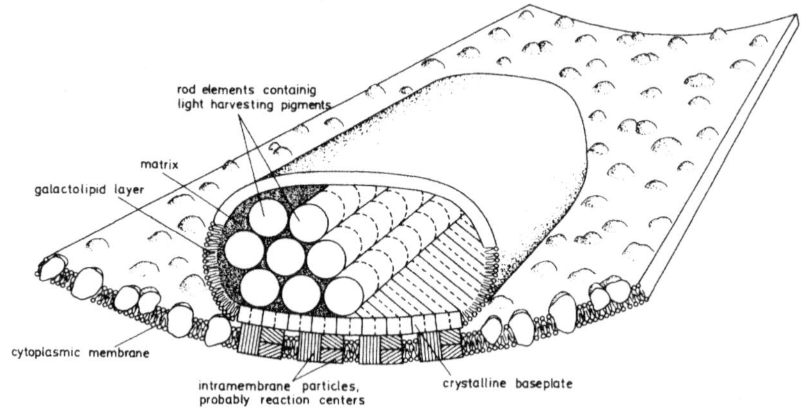

Figure 8. Model of a chlorosome and the associated cytoplasmic membrane in green sulfur bacteria (Redrawn form Staehlin *et al*. 1980).

3.3.2. The photosynthetic apparatus. The reaction center of *Chlorobium* contains a dimer of Bchl *a* P840, which functions as the primary electron donor. The initial electron acceptor is *not* bacteriopheophytin as in purple bacteria, but a special form of Bchl *c*, Bchl 663 (for review see Pierson and Olson, 1986). The second acceptor is an ironsulfur protein with a redox potential of-0.54 V, which is capable of reducing ferredoxin and thus NAD^+ directly. The electron transport in the reaction center is illustrated in Fig. 5b. The resemblance of the reaction center in *Chlorobium* to that of photosystem I of cyanobacteria and chloroplasts is striking, whereas there is little or no resemblance to that of purple bacteria. When the electróns are utlized for the reduction of ferrodoxin and NAD^+ in green sulfur bacteria, sulfide, thiosulfate or molecular hydrogen act as electron donors for the reaction center, *via* cytochrome c. As is indicated in Fig. 5b., there is only indirect evidence for cyclic electron flow in green sulfur bacteria (Pierson and Olson, 1986). Analogous to what takes place in the purple bacteria however, the reaction center, the iron-sulfur complex and the cytochromes are assumed to be localized in the membrane in such a way that proton translocation occurs and a proton motive force is developed which can be utilized for ATP production.

The reaction center of *Chloroflexus aurantiacus* resembles that of purple bacteria (for review see Pierson and Olson, 1986). The primary electron donor is a dimer of Bchl a, the primary electron acceptor is bacteriophaeophytin a, and the secondary acceptor is a quinone. Thus it reasonable to assume that *Chloroflexus* reduces NAD^+ by reversed electron flow like the purple bacteria (Fig. 5c.).

Recently, green phototrophic bacteria (*Heliobacterium*) with a new type of bacteriochlorophyll (Bchl *g*) and lacking chlorosomes have been isolated. These are strict anarobes which grow on organic substrates in the light.

3.4. Molecular biology and genetics

The molecular biology and genetics of anoxygenic phototrophic bacteria have recently been reviewed by Marrs (1983). There is no evidence that proteins or nucleie acids of these bacteria are synthesized in manner basically different from that of other bacteria.

However, for studying metabolic regulatory mechanisms and differentiation of membranes at the molecular level, both the purple nonsulfur and the green gliding bacteria are better suited than organisms like *Escherichia coli* because of their metabolic versatility. This is illustrated by a recent investigation of the effect of oxygen on the development of chlorosomes in *Chloroflexus* (Fuller *et al*., 1985). It has been known for a long time that in certain species of facultatively phototrophic purple bacteria, development of the photosynthetic apparatus is affected by molecular oxygen (Cohen - Bazire and Sistrom, 1966). In *Chloroflexus*, changes in oxygen concentration were followed in aerobic, darkgrown cultures and the cells analyzed for bacteriochlorophylls and chlorosome polypeptides during development of the photosynthetic apparatus. Only when the oxygen concentration fell below a certain critical level, were Bchl *a* and *c* detected, whereas some, at least of the chlorosome polypeptides which were identified, were present under both conditions. These polypeptides became functional when required for the assembly of the chlorosomes.

To fully understand the functioning of living organisms, it is necessary to study the arrangement and regulation of genes within the cells. The development in recent years of the techniques of genetic recombination, provide powerful approaches to the dissection of gene structure and function. What is known so far about gene organization and regulation in the phototrophic bacteria is based for a large part on recombinant-DNA methods. Thus the genes for RuBisCO (Tabita, 1985) and for the light harvesting pigment protein complexes (Kiley _et al_., 1985 from _R. sphaeroides_ have recently been cloned in _E. coli_.

Using cloned genes for sequencing and hybridization analysis of transcription have yielded a certain amount of biochemical information. However, proper genetic analysis requires that the cloned genes are returned to the organism from which they were obtained, in order to determine the properties of specially constructed mutants, or to determine whether a given mutant can be complemented. A system for gene transfer from one organism to another is thus a prerequisite.

In bacteria the classical methods for gene transfer are _transduction_ (transfer of bacterial DNA within a virus particle from one bacterium to another), _conjugation_ (transfer of DNA by cell to cell contact) and _transformation_ (incorporation of free DNA into a recipient cell). The development of these methods in phototrophic bacteria has been slow compared to the progress made with bacteria like _E. coli_ (for reviews se Marrs, 1983; Saunders, 1984).

Although a number of bacteriophages infect purple bacteria specifically only the gene transfer agent, GTA, of _R. capsulata_ is known to mediate genetic exchange. In a process termed "capsduction" this takes place between diffferent strains of this organism. The GTA resembles a small bacteriophage, and superficially "capsduction" resembles generalized transduction in that gene transfer from any region of the genome can take place. However, in contrast to transduction, no GTA-specific DNA has been formed in association with the production of GTA particles. Using GTA, the genes coding for nitrogen fixation in _R. capsulta_ have been manipulated (Wall _et al_., 1975), and a genetic map of the photopigment region of the organism has been constructed (Yen and Marrs, 1976).

When bacterial chromosomal DNA is transferred by conjugation, the chromosome in the donor cell is mobilized by a plasmid. Plasmids are extrachromosomal, circular DNA-molecules which can replicate autonomously in the cell. In addition to mobilizing the chromosome, plasmids can code for properties like drug resistance, tumor induction, nitrogen fixation and the metabolism of rare, natural compounds. So far, among anoxygenic phototrophic bacteria, intrinsic (indigenous) plasmids have been isolated only from purple nonsulfur bacteria, where they have been shown to code for resistance to antibiotics. It has however been suggested that plasmid-DNA in these bacteria is responsible for resistance to heavy metals, production of toxins, photoproduction of hydrogen and genes specifying the photosynthetic apparatus (Saunders, 1984). No report exists of conjugation mediated by such plasmids in phototrophic bacteria.

Exogenous (promiscuous) plasmdis are capable of promoting their own transfer to a wide range of bacteria and of being stably maintained in various hosts. Such plasmids have been transferred to several purple nonsulfur bacteria, where they have been reported to mediate low frequencies of chromosome transfer in some species. However, high frequencies of chromosome transfer have been obtained using hybrid plasmids carrying transposons. Transposons are mobile genetic elements which can insert at random into plasmids or the bacterial chromosome. Using a plasmid carrying a transposon, Pemberton and Bowen (1981) were able to promote high frequency chromosome transfer in R. sphaeroides, which made it possible to construct genetic maps of two regions of the genome of this organism.

More recently, a generalized transformation system has been developed, where plasmid-DNA is taken up in the presence of $CaCl_2$ (Fornari and Kaplan, 1983; Saunders, 1984).

Although gene transfer systems for the anoxygenic phototrophic bacteria are poorly known at this stage, new and powerful genetic tools are becoming available by a combination of recombinant DNA-tecniques and traditional genetic systems. Thus, our knowledge of these bacteria is expected to increase enormously in the near future.

3.5. Possible practical applications of anoxygenic phototrophic bacteria

The possibility of taking practical and industrial advantage of the energetic potential of phototrophic bacteria has long been a subject of discussion. In only a few cases however, has this become a reality. In Japan, anoxygenic phototrophic bacteria have been used successfully in the treatment of various industrial waste waters (Kobayashi and Tchan, 1973). During growth of the phototrophic organisms the waste water was purified without the production of sludge, and the by-products of the treatment, the phototrophic bacteria, were harvested, dried and used as food supplement for fish and poultry.

More recently, the green sulfur bacterium, Chlorobium was used as a biocatalyst in the removal of acid components (H_2S and CO_2) from fuel processing gases supplementd with N_2 (for nitrogen fixation) (Cork et al., 1983). The gases were led into a reactor containing Chlorobium, which under the conditions employed fixed 28% of the CO_2 into organic carbon and converted 67% of the H_2S in to insoluble sulfur, which was removed by continuous centrifugation. Since it is known that part of the CO_2 fixed is converted into endogenous polysaccharide in Chlorobium, a scheme was proposed whereby the cells were allowed to ferment the polysaccharide to acetate. After the fermentation process was completed, the intact cells were recycled back to the reactor.

In addition to giving basic information about the organisms as such, recombinant-DNA techniques are expected to provide opportunities for genetic engineering of phototrophic organisms directed towards biotechnology. Such possibilities are discussed in general terms by Marrs (1983), who points to the possibility of exploring the ability of the organisms to produce carotenoids (polyisoprene). By manipulating the genes involved in the synthesis of carotenoids, one could cause the cells to accumulate precursors of carotenoids, which in turn could be converted into a variety of useful compounds.

4. REFERENCES

Andrews, T.J., Abel, K.M., Menzel, E. and Badger, M.R. (1981). Molecular weight and quaternary structure of ribulose bisphosphate carboxylase from the cyanobacterium, Synechococcus sp. Arch. Microbiol. **130,** 344-348.

Beuscher, N. and Gottschalk, G. (1972). Lack of citrate lyase - the key enzyme of the reductive carboxylic acid cycle - in Chlorobium thiosulfatephilum and Rhodospirillum rubrum. Z. Naturforsch. **27b,** 967-973.

Buchanan, B.B., Sirevåg, R. (1976). Ribulose, 1.5-diphosphate carboxylase and Chlorobium thiosulfatophilum. Arch. Microbiol. **109,** 15-19.

Codd, G.A. (1984). Aspects of carbon dioxide assimilation by autotrophic procaryotes. In: Aspects of microbiol metabolism and ecology (Ed. G.A. Codd), pp. 129-173. Academic press.

Cohen-Bazire, G. and Sistrom, W.R. (1966). The prokaryotic photosynthetic apparatus. In: The chlorophylls (Eds. L.P. Vernon and G.R. Seely) pp. 313-341. Academic Press.

Cook, C.M. (1980). The characterization of ribulose bisphosphate carboxylase from cyanobacteria. Ph.D. Thesis, University of Dundee.

Cork, D.J., Garunas, R., Saggad, A. (1983). Chlorobium limicola forma thiosulfatophilum: Biocatalyst in the production of sulfur and organic carbon from a gas stream containing H_2S and CO_2. Appl. Environ. Microbiol. **45,** 913-918.

Deisenhofer, J., Michel, H. and Huber, R. (1985). The structural basis of photosynthetic light reactions in bacteria. Trends Biochem. Sci. **10,** 243-248.

Evans, M.C.W., Buchanan, B.B., Arnon, D.I. (1966). A new ferredoxin dependent carbon reduction cycle in a photosynthetic bacterium. Proc. nat. Acad. Sci. **55,** 928-934.

Evans, M.C.W. and Heathcote, P. (1983). The early photochemical events in bacterial photosynthesis. In: The phototrophic bacteria: anaerobic life in the light. (Ed. J.G. Ormerod) pp. 35-60. Blackwell scientific publications .

Fornary, C.S. and Kaplan, S. (1982). Genetic transformation of Rhodopseudomonas sphaeroides by plasmid DNA. J. Bacteriol. **152,** 89-97.

Fuchs, G., Stupperich, E. and Eden, G. (1980). Autotrophic CO_2 fixation in Chlorobium limicola. Evidence for the operation of a reductive carboxylic acid cycle in growing cells. Arch. Microbiol. **128,** 64-71.

Fuchs, G. and Stupperich, E. (1985). Evolution of autotrophic CO_2 fixation. In: Evolution of procaryotes (Eds. Schleifer, K. and Stackebrandt, E.) pp. 235-251. Academic press.

Fuller, R.C., Foster, J.M. and Redlinger, T.E. (1985). Oxygen regulation of pigment-protein synthesis in Chloroflexus. V. Int. Symp. on Photosynthetic Prokaryotes, Grindelwald, Switzerland.

Gibson, J.L. and Tabita, F.R. (1977a). Different molecular forms of

D-ribulose-1.5-bisphosphate carboxylase from Rhodopseudomonas sphaeroides. J. Biol. Chem. **252,** 943-949.

Gibson, J.L. and Tabita, F.R. (1977b). Isolation and characterization of two forms of ribulose 1.5-bisphosphate carboxylase from Rhodopseudomonas sphaeroides. J. Biol. Chem. **252,** 943-949.

Gottschalk, G. (1986). Bacterial metabolism Springer-Verlag.

Holo, H. and Sirevåg, R. (1986). Autotrophic growth and CO_2-fixation in Chloroflexus aurantiacus. Arch. Microbiol. (in press).

Holo, H. (1986). Autotrophic growth and CO_2-fixation in Chloroflexus aurantiacus. Dr.scient. thesis, University of Oslo.

Ivanovsky, R.N., Sinstov, V.N. and Kondratieva, E.N. (1980). ATP-linked citrate lyase activity in the green sulfur bacterium Chlorobium limicola forma thiosulfatophilum. Arch. Microbiol. **128,** 239-241.

Johansson, B.C., Nordlund, S. and Baltscheffsky, J. (1983). In: The phototrophic bacteria: anaerobic life in the light (Ed. J.G. Ormerod) pp. 120-145. Blackwell scientific publications.

Kiley, P.J. (1985). Analysis of the light-harvesting genes from Rhodopseudomonas sphaeroides. V. Int. Symp. on Photosynthetic Prokaryotes, Greendelwald, Switzerland.

Kobayashi, M., Tchan, Y.T. (1973). Treatment of industrial waste solutions and production of useful by-products using a photosynthetic bacterial method. Water Res., **8,** 1219-24.

Marrs, B.L. (1983). Gentetics and molecular biology. In. The phototrophic bacteria: anaerobic life in the light (Ed. J.G. Ormerod) pp. 186-214. Blackwell scientific publications.

Michel, H. (1982). Three-dimentional crystals of a membrane protein complex. The photosynthetic reaction centre from Rhodopseudomonas viridis. J. Mol. Biol. **158,** 567-572.

Oelze, J. (1983). Structure of phototrophic bacteria, development of the photosynthetic apparatus. In: The phototrophic bacteria: anaerobic life in the light (Ed. J.G. Ormerod) pp. 8-34. Blackwell scientific publications.

Okabe, K. and Godd, G.A. (1980). Structural and regulatory studies on the ribulose bisphosphate carboxylase-oxygenase of Anabaena cylindrica. Plant Cell Physiol. **21,** 1117-1127.

Ormerod, J.G. and Sirevåg, R. (1983). Essential aspects of carbon metabolism. In. The phototrophic bacteria: anaerobic life in the light. (Ed. J.G. Ormerod) pp. 100-119. Blackwell scientific publications.

Pemberton, J.M. and Bower, A.R. St. G. (1981). High frequency chromosome transfer in Rhodopseudomonas sphaeroides promoted by the broad host range plasmid RPI carrying the mercury transposon Tu 501. J. Bacteriol. **147,** 110-117.

Pierson, B.K. and Castenholz, R.W. (1974). A phototrophic gliding, filamentous bacterium of hot springs, Chloroflexus aurantiacus, gen. and Sp. nov. Arch. Microbiol. **100,** 5-24.

Pierson, B.K. and Olson, J.M. (1986). Photosynthetic bacteria. In: New comprehensive Biochemistry. Photosynthesis (Ed. J. Amesz) in press Elsevier.

Purohit, K., McFadden, B.A. and Lawlis, V.B. (1979). Ribulose biphosphate carboxylase/oxygenase from Thiocapsa rosepersicinor. Arch. Microbiol. **121,** 79-82.

Reed, D.W. and Clayton, R.K. (1968). Isolation of a reaction center fraction from Rps. sphaeroides. Biochem. Biophys. Res. Commun. **30,** 471-475.

Saunders, V. (1984). Molecular biology of phototrophs. In: Aspects of microbial metabolism and ecology (Ed. G.A. Codd), pp. 241-276. Academic Press.

Sirevåg, R., Ormerod, J.G. (1970). Carbon dioxide fixation in green sulphur bacteria. Biochem. J. **120,** 399-408.

Sirevåg, R. (1974). Further Studies on carbon dioxide fixation in Chlorobium. Arch. Microbiol. **98,** 3-18.

Sirevåg, R., Buchanan, B.B., Berry, J.A., Troughton, J.H. (1977). Mechanisms of CO_2 fixation in bacterial photosynthesis studied by the carbon isotope fractionation technique. Arch. Microbiol. **112,** 35-38.

Smillie, R.M., Rigopolulos, N. and Kelly, H. (1962). Enzymes of the reductive pentose phosphate cycle in the purple and in the green photosynthetic sulphur bacteria. Biochim. Biophys. Acta (Amst.) **56,** 612-614.

Staehelin, L., Golecki, J.R. and Drews, G. (1980). Supramolecular organization of chlorosomes (Chlorobium vesicles) and of their membrane attachment sites in Chlorobium limicola. Biochem. Biophys. Acta. **589,** 30-45.

Stewart, R., Auchterlonie, C.C. and Codd, G.A. (1977). Studies on the subunit structure of ribulose 1.5-bisphosphate carboxylase from the blue-green alga Microcystis aeruqenosa. Planta **136,** 61-64.

Tabita, F.R. and McFadden, B.A. (1974). D-ribulose 1.5-diphosphate carboxylase form Rhodospirillum rubrum. J. Biol. Chem. **249,** 3453-3459.

Tabita, F.R., McFadden, B.A. and Pfennig, N. (1974). D-ribulose-1.5-biphosphate carboxylase in Chlorobium thisoulfatophilum, Tassajara. Biochim. Biophys. Acta (Amst.) **341,** 187-194.

Tabita, F.R. and McFadden, B.A. (1976). Molecular and catalytic properties of ribulose 1.5-bisphosphate carboxylase from the photosynthetic extreme halophile Ectothiorhodospira halophila. J. Bacterial. **126,** 1271-1277.

Tabita, F.R. (1985). Molecular and enzymological studies of recombinant ribulose 1.5-bisphosphate carboxylase/oxygenase. V. Int. Symp. on Photosynthetic Prokaryotes, Grindelwald, Switzerland.

Takabe, T. and Akazawa, T. (1977). A comparative study of the effect of O_2 on photosynthetic carbon metabolism by Chlorobium thiosulfatophilum and Chromatium vinosum. Plant Cell Physiol. **18,** 753-756.

Takabe, T. (1977). Ribulose 1.5-bisophosphate carboxylase from the blue-green alga Anabaena cylindrica. Agric. Biol. Chem. **41,** 2255-2260.

Taylor, S.C. and Dow, C.S. (1980). Ribulose-1.5-bisphosphate carboxylase from Rhodomicrobium vannieli. J. Gen. Microbiol. **116,** 81-87.

Wall, J.D., Weaver, P.F. and Gest, H. (1975). Gentic transfer of nitrogenase-hydrogenase activity in Rhodopseudomonas capsulata. Nature (London) **258,** 360-361.

Wechsler, T., Suter, F., Fuller, R.C. and Zuber, H. (1985). The complete amino acid sequence of the bacteriochlorophyll c binding polypeptide from chlorosomes of the green photosynthetic bacterium Chloroflexus

aurantiacus. FEBS Lett. **181** ,173-178.
Yen, H.C. and Marrs, B. (1976). Map of genes for carotenoid and bacteriochlorophyll biosynthesis in Rhodopseudomonas capsulata. J. Bacteriol. **126,** 619-629.

CARBON DIOXIDE REDUCTION BY ANAEROBIC BACTERIA

Georg Fuchs
Abteilung Angewandte Mikrobiologie
Universität Ulm
D-7900 Ulm
Federal Republic of Germany

ABSTRACT. Anaerobic bacteria exhibit a large variety of enzymes catalyzing carboxylation or reductive carboxylation reactions. In some of these organisms CO_2 functions as the principal electron acceptor in energy metabolism: CO_2 is reduced with hydrogen to CH_4 (methanogenic bacteria) or to acetate (acetogenic bacteria). These exergonic reactions serve as the sole energy yielding processes for life of these microorganisms and are coupled to a synthesis of ATP from ADP and P. In addition, many anaerobic bacteria can synthesize all cellular compounds from CO_2 (autotrophic anaerobic bacteria). Even when a preformed organic compound is used as the main source of cell carbon (heterotrophic bacteria), CO_2 is also required for biosynthetic reactions. A brief overview is given, why these organisms use CO_2 in metabolism, how they reduce CO_2 or incorporate it into cellular compounds, and what are the strategies used compared to known chemical processes.

INTRODUCTION

In strict anaerobic bacteria CO_2 plays a dual function in metabolism. CO_2 can be used in energy metabolism as an oxidant, which accepts reducing equivalents either derived from molecular hydrogen or from the dehydrogenation of organic compounds. Most methanogenic bacteria can reduce CO_2 with H_2 to methane according to

$$CO_2 + 4\ H_2 \longrightarrow CH_4 + 2\ H_2O;\ \Delta G^{o\prime} = -131\ kJ/Mol\ CH_4.$$

Analogously most acetogenic bacteria can reduce CO_2 with H_2 to acetate according to

$$2\ CO_2 + 4\ H_2 \longrightarrow CH_3COO^- + H^+ + 2H_2O;\ \Delta G^{o\prime} = -95\ kJ/Mol\ acetate.$$

These specialized anaerobic microorganisms obtain all their energy for biosynthesis and other energy consuming cellular processes from these exergonic redox reactions. In addition, many of them can use

M. Aresta and G. Forti (eds.), Carbon Dioxide as a Source of Carbon, 263–273.

CO_2 as the sole source of carbon for biosynthetic reactions according to

$$nCO_2 + 4n[H] \longrightarrow [CH_2O]n + n\ H_2O.$$

The mechanisms of CO_2 incorporation into cellular compounds are totally different from those encountered in green plants and in most aerobic autotrophic bacteria. Besides these CO_2 metabolizing processes all living organism use CO_2 for carboxylation reactions according to

$$[C_n]\ H + CO_2 \longrightarrow [C_n]\text{-}\ COO^- + H^+.$$

Consequently, large a variety of CO_2 metabolizing enzymes occurs in anaerobic bacteria most of which have been discovered or studied in more detail only very recently. Among them are metallo-enzymes containing one or two metals of various oxidation states (Ni, Mo, W, Fe, Co), but also non-metallo-enzymes, which may or may not contain an organic prosthetic group. An incomplete list of these enzymes is given in Table I.

This contribution intends (1) to give an overview of CO_2 metabolizing processes and enzymes in anaerobic bacteria, and (2) to serve as a guide to the literature for non-microbiologists.

ROLE OF CO_2 REDUCTION TO CH_4 AND CH_3COOH

In non-marine anaerobic environments, organic compounds are decomposed by a consortium of anaerobic bacteria basically to CO_2 and CH_4, which is called biogas. The disproportionation of organic carbon to the most reduced and most oxidized one-carbon compounds is associated with a maximum free energy change. The overall process is schematically presented in Fig. 1 using cellulose as the most predominant organic matter on earth. Acetogenic and methanogenic bacteria as the last members in this chain metabolize acetate, CO_2, and H_2, the metabolic end products of various fermenting bacteria. They obtain all their energy from these CO_2 reduction processes which are coupled to a net synthesis of ATP. For literature on the biology of these bacteria see [1, 2].

CO_2 REDUCTION TO CH_4 CATALYZED BY METHANOGENIC BACTERIA

Biological methane formation from CO_2 and H_2 does not proceed via the free one-carbon compounds HCOOH, HCHO, CH_3OH. Rather, methanogenic bacteria use a variety of novel coenzymes which carry these C_1-compounds. Similarly, the electrons from hydrogen oxidation (catalyzed by the enzyme hydrogenase) are transferred by several different electron carriers to the four $2e^-$ reduction steps (Fig. 2) required for CO_2 reduction to CH_4. The CO_2 fixation reaction in methanogenesis is reaction 3 in Table I. The last strongly exergonic reduction of CH_3-coenzyme M to CH_4 is coupled to ATP synthesis by an unknown

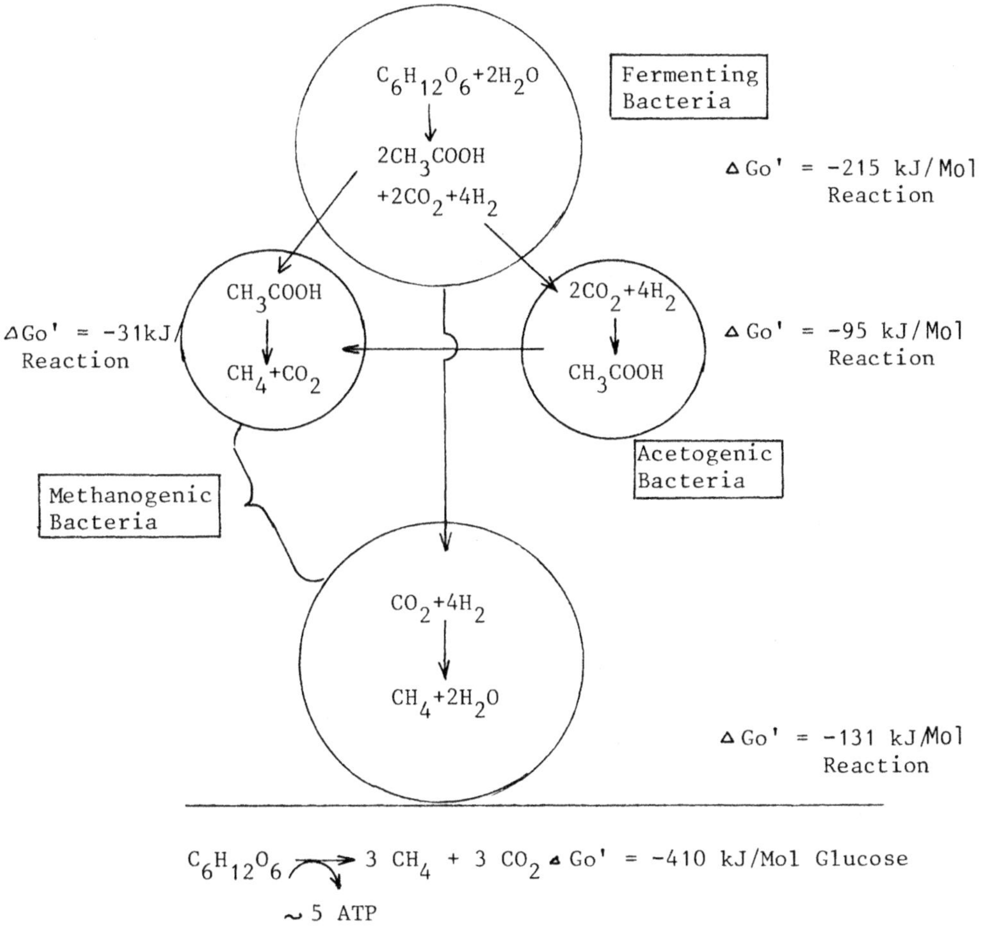

Fig. 1 Scheme of the role of CO_2 reduction(for energy metabolism) in the anaerobic decomposition of organic material (non-marine environment). As an example, the degradation of cellulose $(C_6H_{12}O_6)_n$ is shown. Several types of interdependent anaerobic bacteria symbolized by circles are involved in this complex process. No electron acceptors are available except for H^+ and CO_2, resulting in intermediary H_2 and final CH_4 gas production (CH_4 + CO_2 = biogas).

TABLE I Important CO_2 fixation reactions in chemotrophic anaerobic bacteria. The table does not include all CO_2 fixation reactions, especially those common in aerobic bacteria. Note, that not all enzymes are present in one organism.

[H] = reducing equivalent; [CO] = enzyme-bound CO; P_i = inorganic phosphate; PP_i = inorganic pyrophosphate; CO_2 (g) is used for calculations of ΔGo' values

Reaction	ΔGo' (2[H] = H_2) (kJ/Mol CO_2)	CO_2 Acceptor	Enzyme
1. CO_2 + 2 [H] ⇌ $HCOO^-$ + H^+	+ 3.4	Mo or W enzyme	Formate dehydrogenase
2. CO_2 + 2 [H] ⇌ [CO] + H_2O	< + 20.1	Ni-enzyme	CO dehydrogenase
3. CO_2 + Methanofuran ⇌ Carboxy-methanofuran	?	Methanofuran	not yet characterized
4. CH_3-CO-SCoA + CO_2 + 2[H] ⇌ CH_3-CO-COO^- + H^+ + CoASH	+ 16.3	Acetyl-coenzyme A	Pyruvate synthase
5. ^-OOC-CH_2-CH_2-CO-SCoA+CO_2+2[H] ⇌ ^-OOC-CH_2-CH_2-CO-COO^-+H^++ CoASH	+ 14.8	Succinyl-coenzyme A	2-Oxo-glutarate synthase
6. ^-OOC-CH_2-CH_2-CO-COO^-+CO_2+2[H] ⇌ ^-OOC-CH_2-CH(COO^-)-CHOH-COO^-+ H^+	- 9.6	2-Oxo-glutarate	Isocitrate dehydrogenase
7. CH_3-CO-COO^-+CO_2+ATP ⟶ ^-OOC-CH_2-CO-COO^-+H^++ADP+P_i	0.1	Pyruvate	Pyruvate carboxylase
8. CH_2= C(O-PO_3^{2-})-COO^- + HCO_3^- ⟶ ^-OOC-CH_2-CO-COO^-+P_i	- 19.7	Phosphoenol-pyruvate (PEP)	PEP carboxylase
9. PEP + HCO_3^- + P_i ⟶ ^-OOC-CH_2-CO-COO^-+ PP_i	+ 2.3	Phosphoenol-pyruvate (PEP)	PEP carboxykinase (PP)

chemiosmotic mechanism. It is mostly unknown how the fixation of CO_2 and the endergonic reductions to the formaldehyde oxidation state are brought about. Only the enzyme system catalyzing reaction 8 has been studied in more detail. For a review on the biochemistry of CO_2 reduction to CH_4 see [3].

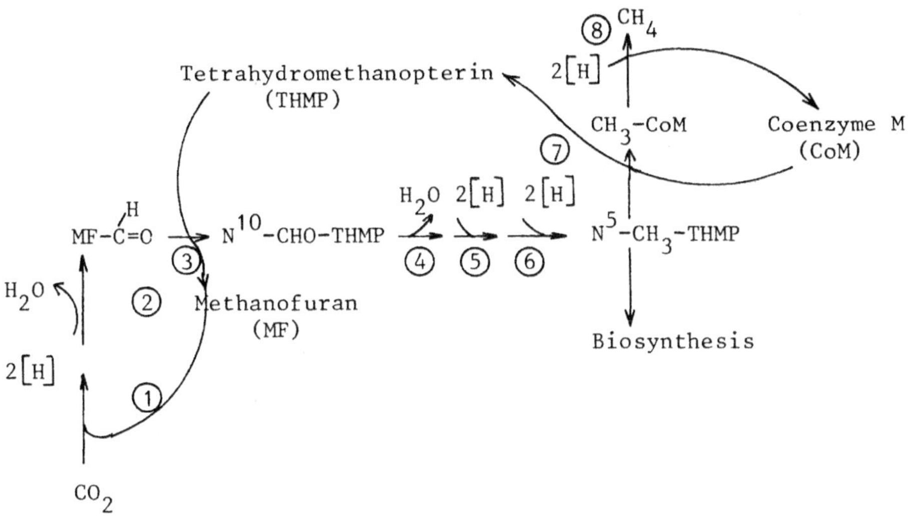

Fig. 2 Proposed reactions ① - ⑧ of microbial CO_2 reduction to CH_4 catalyzed by methanogenic bacteria. The reducing equivalents [H] for methane formation come from H_2, which is an important intermediate in the anaerobic breakdown of organic compounds (see Fig. 1). It is unknown, how the endergonic reactions ① - ③ are coupled with the exergonic reactions ⑥ - ⑧, and how energy conservation (net ATP synthesis) is achieved. Biosynthesis of cell compounds starts from N^5-CH_3-THMP and uses the same principle as in acetogenic bacteria (Fig. 3).

CO_2 REDUCTION TO CH_3COOH CATALYZED BY ACETOGENIC BACTERIA

Biological acetate formation from CO_2 and H_2 uses two different CO_2 fixation and reduction mechanisms (Fig. 3). The carboxyl of acetate is derived from one CO_2 wich is reduced to an enzyme-bound carbon monoxide. The highly interesting enzyme CO dehydrogenase catalyzing this reaction contains nickel and iron in the active

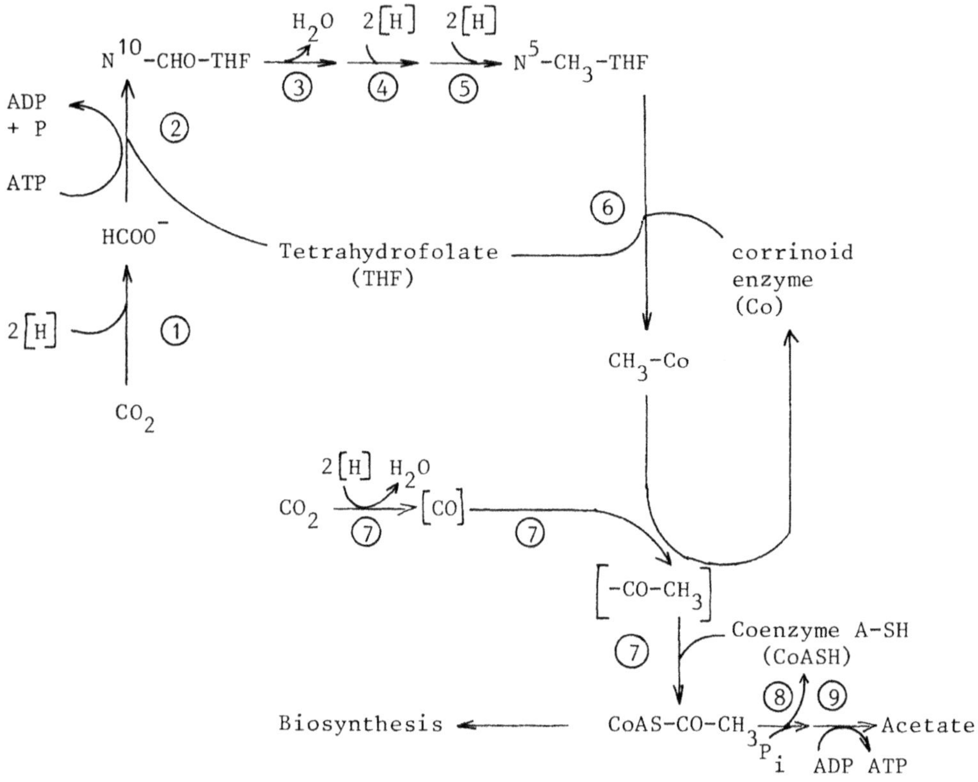

Fig. 3 Proposed reactions ①- ⑨ of microbial CO_2 reduction to CH_3COOH catalyzed by acetogenic bacteria. The reducing equivalents for acetate formation either come from H_2 or from the oxidation of e.g. glucose to 2 acetate, 2 CO_2, and 8 reducing equivalents performed by the acetogenic bacterium itself. It is unknown how energy conservation (net ATP synthesis) is achieved.

center and is presently being studied in several laboratories (see reaction 2 in Table I). This is the first example of carbon monoxide being a natural intermediate in biosynthesis. The methyl of acetate is derived from another CO_2. CO_2 is reduced to free formate, catalyzed by the Mo, Fe/S-or W, Fe/S-enzyme formate dehydrogenase (reaction 1 in Table I). Formate then is added to a tetrahydropteridin in an endergonic reaction which is stoichiometrically coupled with exergonic ATP hydrolysis to ADP + P. The reduction of the formyl to the methyl requires two $2e^-$ reductions. The methyl of CH_3-tetrahydropteridin is transfered via a CH_3Co-enzyme to the carbonyl-carrying nickel enzyme CO dehydrogenase and combined to an acetyl. The acetyl is released by thiolysis (rather than hydrolysis) using coenzyme A to give the thioester acetyl CoA. The whole enzyme complex consists of several peptides which remain to be studied in detail. Biosynthesis starts from acetyl CoA. However, most of the acetyl CoA is converted via the intermediate acetyl phosphate to acetate which is stoichiometrically coupled with ATP synthesis. Since one ATP is required for formate activation, no net ATP gain would be possible. Probably, methylene reduction to methyl is coupled to additional ATP synthesis via a chemiosmotic mechanism. For reviews on the biochemistry of CO_2 reduction to CH_3COOH see [2, 4-7]. For energy conservation in anaerobic bacteria see [8].

AUTOTROPHIC CO_2 FIXATION AND HETEROTROPHIC CO_2 FIXATION

Many anaerobic bacteria can use CO_2 as the sole or principal carbon source for biosynthesis, irrespective of their mode of energy conservation (autotrophic bacteria). None of these organisms uses the CO_2 fixation mechanism of green plants and of most autotrophic aerobic bacteria, ribulose 1,5-bisphosphate carboxylase reaction. Two different pathways both leading to a synthesis of acetyl CoA from 2 CO_2 are found. Acetyl CoA may be formed in a reductive citric acid cycle (see the contribution by Sirevåg, this volume). A second possibility to synthesize acetyl CoA is via the acetyl CoA pathway. In these bacteria 2 CO_2 are assimilated via a bound carbon monoxide and a methyl tetrahydropteridin, as indicated in Fig. 3. For reviews on autotrophic CO_2 fixation by anaerobic bacteria see [5, 9].

Heterotrophic anaerobes depend on a preformed organic compound as the main source of cell carbon, e.g. acetate. In order to assimilate these compounds additional CO_2 fixation reactions are required. Also, the autotrophic anaerobes require carboxylations for biosynthesis starting from acetyl CoA. A selection of these reactions is given in Table I, reactions 4-9 [10]. Among them, the reductive carboxylation of coenzyme A thioesters of carboxylic acids to the corresponding 2-oxoacids and coenzyme A are noteworthy [11, 12].

COMPARISON OF CHEMICAL AND BIOLOGICAL SYNTHESIS OF ACETATE FROM CO_2 AND H_2

The last steps of CO_2 reduction with H_2 to acetate by acetogenic and autotrophic bacteria has very much in common with industrial acetate synthesis, e.g. by the Monsanto-process (Fig. 4), or with other Reppe syntheses [13, 14 .

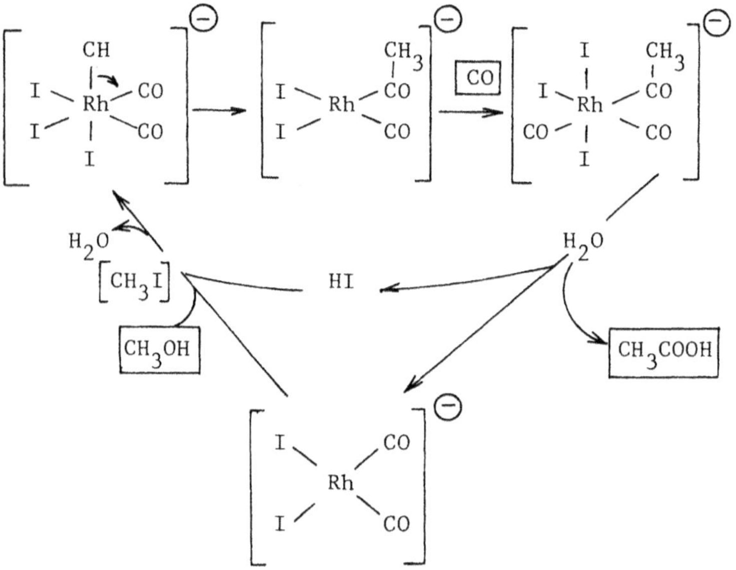

Fig. 4 Proposed catalytic reaction cycle in rhodium-catalyzed industrial acetate synthesis by methanol carbonylation (from [13], modified). The reaction is promoted by catalytic amounts of HI reacting with methanol to methyl iodide. Methyl iodide generates an initial metal-carbon bond with a metal carbonyl-containing species ('oxidative addition'), immediately followed by formation of an ('energy-rich') acetyl complex which decomposes to acetyl iodide. With hydroxylic solvents, acetyl iodide is rapidly solvolyzed to acetic acid, regenerating HI. For comparison see Fig. 3.

Both processes start from methanol and carbon monoxide, make use of transition metals as catalytic principle, form a metal carbonyl, and involve catalytic amounts of compounds producing a CH_3^+. It will be extremely interesting to understand the catalytic mechanism of the enzyme complex CO dehydrogenase involved in this reaction. Possibly this will give new impetus in the search for improved chemical catalysts. The overall synthesis from CO_2 and H_2 involves a dozen or so enzymes, many coenzymes, the expenditure of energy by ATP hydrolysis, and bound intermediates. By these means the endergonic CO_2 reduction to the formaldehyde and CO level is overcome, and the whole process is conducted such that even net ATP is synthesized. In contrast, the chemical approach uses one catalyst, free intermediates and extreme reaction conditions to facilitate the endergonic CO_2 reduction to the formaldehyde and CO level (Fig. 5). Aigain, the study of the biological system may give new directions to the strategy of chemical CO_2 reduction processes.

Chemical Synthesis	Eo' (mV)	Biological Synthesis	Eo' (mV)
$CO_2 \xrightarrow{H_2} HCOOH$	-420	$CO_2 \xrightarrow{H_2} \xrightarrow[R]{ATP} R\text{-}C(=O)H$	-420
$HCOOH \xrightarrow{H_2} HCHO$	-540	$R\text{-}C(=O)H \xrightarrow{H_2} R\text{-}CH_2OH$	-295
$HCHO \xrightarrow{H_2} CH_3OH$	-185	$R\text{-}CH_2OH \xrightarrow{H_2} R\text{-}CH_3$	-120
$CO_2 \xrightarrow{H_2} CO$	-540	$CO_2 \xrightarrow[Energy]{H_2} CO\text{-}[Ni]$	< -540
$CO + CH_3OH \longrightarrow CH_3COOH$		$[Ni]\text{-}CO + R\text{-}CH_3$ $\longrightarrow [\text{-}CO\text{-}CH_3] \xrightarrow{CoASH}$ $CoAS\text{-}CO\text{-}CH_3$ (→ ATP) CH_3COOH Biosynthesis	

Fig. 5 Comparison of chemical and biological synthesis of acetate from CO_2 and H_2.

References

[1] Balch, W.E., Fox, G.E., Magrum, L.J., Woese, C.R., Wolfe, R.S. (1979) Methanogens: Reevaluation of a Unique Biological Group. Microbiol.Rev., 43, 260-296.

[2] Ljungdahl, L.G., Wood, H.G. (1982) Acetate Biosynthesis. In: B_{12} (D. Dolphin, ed.), Vol. 2, pp. 165-202. Wiley & Sons, New York.

[3] Wolfe, R.S. (1985) Unusual coenzymes of methanogenesis. TIBS, 10, 396-399.

[4] Wood, H.G., Ragsdale, S.W., Pezacka, E. (1986) The acetyl-coA pathway of autotrophic growth.(CO dehydrogenase, acetogenesis, methanogenesis, sulfate-reducers). FEMS Microbiol.Rev., in press.

[5] Fuchs, G. (1986) CO_2 fixation in acetogenic bacteria: variations on a theme. FEMS Microbiol.Rev. 39, 181-213.

[6] Diekert, G., Fuchs, G., Thauer, R.K. (1985) Properties and function of carbon monoxide dehydrogenase form anaerobic bacteria. In: Microbial gas metabolism: mechanistic, metabolic and biotechnological aspects (R.K. Poole, C.S. Dow, eds.) pp. 115-129. Academic Press London.

[7] Zeikus, J.G. (1983) Metabolism of one-carbon compounds by chemotrophic anaerobes. Adv. Microb. Physiol. 24, 215-299.

[8] Thauer, R.K., Jungermann, K., Decker, K. (1977) Energy conservation in chemotrophic anaerobic bacteria. Bacteriol.Rev. 41, 100-180.

[9] Fuchs, G., Stupperich, E. (1986) Carbon assimilation pathways in archaebacteria. System.Appl.Microbiol. 7, 364-369.

[10] Boyer, P.D. (ed) The Enzymes, 3. edition. Academic Press New York (1974).

[11] Kerscher, L., Oesterhelt, D. (1982) Pyruvate ferredoxin oxido-reductase - new findings on an ancient enzyme. Trends Bioch. Sci. 7, 371-374.

[12] Evans, M.C.W., Buchanan B.B., Arnon, D.I. (1966) A new ferredoxin-dependent carbon reduction cycle in a photosynthetic bacterium. Proc.Natl.Acad.Sci. (USA) 55, 928-933.

[13] Forster, D. (1979) Mechanistic pathways in the catalytic carbonylation of methanol by rhodium and iridium complexes. Adv. Organomet. Chem. 17, 255-267.

[14] Mullen, A. (1980) Carbonylations catalyzed by metal carbonyls-Reppe reactions. In: New synthesis with carbon monoxide (Falbe, J. ed) pp. 286-290. Springer Berlin.

STOICHIOMETRIC REACTIONS OF C-C BOND FORMATION PROMOTED BY METAL SYSTEMS

Heinz Hoberg
Max-Planck-Institut für Kohlenforschung
Kaiser-Wilhelm-Platz 1
D-4330 Mülheim a. d. Ruhr, Germany

ABSTRACT. Carbon dioxide can be so simply activated by both (Lig)Ni(O) and (Lig)Fe(O) complexes that it already undergoes coupling reactions with π-systems under mild conditions. Metalla-five-membered rings are obtained in good yields by 1,1 coupling. Of particular interest are the C-C coupling products obtained by reaction of CO_2 with alkynes, alkenes and 1,2- or 1,3-dienes. Metallorganic complexes are obtained in this novel route from CO_2 and π-systems. This means that these compounds contain a metal- (Ni-C or Fe-C) σ-bond. Some properties of five-membered ring complexes are discussed and their use for preparative chemistry is illustrated. Further, the numerous ways in which the P- or N-ligands (mono- and bidentate) can effect the reactivity of the metal C-σ bond. Some model reactions for catalyses which have not yet been realized are also shown.

In view of its ready availability, there has been a great increase in recent years in the use of CO_2 as a starting material in preparative chemistry. Above all, reactions involving transition metal compounds have gained in significance, since in contrast to compounds of the main group metals, transition metal complexes can also act as catalysts. Some important review articles on this topic are to be found in (1).

It has been known for a long time that the electrophilic molecule CO_2 can react with metal carbon σ-bonds <u>via</u> insertion with C-C coupling to form the corresponding carboxylates. This reaction can take place with organometallic compounds of both the main group metals and transition metals.

M. Aresta and G. Forti (eds.), Carbon Dioxide as a Source of Carbon, 275–293.

$$\text{M-R} + CO_2 \longrightarrow \text{M-O-CO-R}$$

Those reactions known since the introduction of Grignard reagents will not be discussed further here.

Carbon dioxide is linear in its ground state with different possible forms arising from mesomerism. The electrophilic nature of CO_2 results from the difference between the electronegativities of oxygen and carbon. These factors also lead to the different modes of complexation of CO_2 in transition metal complexes (Fig. 1).

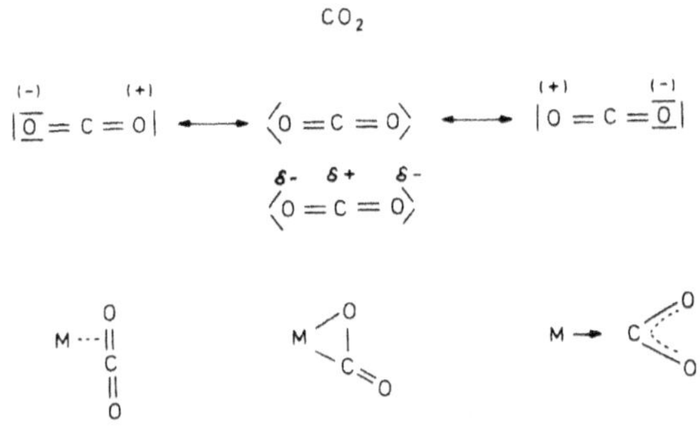

Fig. 1: Mesomerism and different modes of complexation of CO_2 in transition metal complexes

In the last few years a number of compounds have been prepared showing that CO_2 can form stable complexes with transition metals. In some cases the complexes have been characterized by X-ray crystallography. In most of these compounds the CO_2 is not present in the linear form of the ground state but is found in a bent or 'activated' form (Fig. 2).

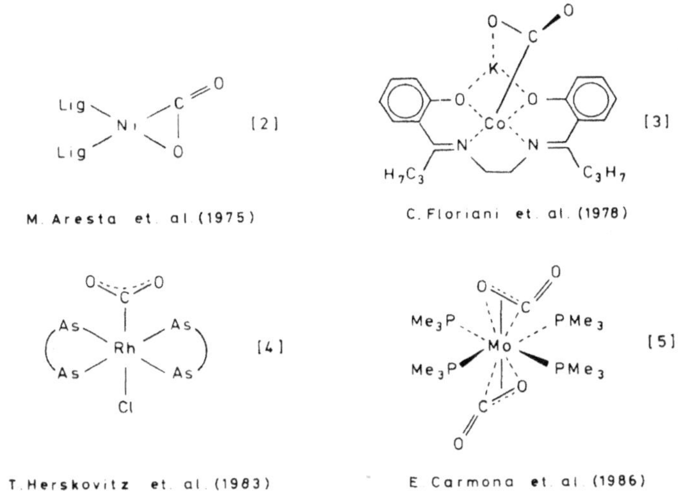

Fig. 2: Some CO_2 transition metal complexes

Depending upon the particular transition metal involved, its oxidation state and the type of stabilizing ligands to which it is bonded (e.g., mono- or bidentate, etc.), the complexed CO_2 can be activated such that it undergoes further reaction with suitable substrates.

Reaction of Ir(I) complexes with two molecules of CO_2 leads to C-O coupling to form the very stable five-membered metallacyclic iridia-complex; the structure of this compound has been confirmed by X-ray analysis (6).

(Lig)Ni(O) complexes (e.g., Lig = BIPY, TMEDA) are able to activate the CO_2 to such an extent that the intermediate pentacyclic compound is not isolated and its immediate 'disproportionation' to form the carbonate and carbon monoxide takes place (7) (Fig. 3).

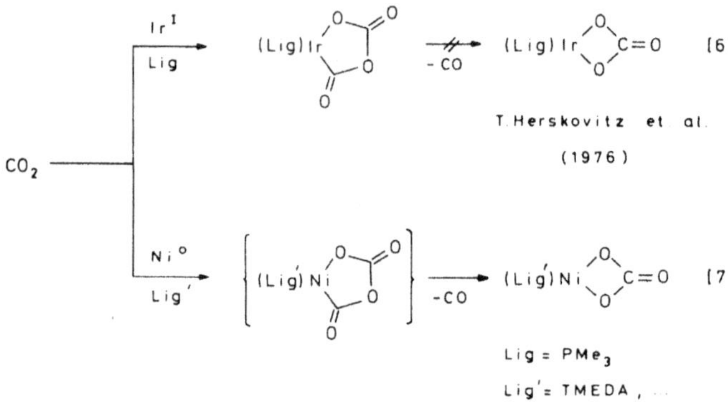

Fig. 3: Reaction of CO_2 on different transition metal complexes

The five-membered ring is apparently formed by the strongly activated molecule of carbon dioxide complexed to Ni(O) reacting with a second molecule of CO_2 via electron shifts and simultaneous oxidation of the metal.

Based on these findings, a series of novel metalla-5-ring complexes can be prepared in good yield by the reaction of CO_2 with C=N, C=O and N=N-systems on (Lig)Ni(O) compounds. However, it should be emphasized that in these cases the coupling reaction results in the formation of a C-X and not a C-C bond (Fig. 4).

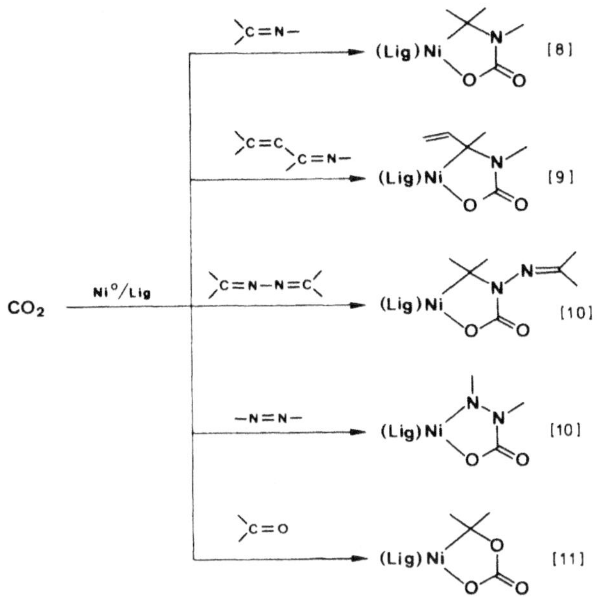

Fig. 4: Nickela-5-ring complexes from CO_2 and C-X π-systems

The discovery of novel types of C-C coupling reactions, particularly those induced by transition metal compounds is, however, the domain of organometallic chemistry. In this context, we have succeeded in proving that it is perfectly possible to bring CO_2 to reaction in this way with C-C multiple bond systems on transition metal complexes such that a 1:1 C-C coupling occurs. Some examples are collected in the following scheme (Fig. 5).

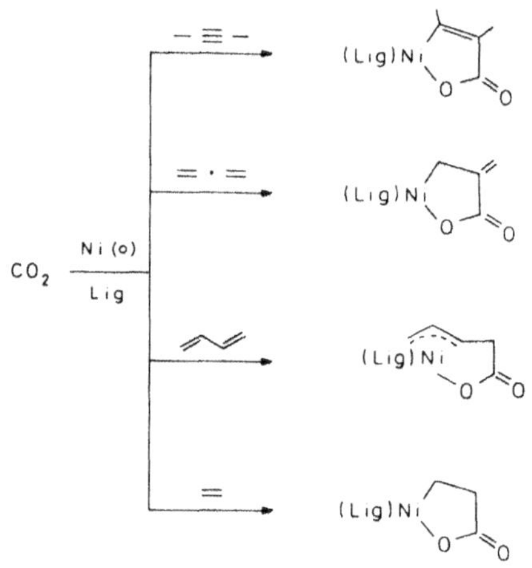

Fig. 5: 1:1 C-C coupling complexes from CO_2 and different C-C π-systems

The starting materials are nickel(O) complexes stabilized by various ligands. In the course of a C-C coupling reaction, a nickela-system with three new σ-bonds is the formed - a Ni-C, a Ni-O and a C-C bond. The two electrons that are needed to produce these bonds are supplied by the metal

$$Ni(O) \longrightarrow Ni^{2+} + 2e.$$

As far as alkynes are concerned, it can be shown that when starting with defined alkyne-Ni(O) or with CO_2-Ni(O) complexes, it is possible to form complexes with five-membered rings. Thus we cannot yet determine whether CO_2 first becomes complexed to the nickel and then reacts with the alkyne to form the metallacycle or whether the sequence of reactions proceeds the other way round (12) (Fig. 6).

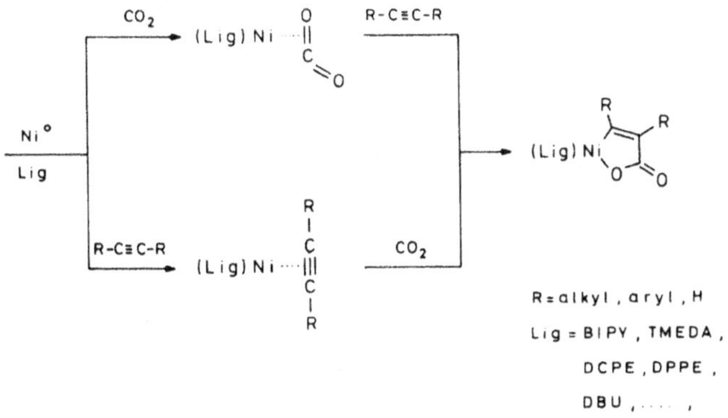

Fig. 6: Different reaction pathways to nickela-5-ring complexes.

These nickela-5-ring complexes can be prepared easily and in good yield from alkynes and CO_2 on (Lig)Ni(O) and are outstanding educts for preparative chemistry (13) (Fig. 7).

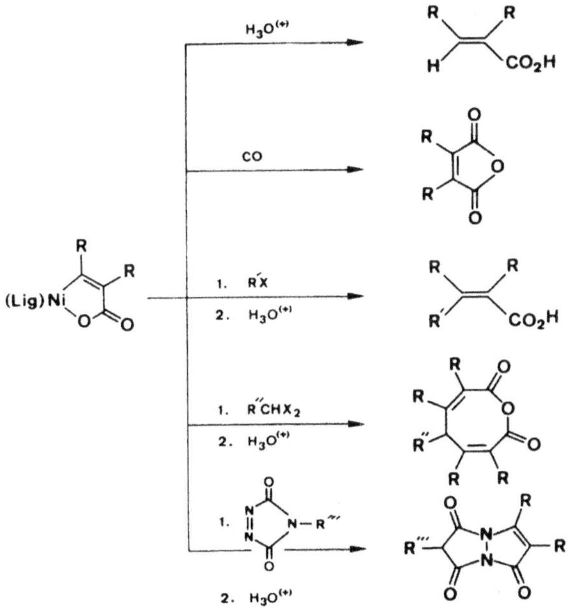

Fig. 7: Nickela-ring complexes are educts for preparative chemistry

In 1977, Inuoe et al. (14) described the Ni(O)-catalyzed preparation of 2-pyrones from alkynes and CO_2. The authors formulated a reaction sequence in which in the first step two alkynes undergo coupling on the metal complexes to produce a nickelacyclopentadiene, into which CO_2 is inserted to give a nickela-7-ring. Reductive elimination then forms the 2-pyrone and releases the (Lig)Ni(O) starting material.

We have been able to show that the oxanickela-5-ring complex formed from alkyne and CO_2 undergoes an insertion reaction with activated alkynes ($R-C_2-R$, R = CF_3 or CO_2CH_3) to form a complex with the seven-membered ring structure suggested by Inuoe (confirmed by X-ray crystallography). Here too, the 2-pyrone is formed on heating this compound, regenerating the starting material. In these studies, we have used the catalytically inactive chelate-forming ligands TMEDA or BIPY. The kinetic studies that are required to confirm this formulated reaction sequence have not yet been carried out (15) (Fig. 8).

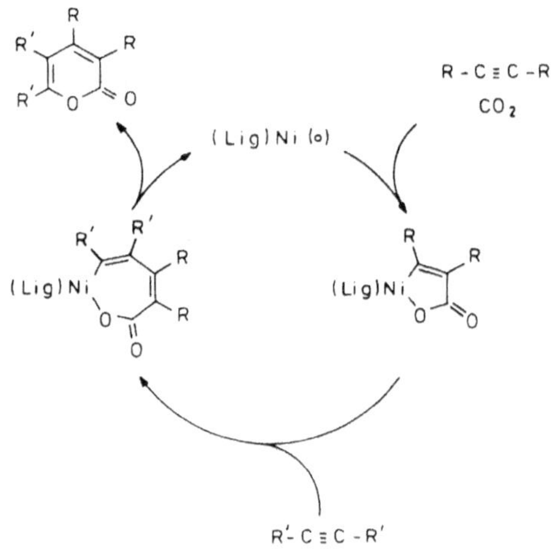

Fig. 8: Reaction pathway for the nickel-catalyzed formation of pyrones

1,2-Dienes also undergo a 1:1 coupling reaction with CO_2 on (Lig)Ni(O) to afford nickela-5-ring compounds in good yield. The coupling reaction always proceeds in such a way that the product has a conjugated system. However, when the 1,2-dienes are substituted, two different 5-ring complexes are produced (16) (Fig. 9).

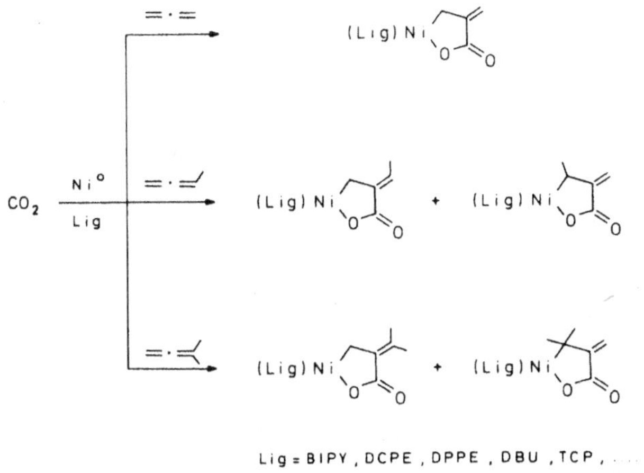

Fig. 9: Different 1:1 C-C coupling products from CO_2 and 1,2-dienes

These novel nickela-5-ring molecules are very inactive towards electrophiles. Obviously the nucleophilicity of the α-C-atom is too low for reaction to occur. However, the reactivity of the Ni-C σ-bond is strongly influenced by the ligands present and by the introduction of Cu(I) salts. Under these conditions, γ-lactones are obtained in good yield on addition of aldehydes (18) (Fig. 10).

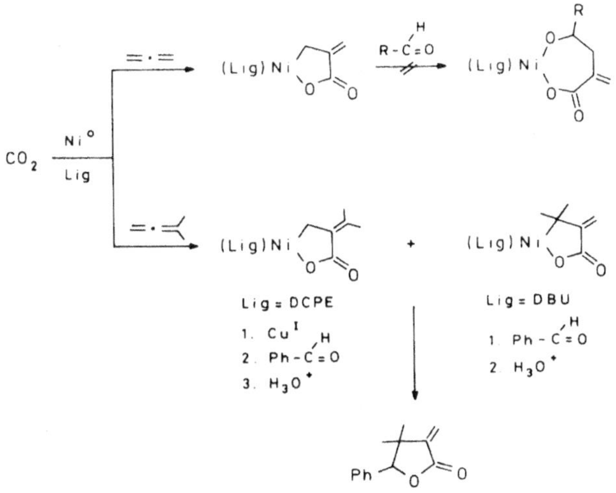

Fig. 10: Formation of γ-lactones from nickela-5-rings and aldehydes

Particularly interesting reactions occur when 1,3-dienes are brought to reaction with CO_2 on (Lig)Ni(O)-systems. It should be emphasized that the 1:1 coupling reaction always proceeds such that the thermodynamically more stable η^3-allyl system is formed. The C-C coupling step preceeds in the same way in the case of substituted 1,3-dienes (18). The presence of a η^3-allyl system on reaction of 2,3-dimethyl-1,3-butadiene was confirmed by an X-ray study by D. Walther et al. (19).

The hydrolysis of these novel products of C-C coupling leads to the expected formation of mono-unsaturated carboxylic acids, whose structures, of course, can be deduced from the structures of the η^1-allyl complexes with which coupled products are in equilibrium.

On reaction with further CO_2 under more drastic conditions or on addition of pyridine, α,ω-diacids are produced in good yield. No evidence is found of an insertion of CO_2 into the Ni-C σ-bond of the η^1-compound. In that case, 1,2-diacids would be formed. Highly interesting reactions occur when the product of the coupling reaction between the 1,3-diene and CO_2 is treated with maleic anhydride. A reductive elimination takes place without hydrolysis with the formation of conjugated diene carboxylic acids (20) (Fig. 11).

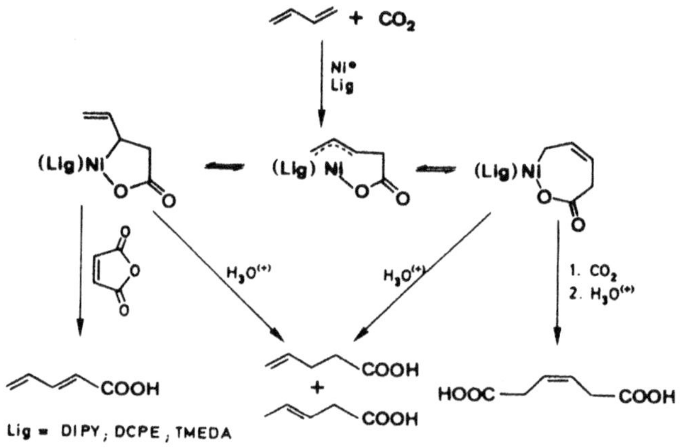

Fig. 11: Formation of mono- and diacids from CO_2 and 1,3-dienes on (Lig)Ni(O)-systems

This reaction is particularly significant since we found that sorbic acid can be prepared this way from piperylene and CO_2. Sorbic acids is, of course, important in the food industry as a preservative (21).

This nickel-induced reaction of piperylene and CO_2 to give sorbic acid is thus the basic reaction of a potential transition metal catalysis, but unfortunately we have not yet succeeded in achieving this catalytic process. We therefore now need to find a (Lig)metal system on which this reaction, which only proceeds stoichiometrically on Ni(O), can be turned into a catalysis. In the case of the reaction of isocyanates with piperylene this has already been achieved (22).

The sequence of reactions can be formulated in two quite different ways (Fig. 12):

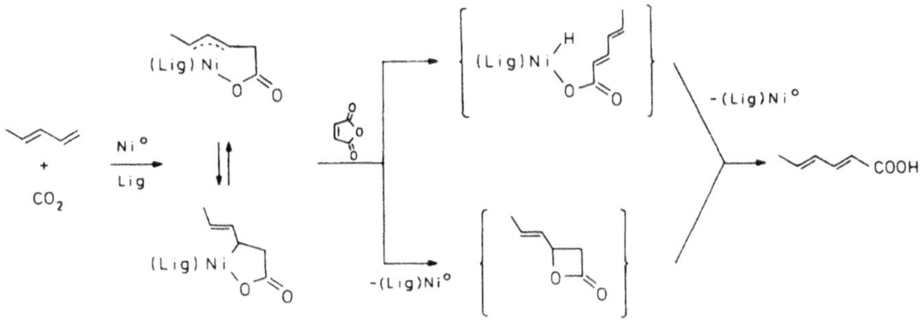

Fig. 12: Different pathways for the formation of sorbic acid

a) First of all a ß-H elimination occurs to give the carboxylate-Ni-H species. Then a reductive elimination liberates the sorbic acid. Thus the (Lig)Ni(O) starting material is regenerated.
b) A reductive elimination takes place to regenerate the (Lig)Ni(O) and form the ß-lactone, which then isomerizes to yield the sorbic acid.

It is not yet possible to say which of these mechanisms is correct.

Alkenes also react with CO_2 on (Lig)Ni(O) to give products of 1:1 C-C coupling reactions. It can be shown for systems containing C-C multiple bonds that the C-C coupling reaction is reversible for ligands like TMEDA, BIPY, DCPE, DPPE and is irreversible for ligands like (alkyl)$_2$N-C_2H_4-P-(alkyl)$_2$. The formation of the cyclic compounds proceeds under kinetic control and the rate of reaction to the thermodynamically more stable final product depends upon the temperature and the reaction time. Complexes with a $Ni-C_{prim}$ bond are thermodynamically more stable than those with a $Ni-C_{sec}$ bond.

In agreement with the reversibility of the reactions of formation discussed above, it is possible to replace the alkene which has undergone coupling to form a 5-membered ring metallacycle by other alkenes or alkynes.

The 5-ring complexes derived from alkenes and CO_2 eliminate CO_2 on heating. The corresponding (Lig)Ni(O)-alkene complex can be isolated as crystals. This means that the alkene is more tightly bound to the Ni than CO_2 (23) (Fig. 13).

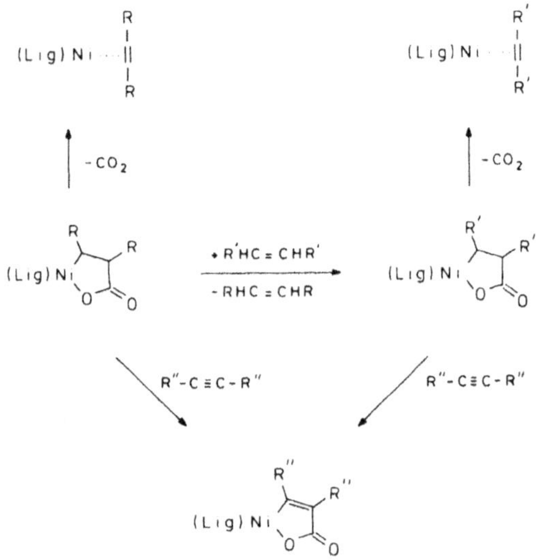

Fig. 13: Reversibility of the C-C coupling reaction between CO_2 and C-C π-systems on (Li)Ni(O)-complexes

Before turning to some very interesting reactions of alkenes with CO_2, it should be mentioned that it is also possible to react polyenes such as cyclooctatetraene and bicycloheptadiene with CO_2 on Ni(O) to form carboxylic acids (24).

The reaction of styrene with CO_2 on (Lig)Ni(O) can be carried out such that the 1:1 C-C coupling is followed by a ß-elimination. On hydrolysis, cinnamic acid is then produced in good yield (25) (Fig. 14). E. Carmona et al. have described an analogous coupling reaction between ethylene and CO_2 which takes place on W(O) complexes (26).

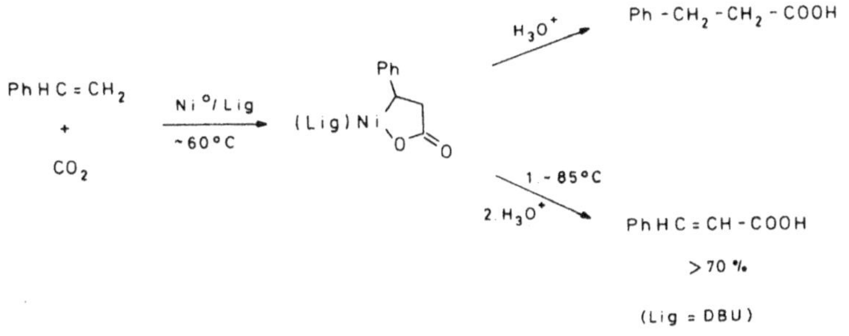

Fig. 14: Cinnamic acid from CO_2 and styrene on (Lig)Ni(O)

Two alternative reaction mechanisms leading to cinnamic acid are given in the next reaction scheme. At the present time it is not possible to decide which is the correct one (25).

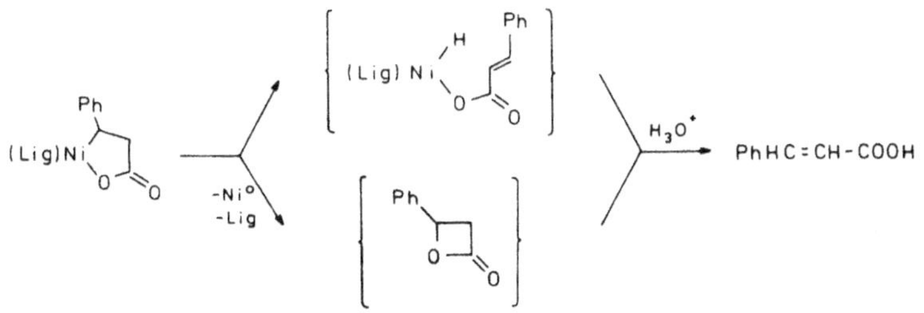

Fig. 15: Alternative reaction mechanism to cinnamic acid

With certain (Lig)Ni(O) systems ethylene can be brought to a stoichiometric reaction with CO_2 to produce unsaturated C_5-acids in over 90 % yield (relative to Ni(O)) (27) (Fig. 16).

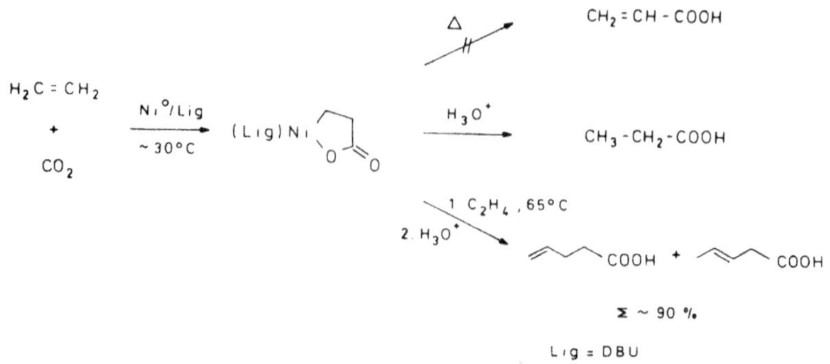

Fig. 16: Formation of n-pentenoic acids from ethene and CO_2

We propose the following sequence of reactions to explain the formation of C_5-acids. First of all, a 1:1 C-C coupling reaction takes place to produce a nickela-5-ring (these compounds have been isolated). Under more drastic conditions, insertion of an additional molecule of ethylene occurs to form the nickela-7-ring. This can then yield the unsaturated carboxylic acid in two different ways (Fig. 17).

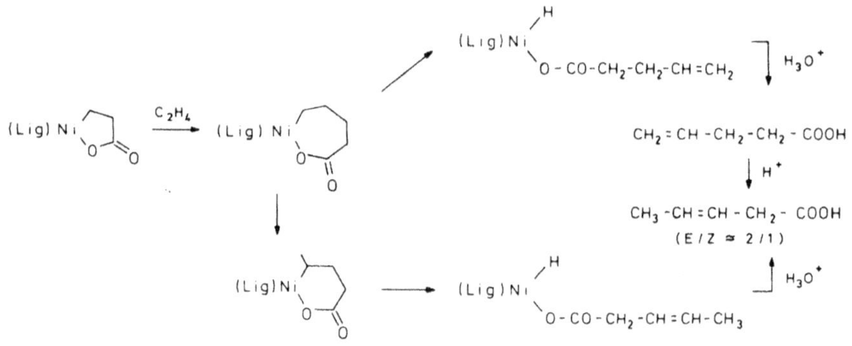

Fig. 17: Different reaction pathways for the formation of C_5-acids

The reaction of ethylene with isocyanates can be catalyzed by (Lig)Ni(O) to produce acrylic or n-pentanoic amides (28).

The results discussed above show that CO_2 can undergo C-C coupling reactions with C-C multiple bond systems on (Lig)Ni(O)-complexes. We have extended our studies of these novel C-C coupling reactions by examining the effect of Fe(O) complexes; we report here the first results.

1,3-Butadiene coordinated to Fe(O) reacts cleanly with CO_2 to form a carboxylate. The X-ray structural analysis shows that this compound, like the corresponding nickel carboxylates, contains a η^3-allyl structure (Fig. 18). The geometry around the metal is a trigonal bipyramid. The NMR spectra indicate that in solution a dynamic system is present, which becomes 'frozen out' at -80°.

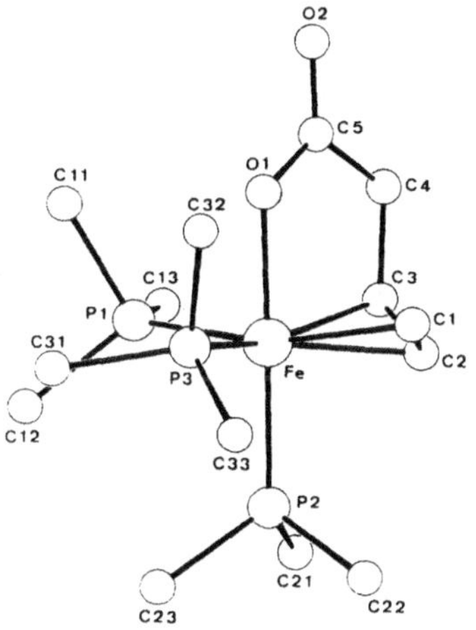

Fig. 18: X-ray structure of the Fe-carboxylate complex from 1,3-diene and CO_2

The further reactions of this Fe-carboxylate complex can be predicted, as expected, on the basis of the ^{1}-allyl structure. On hydrolysis the unsaturated carboxylic acid is produced and reaction with CO_2 leads to insertion in the Ni-C -bond to form the dicarboxylate. Hydrolysis of the compound gives , -dicarboxylic acids.

An unexpected reaction, which differs from that on the corresponding nickel complexes, is found when monocarboxylates react with $FeCl_3$. On hydrolysis an α,ω-dicarboxylic acid is formed in which two moles of the 1,3-diene have undergone coupling with two moles of CO_2 (29) (Fig. 19).

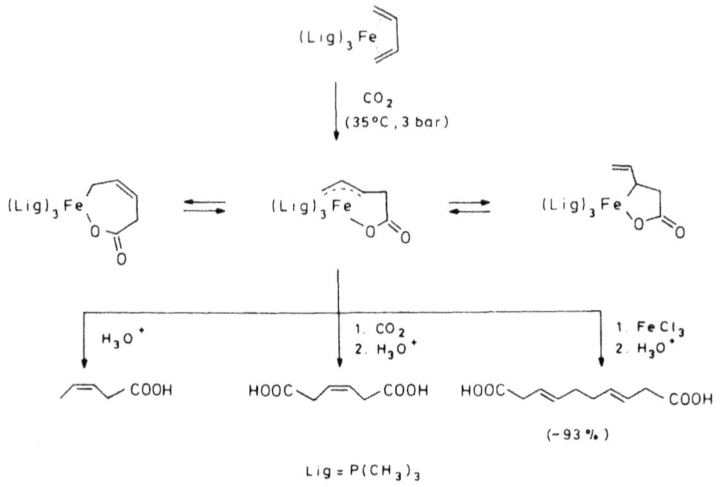

Fig. 19: Several mono- and dicarboxylic acids from 1,3-diene and CO_2 with Fe-complexes

These C-C coupling reactions between CO_2 and C-C multiple bond systems induced by Fe(O), which are reported here for the first time, differ from the corresponding reactions on Ni(O) in the following ways:

a) 1,3-Butadiene reacts with CO_2 on (Lig)Fe(O) with $P(CH_3)_3$ and $P(C_2H_5)_3$ ligands to give the mono-carboxylate. With excess CO_2 the dicarboxylates are then formed. However, with $P(CH_3)_3$ the predominant product is the 1,4-dicarboxylate, while with $P(C_2H_5)_3$ more of the 1,2-dicarboxylate is formed (30) Fig. 20).

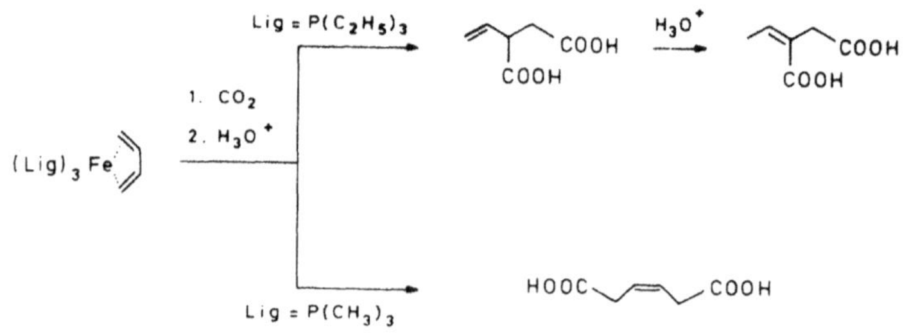

Fig. 20: Different carboxylic acids from CO_2 and 1,3-diene induced by (Lig)Fe(O)-systems

b) In contrast to the corresponding Ni(O) reactions, the C-C coupling reactions on Fe(O) are irreversible. Treatment of piperylene with CO_2 in the presence of R_3P leads to the formation of both carboxylates in a molar ratio of 1:1 in a kinetically controlled reaction (30) (Fig. 21).

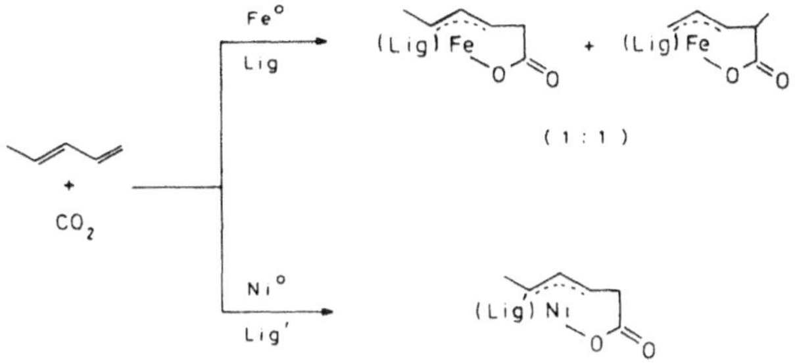

Fig. 21: Reversible and irreversible C-C coupling reactions on Ni(O) and Fe(O) systems

The examples discussed here demonstrate impressively that CO_2 can undergo C-C coupling reactions with various C-C multiple bond systems on both Ni(O) and Fe(O) systems. The novel metal-carboxylate complexes that are thus formed are obtained in good yield and are suitable for a variety of applications in preparative chemistry.

Furthermore, a series of well characterized C-C coupling reactions has been presented which can serve as the basic reactions of catalyses. Based upon these findings, metal-ligand systems now have to be developed which can achieve this in practice.

In my lecture I presented examples of interest as well as developments originating from my group. Many new organometallic pathways have been detected and synthetic application demonstrated. Our research is now developing along the lines which have emerged from testing working hypotheses experimentally. In particular C-C bond formation via catalytic CO_2 activation and syntheses of carbonic acids are being actively investigated.

The aim of this review was to show the organic chemists participating in this meeting, who are not so familiar with organic transition metal chemistry, that homogeneous transition metal complexes can be extremely useful for organic synthesis.

REFERENCES

1) S. Inoue, N. Yamazaki, Organic and Bio-organic Chemistry of Carbon Dioxide, John Wiley & Sons, New York 1982;
R.P.A. Sneeden, in G. WIlkinson (ed.) Comprehensive Organometallic Chemistry, Vol. 8, Pergamon Press, Oxford 1982, p. 225;
D. Walther, E. Dinjus, Zeitschrift für Chemie 23 (1983) 237;
A. Behr, Chem. Ing. Techn. 57 (1985) 839.
2) M. Aresta, C.F. Nobile, V.G. Albano, E. Forni, M. Manessero, J. Chem. Soc., Chem. Commun. 1975, 636;
M. Aresta, C.F. Nobile, J. Chem. Soc. Dalton Trans., 1976, 708.
3) G. Facchinetti, C. Floriani, J. Am. Chem. Soc. 100 (1978) 7405.
4) J.C. Calabrese, T. Herskovitz, J.B. Kinney, J. Am. Chem. Soc. 105 (1983) 5914.
5) R. Alvarez, E. Carmona, J.M. Marin, M.L. Poveda, E. Gutierrez-Puebla, A. Monge, J. Am. Chem. Soc. 108 (1986) 2286.
6) T. Herskovitz, L.J. Guggenberger, J. Am. Chem. Soc. 98 (1976) 1815.
7) G. Burkhart, H. Hoberg, Angew. Chem. 94 (1982) 75; Angew. Chem. Int. Ed. Engl. 21 (1982) 76; Angew. Chem. Suppl. 1982, 147.
8) K. Sümmermann (H. Hoberg), Diplom-Arbeit, Universität Bonn, 1982;
D. Walther, E. Dinjus, E. Sieler, J. Kaiser, J. Organomet. Chem. 240 (1982) 289.
9) K. Sümmermann (H. Hoberg), Diplom-Arbeit, Universität Bonn, 1982;
D. Walther, E. Dinjus, Z. Chemie 24 (1984) 298.
10) K. Sümmermann (H. Hoberg), Diplom-Arbeit, Universität Bonn, 1982.
11) K. Sümmermann (H. Hoberg), Diplom-Arbeit, Universität Bonn, 1982;
E. Dinjus, J. Kaiser, J. Sieler, D. Walther, Z. Anorg. Allg. Chem. 483 (1981) 63.
12) D. Schaefer (H. Hoberg), Dissertation, Ruhr-Universität Bochum, 1984.
13) H. Hoberg, D. Schaefer, G. Burkhart, J. Organomet. Chem. 228 (1982) C 21.
14) S. Inoue, Y. Itoh, H. Hashimoto, Chem. Lett. 1977, 855;
S. Inoue, Y. Itho, H. Kazama, H. Hashimoto, Bull. Chem. Soc. Japan 53 (1980) 3329.
15) H. Hoberg, D. Schaefer, J. Organomet. Chem. 238 (1982) 383.
16) H. Hoberg, B.W. Oster, J. Organomet. Chem. 266 (1984) 321.

17) H. Hoberg, S. Groß, unbpulished results, 1986.
18) H. Hoberg, D. Schaefer, B.W. Oster, J. Organomet. Chem. 266 (1984) 313.
19) D. Walther, E. Dinjus, J. Sieler, N.N. Thanh, W. Schade, J. Lehn, Z. Naturforsch. 38b (1983) 835.
20) H. Hoberg, D. Schaefer, B.W. Oster, J. Organomet. Chem. 266 (1984) 313;
H. Hoberg, B. Apotecher, J. Organomet. Chem. 270 (1984) C 15.
21) H. Hoberg, D. Schaefer, J. Organomet. Chem. 255 (1983) C 15.
22) H. Hoberg, H. Hernandez, Angew. Chem. 97 (1985) 987; Angew. Chem. Int. Ed. Engl. 24 (1985) 961.
23) H. Hoberg, D. Schaefer, G. Burkhart, C. Krüger, M.J. Romao, J. Organomet. Chem. 266 (1984) 203.
24) A. Behr, G. Thelen, C 1 Mol. Chem. 1 (1984) 137.
25) H. Hoberg, Y. Peres, A. Milchereit, J. Organomet. Chem. 307 (1986) C 38.
26) R. Alvarez, E. Carmona, D.J. Cole-Hamilton, A. Galindo, E. Guiterrez-Puebla, A. Monge, M.L. Poveda, C. Ruiz, J. Am. Chem. Soc. 107 (1985) 5529.
27) H. Hoberg, Y. Peres, A. Milchereit, J. Organomet. Chem. 307 (1986) C 41.
28) H. Hoberg, E. Hernandez, J. Chem. Soc., Chem. Commun. 1986, 544.
29) H. Hoberg, K. Jenni, C. Krüger, E. Raabe, Angew. Chem. 98 (1986) in press; Angew. Chem. Int. Ed. Engl. 25 (1986) in press.
30) H. Hoberg, K. Jenni, unpublished results, 1986.

CATALYTIC REACTIONS OF CARBON DIOXIDE WITH HYDROCARBONS

Arno Behr
Institute of Technical Chemistry and Petrochemistry
Aachen University of Technology, Worringer Weg 1,
D-5100 Aachen
Federal Republic of Germany

ABSTRACT. Carbon dioxide can be reacted with a great number of hydrocarbons in the course of which a new carbon-carbon bond is formed. These reactions can in the broadest sense be called carboxylations, in which the whole CO_2 molecule is passed into products such as carboxylic acids, esters, lactones and pyrones. The carboxylation of aromatic compounds is the well-known Kolbe-Schmitt process and a completely analogous reaction is possible using saturated CH-acidic hydrocarbons. Most promissing syntheses, however, are the homogeneous catalyzed reactions of carbon dioxide with unsaturated hydrocarbons, that is, with alkynes, alkenes, methylenecyclopropanes, allenes and 1,3-dienes. These conversions are well controlled by transition metal catalysts, especially by complexes of palladium, rhodium and nickel. Molecules of high complexity can be built up by simple one-stage processes. For instance, propyne yields dimethylpyrone, methylenecyclopropane yields γ-lactones and 1,3-butadiene yields γ-lactones, δ-lactones and esters. These fine chemicals can become of high interest to the pharmaceutics and perfume industries.

1. INTRODUCTION

The syntheses of chemicals using carbon dioxide as raw material are well known reactions. The production of urea, the reaction with ethylene oxide yielding ethylene carbonate and the use of CO_2 in the synthesis of methanol are important examples. However, the utilization of carbon dioxide by chemical routes is, so far, very limited. A new start in CO_2 chemistry was found in the early seventies when several research groups began to study the activation of carbon dioxide by transition metal catalysis [1]. In the following homogeneous transition metal catalyzed reactions of CO_2 with alkynes, olefins and dienes will be discussed, but also some reactions with aromatic compounds and alkanes, which do not require a metal catalyst, will be included.

M. Aresta and G. Forti (eds.), Carbon Dioxide as a Source of Carbon, 295–320.

2. REACTIONS OF CARBON DIOXIDE WITH AROMATICS

The reaction of carbon dioxide with sodium phenolate, the so-called "Kolbe-Schmitt-process", yields salicylic acid, which is widely used in the preparation of pharmaceutics, dyes and pesticides [2]. In 1980 in the USA the production capacity of salicylic acid was 25 000 t/a. In 1874 it was manufactured first by Kolbe; in 1884 Schmitt raised the yield from ca. 50 % to ca. 90 % when he introduced a carbon dioxide pressure of about 5 bar (Equation 1).

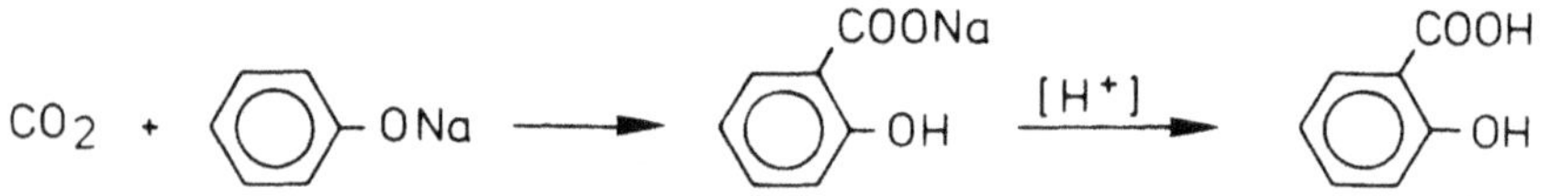

(1)

There are numerous suggestions about the mechanism of the reaction, such as being classified as an electrophilic substitution reaction (Equation 2).

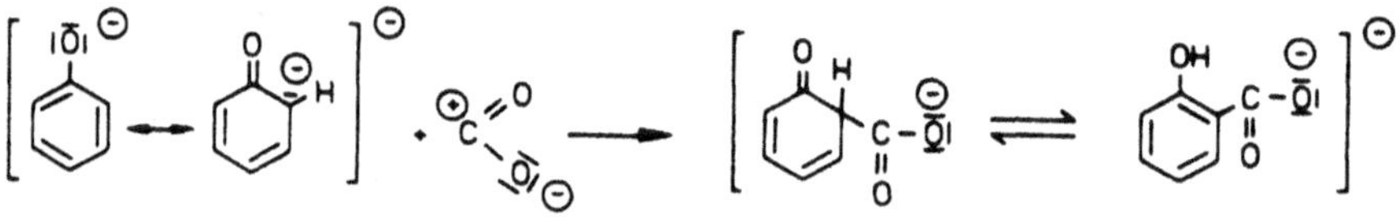
(2)

The reaction conditions and the kind of alkali cation influence production a lot. Sodium phenolate and lithium phenolate produce salicylate at 150°C to 180°C selectively. Potassium phenolate mainly results in p-hydroxybenzoate at 220°C. Hydroxydicarboxylic acids are chiefly produced at a pressure of 100 bar and temperatures above 220°C. Substituted phenolates can also be used in the Kolbe-Schmitt process, substituted groups with electron donor characteristics in particular make carboxylation easier. This means, for example, that aromatic di- and trihydroxy compounds and aminophenols can be transformed particularly easily. The Kolbe-Schmitt reaction is used, too, in the reaction of cresols to cresotic acids.

Although the principle of this reaction has been known for a long time, new research into it is still being published. Particular mention should be made of the numerous and thorough investigations made by Hirao's team [3]. The reaction with 1-naphthol to 1-naphthol-2-carboxylic acid has been described by Yatsuka et al. [4], and the reaction with 2-naphthol to 2-naphthol-3-carboxylic acid by Bachmann et al. [5]. 2-Hydroxy-3-methylpyridine could be reacted into 2-hydroxy-3-methyl-5-pyridine carboxylic acid [6] and 3-hydroxy-pyridine derivatives produced a 70 % yield of the appropriate carboxylic acid [7].

These examples are meant to demonstrate the many ways in which the Kolbe-Schmitt synthesis can be applied. The assumption can therefore be made that this reaction will find further uses in future in chemical technology in producing a variety of fine chemicals in the field of pharmaceutics and plant protection.

3. REACTIONS OF CARBON DIOXIDE WITH ALKANES

Activated alkanes, for instance CH-acidic hydrocarbons react very analogous to aromatic compounds. Bottaccio and Chiusoli [8] reported already in 1966 that such compounds as acetophenone or cyclohexanone react with carbon dioxide in the presence of phenolates to yield carboxylic acids (Equation 3).

$$\text{Ph-}\overset{\text{O}}{\overset{\|}{\text{C}}}\text{-CH}_3 + CO_2 \xrightarrow[\text{DMF}]{\text{PhOK}} \text{Ph-}\overset{\text{O}}{\overset{\|}{\text{C}}}\text{-CH}_2\text{-COOK} \qquad (3)$$

This reaction is of interest because of the moderate reaction conditions and its wide range of application. For example, Equation (4) shows a three-stage preparation of citric acid starting from acetone.

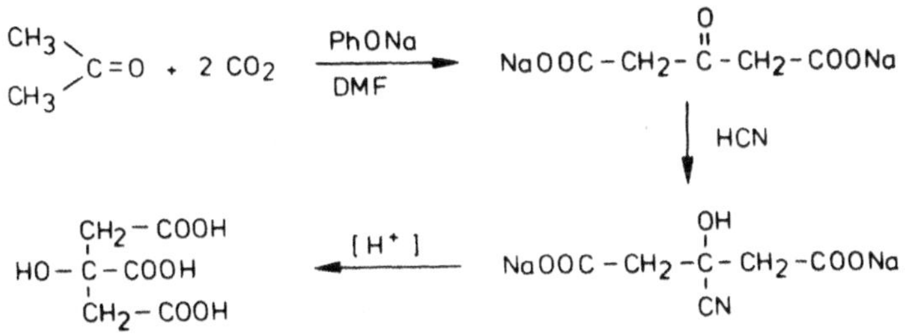

(4)

Non-activated paraffins can be carboxylated only under much more drastic conditions. In a Chevron patent [9], alkanes in the C_{12}- to C_{40}-area are "cracked" in the presence of carbon dioxide at temperatures between 550°C and 650°C and simultaneously converted into short chain carboxylic acids. It is presumed that there is a reaction between CO_2 and alkyl radicals (Equation 5).

$$R-CH_2-CH_3 \xrightarrow[-\ \dot{C}H_2CH_3]{T} R^{\bullet} \xrightarrow[+\ RCH_2CH_3]{+\ CO_2} R-COOH + R\dot{C}HCH_3 \qquad (5)$$

4. REACTIONS OF CARBON DIOXIDE WITH ALKYNES

Already in 1977 Inoue discovered the reaction of hexynes with carbon dioxide catalyzed by nickel complexes [10]. 1-Hexyne reacted with CO_2 in the presence of a nickel/dppe-catalyst to give the butyl-substituted pyrone in a 50 % yield. As by-products the oligomers of 1-hexyne were formed (Equation 6).

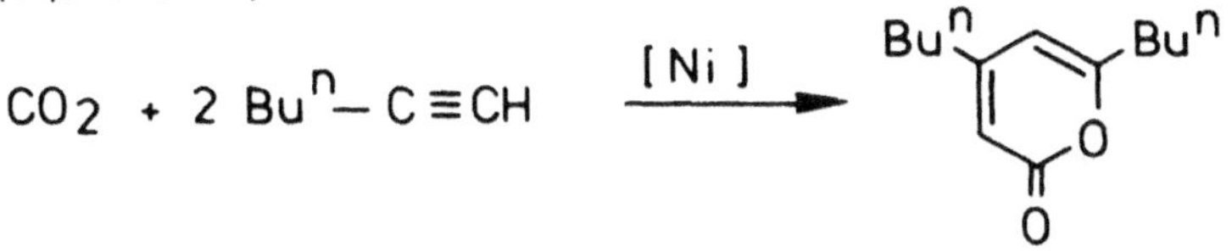

(6)

The mechanism of the nickel-catalyzed reaction was studied intensively by Hoberg [11]. He discussed two different mechanistic pathways which are called cycle A und cycle B in Figure 1. In cycle A the reaction is started by the oxidative coupling of two alkynes yielding the metallacyclopentadiene species into which the CO_2 then inserts. The reductive elimination of the α-pyrone closes the cycle. In the first step of cycle B one alkyne and one CO_2 molecule react with the nickel species, and in step two the second alkyne inserts into the nickel-carbon bond. When the ligand tetramethylethylenediamine is used, a chelate ligand of high basicity, it proved to be possible to isolate both the oxanickelacyclopentene and the oxanickelacycloheptadienone, thus suggesting that cycle B is the catalytic cycle.

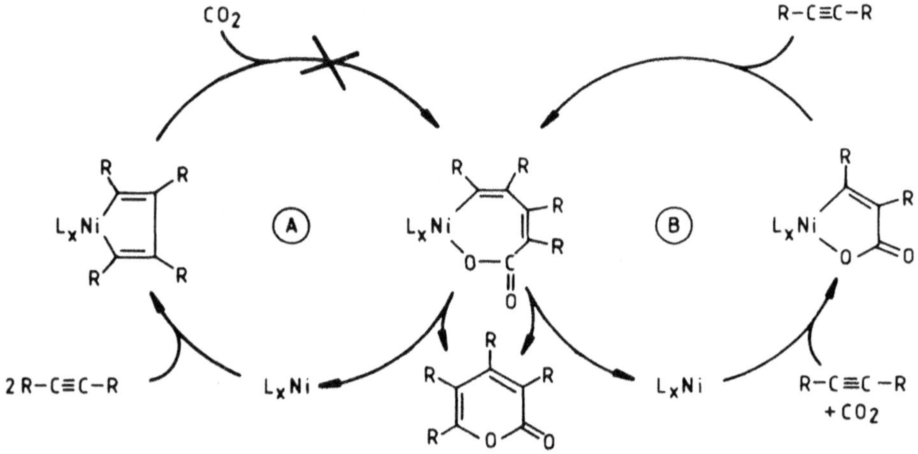

Fig. 1 Discussed mechanism of the nickel-catalyzed reaction of carbon dioxide with alkynes

Aresta and Albano studied the reaction of propyne and CO_2 using the rhodium catalyst Rh(dppe)(η-BPh_4) [12]. They found the formation of the dimethylpyrone, however, only in low yields of about 3 % (Equation 7).

$$2\ CH_3{-}C{\equiv}CH + CO_2 \xrightarrow{[Kat.]}$$ 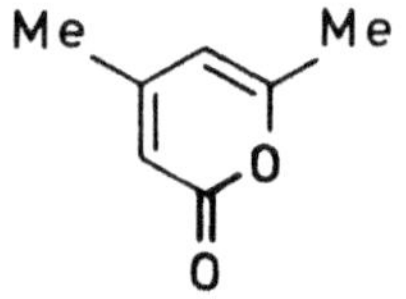 (7)

A mechanism can be envisaged in which non-cyclic intermediates occur (Figure 2).

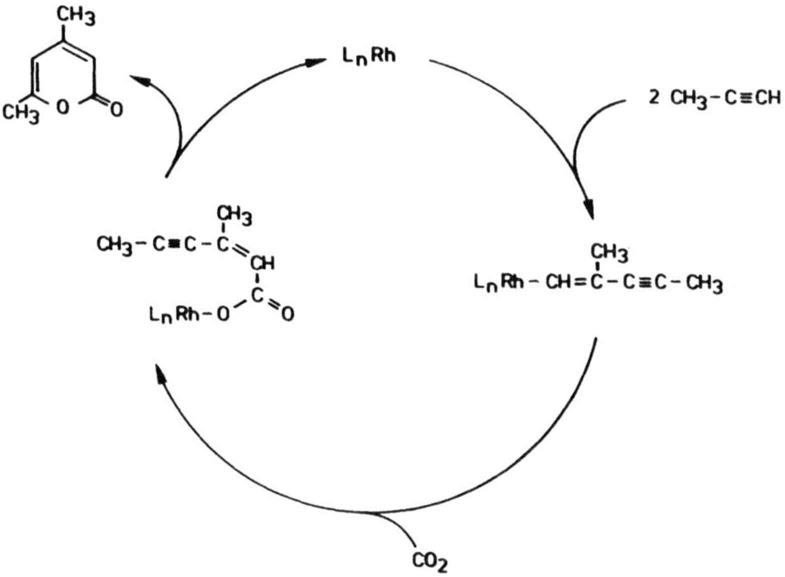

Fig. 2 Discussed mechanism of the rhodium-catalyzed reaction of carbon dioxide with propyne

5. REACTIONS OF CARBON DIOXIDE WITH ALKENES

The reaction between alkenes and carbon dioxide was first mentioned in 1978, when Lapidus discovered the carboxylation of ethene with CO_2 [13]. At high pressures propionic acid and its ethyl ester were formed (Equation 8). In the presence of the Wilkinson catalyst - activated by hydrogen halide promoters - the ethene conversion reached about 90 %. Some typical results are given in Table 1.

$$CH_2{=}CH_2 + CO_2 \xrightarrow{cat.} CH_3CH_2COOH + CH_3CH_2COOEt \quad (8)$$

Table 1 Influence of pressure on the reaction of carbon dioxide with ethene

Pressure (atm)	Conversion (%) relating to		Yields (%)	
	C_2H_4	CO_2	EtCOOH	EtCOOEt
150	36	19	9	20
350	45	38	18	20
600	82	56	25	16
700	92	67	38	11

Another alkene reaction was described by Besecke and Schröder in 1981 [14]. They studied the reaction of propene, carbon dioxide and methanol thus yielding butyric and isobutyric methyl esters. As catalysts they used ruthenium-chloro systems stabilized by triphenylphosphine. However, investigations using labeled methanol proved that the additional carbon of the esters does not originate from carbon dioxide, but from the alcohol [15]. Based on these results, the mechanism shown in Figure 3 is proposed. In the first step, a ruthenium hydride species is formed via reaction of hydrogen - steming from methanol - with the starting complex RuL_x. The catalytic cycle is then initiated by the coordination of propene, which inserts into the metal-hydrogen bond yielding a ruthenium alkyl species. Carbon monoxide, originating from methanol decomposition and probably passing through formaldehyde, initiates the formation of an acyl species, which is than subsequently attacked by further methanol. The catalytic cycle is closed by elimination of the ester, returning to the hydrido-ruthenium complex.

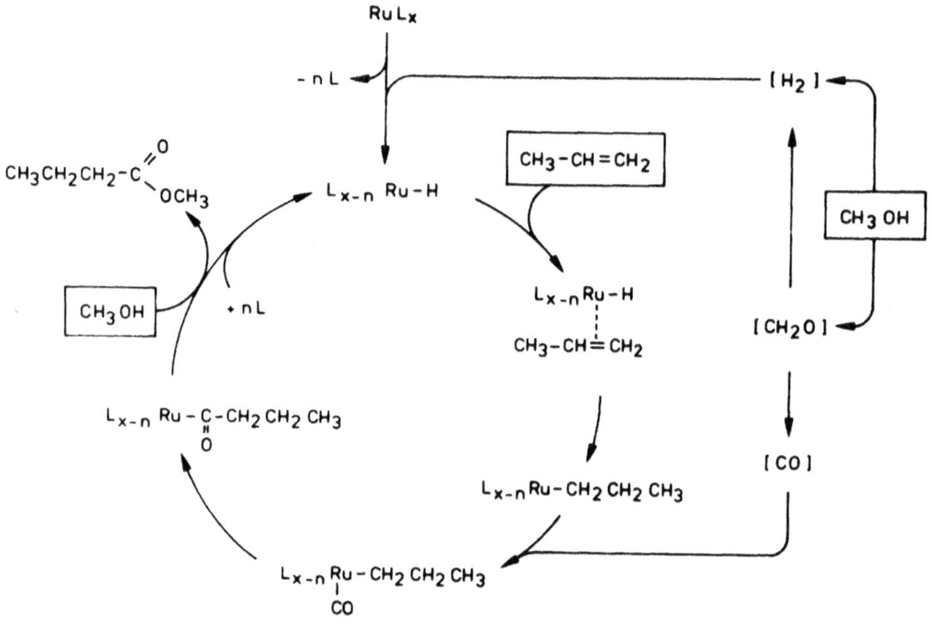

Fig. 3 Discussed mechanism of the ruthenium-catalyzed hydroesterification with methanol as CO/H_2 source

A very promising goal seems to be the connection of a greater number of olefins with carbon dioxide and a first step in this direction has been gone (Equation 9). Using nickel-phenyl complexes, formed by the reaction of nickelbis(cyclooctadiene) and an ylid ligand which are well known catalysts for ethene oligomerization [16], not only the simple insertion of CO_2 into the metal-carbon bond is observed yielding methyl benzoate, but also the insertion of ethene followed by the insertion of CO_2 yielding hydrocinnamic ester could be realized [17]. This reaction could be carried out only stoichiometrically till now. If the catalytic oligomerization of ethene can be combined with the CO_2 insertion step, an interesting synthesis of longer-chained carboxylic acids could be expected.

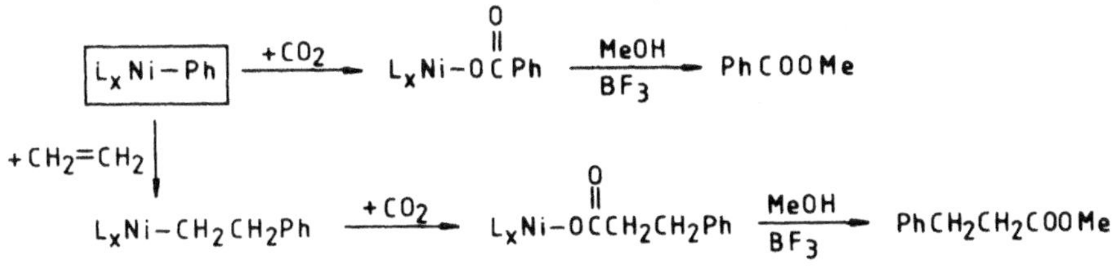

(9)

6. REACTIONS OF CARBON DIOXIDE WITH METHYLENECYCLOPROPANES

In 1979 Y. Inoue et al. found the reaction of CO_2 with substituted methylenecyclopropanes [18]. In the presence of palladium-phosphine catalysts the three-membered ring opened to yield five-membered lactones (Equation 10).

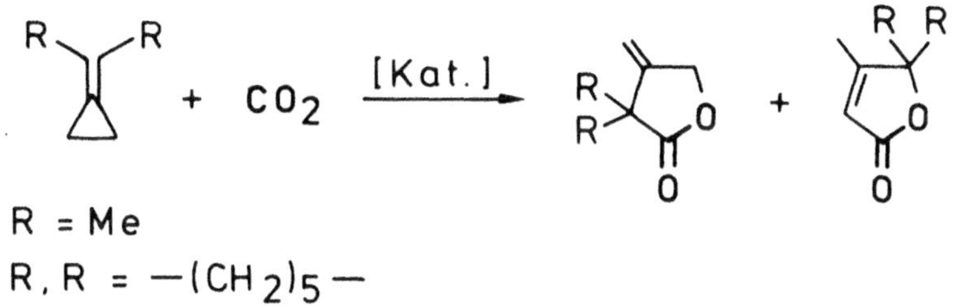

(10)

For instance, isopropylidenecyclopropane reacts with bis(dibenzylideneacetonato)palladium and triphenylphosphine giving the β-methylene lactone in a 69 % yield. When $Pd(dppe)_2$ is used as catalyst the α,β-unsaturated lactone was obtained as the major product, probably derived by an isomerization step. The proposed mechanism contains a trimethylenemethane intermediate which reacts with carbon dioxide yielding a cyclic carboxylate (Figure 4).

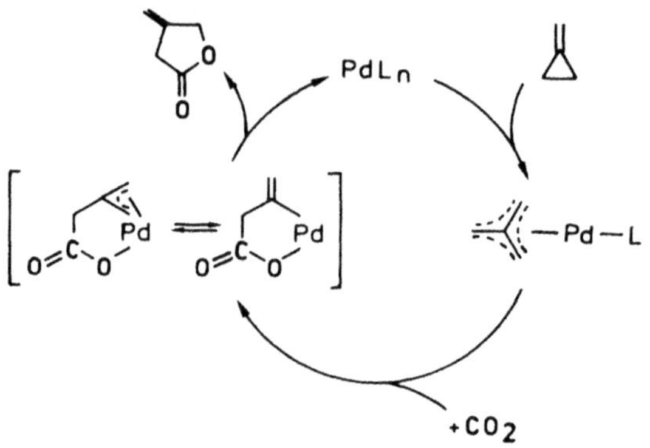

Fig. 4 Discussed mechanism of the palladium-catalyzed reaction of carbon dioxide with methylenecyclopropane

Binger and Weintz [19] investigated the reaction of the unsubstituted methylenecyclopropane with CO_2 and synthesized the 3-methyl-2-buten-4-olide in yields up to 80 %. Besides of this "codimer", also cotrimers, cotetramers and copentamers were formed as by-products (Figure 5).

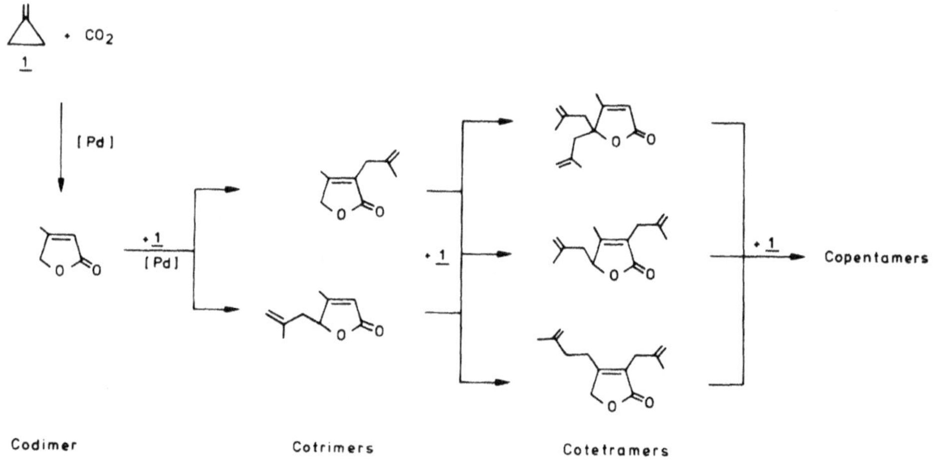

Fig. 5 Products of the reaction of carbon dioxide with methylenecyclopropane

7. REACTIONS OF CARBON DIOXIDE WITH ALLENE

The palladium-catalyzed cooligomerization of allene and CO_2 yields two esters and one pyrone (Equation 11). The pyrone is reminiscent to the product of propyne with carbon dioxide (Equation 7) and suggests that the allene might be initially isomerized to the alkyne. However, comparing experiments with propyne proved that the alkyne does not react in the presence of palladium catalysts [20].

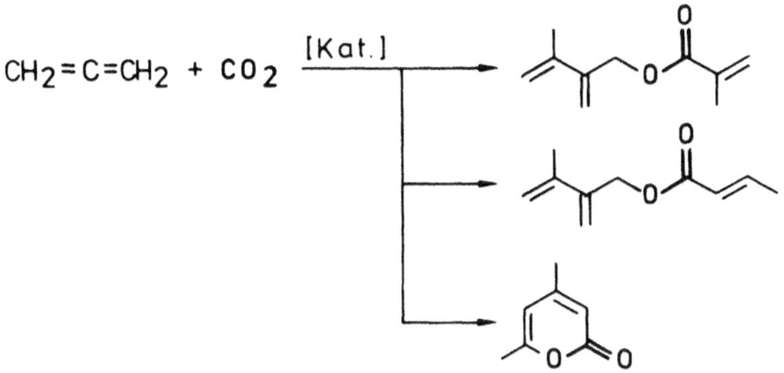

(11)

The most effective catalyst in allene-CO_2-cooligomerization was prepared by treating bis(η^3-allyl)palladium with bis(dicyclohexylphosphino)-ethane, whereas the reaction was less effective in the presence of dppe, PPh_3 or PCy_3. Highest yields (~ 40 %) were obtained with a reaction temperature of 110°C and with toluene as solvent. The reaction proceeds probably via metallacyclic intermediates. Similar cyclic nickelacarboxylate complexes could recently be isolated by Hoberg and Oster [21] in stoichiometric reactions of allenes and CO_2.

8. REACTIONS OF CARBON DIOXIDE WITH BUTADIENE

The reaction of butadiene with CO_2 was already found in 1976 by Inoue who isolated a γ-lactone in yields of about 5 % [22]. In 1978 Musco [23] was able to characterize further reaction products, for instance a δ-lactone and two esters (Equation 12).

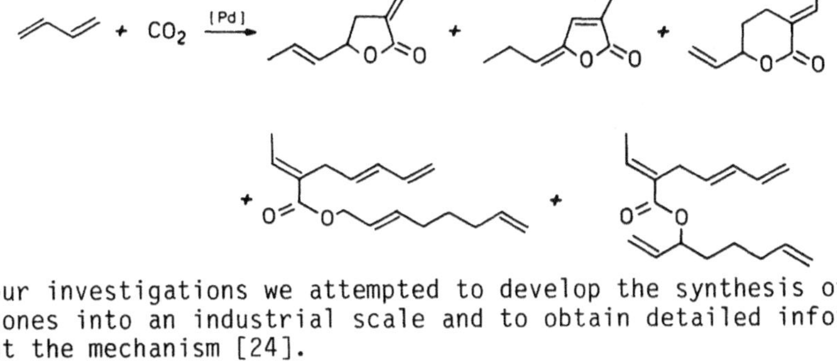

(12)

In our investigations we attempted to develop the synthesis of the lactones into an industrial scale and to obtain detailed information about the mechanism [24].

In a typical reaction the catalyst is composed in situ by palladium-bis(acetylacetonate) and a phosphorous ligand. Table 2 illustrates the strong controlling effect of the ligand. Without ligand only negligible amounts of lactones could be observed. Phosphines with linear alkyl substituents such as tri-n-propylphosphine or tri-n-butylphosphine yielded a mixture with predominant amounts of the two esters, whereas phosphines with alkyl groups branched at the α-carbon atom, such as triisopropylphosphine or tricyclohexylphosphine, gave selectively and in high yields the δ-lactone.

Table 2 Reaction of carbon dioxide with butadiene, influence of the phosphorous ligand

Ligand	Yield (%) δ-Lactone	γ-Lactones	Esters
-	0.6	0.1	-
PEt_3	4.0	0.4	12.2
PPr^n_3	3.1	0.8	14.9
PBu^n_3	3.0	0.1	14.5
PPh_3	1.7	0.7	12.3
PPr^i_3	43.9	0.4	1.5
PCy_3	38.6	0.1	1.4
PCp_3 a)	15.4	1.5	0.4

a) PCp_3 = Tricyclopropylphosphine

In homogeneous catalysis, the catalytic properties of a ligand are often compared with its electronic and steric data. Already at 1970, Tolman [25] introduced the electronic parameter $\Sigma\chi_i$: The most basic phosphine, tris(tertiary-butyl)phosphine, is the starting point of this scale. The steric properties of a ligand are given by the cone angle [26, 27], measured on a model of the ligand molecule. In Figure 6 the yield of the δ-lactone is plotted against the $\Sigma\chi_i$-values and the cone angles of twelve different phosphines.

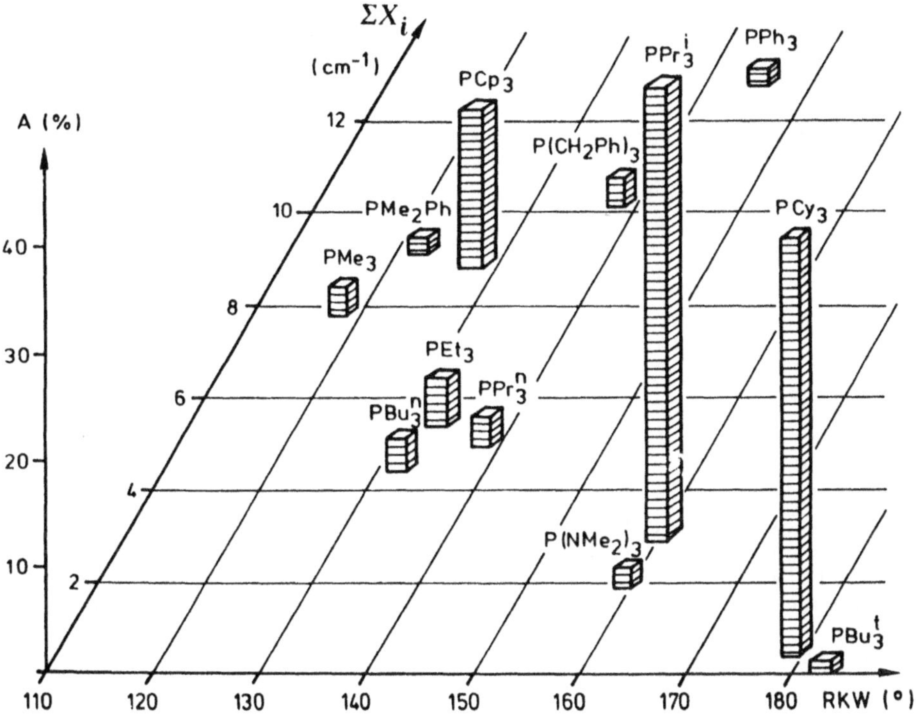

Figure 6 Influence of the ligand parameters on the yield of the δ-lactone

It is remarkable that ligands with similar electronic and steric properties give very different catalytic results. Whereas PPr^i_3 and PCy_3 afford high yields of the δ-lactone, the comparable ligands $P(NMe_2)_3$ and PBu^t_3 yield the lactone only in traces. It is noteworthy that all active ligands contain a CHR -group bound to the phosphorous and it can therefore be assumed that this group activates the carbon dioxide as it is shown in Equation (13).

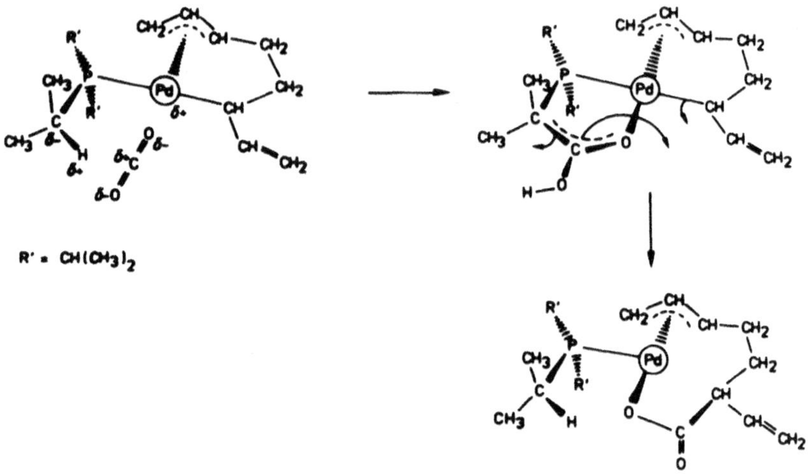

(13)

Carbon dioxide inserts into the C-H bond of the ligand and forms a five-membered ring with formation of new C-C-and Pd-O-bonds. This "activated CO_2" reacts with the C_8-chain built by two butadiene molecules, and a C_9-carboxylate intermediate is formed which is the precursor of the lactones.

When "in situ catalysts" are used, not only the ligand sphere, but also the transition metal component has an important influence on the activity of the catalyst (Table 3). Pd(II)- and Pd(0)-complexes, neutral and ionic palladium species as well as mono- and di-nuclear palladium compounds were applied as catalyst precursors.

Using palladium(II)-complexes, yields of the δ-lactone in the range of 35 to 45 % were obtained, whereas Pd(0)-compounds afforded yields lower than 10 %. Ionic palladium complexes and dimeric palladium compounds gave almost the same results as the neutral monometallic palladium(II) species.

Table 3 Reaction of carbon dioxide with butadiene; influence of the palladium complex

Palladium Complex	Yield (%)		
	δ-Lactone	γ-Lactones	Esters
$[Pd(OAc)_2]_3$	48	0.6	1.3
$Pd(acac)_2$	44	0.4	1.5
$Pd(hfacac)_2$	33	0.1	2.2
$Pd(dba)_2$	38	0.1	7.2
$[Pd(C_3H_5)(cod)]BF_4$	40	0.1	0.9
$[Pd(C_3H_5)(OAc)]_2$	44	0.5	0.9
$Pd_2(OAc)_2C_8H_{14}$	39	0.8	5.2

It is remarkable, that the reaction of butadiene and CO_2 proceeds only in solvents containing a nitrile group such as acetonitrile, propionitrile or benzonitrile. In other solvents such as toluene, acetone or tetrahydrofurane only oligomers of butadiene are formed. This effect is so evidently that it can be presumed that the solvent is involved in the catalytic cycle of the reaction.

The conversion of butadiene with carbon dioxide to the δ-lactone is very sensitive against the reaction conditions applied. In order to optimize the reaction various parameters have been investigated.

Of great importance proved to be the ratio of the phosphine to the palladium-compound as shown in Figure 7. At a phosphorus/palladium ratio of 3:1 a maximum of yield and selectivity of the δ-lactone was observed, whereas an excess of phosphine blocked the free coordination sites of the catalytic intermediate. At ratios lower than 2:1 yield and selectivity decrease evidently thus proving that a little excess of ligand is needed to form the catalytic active species.

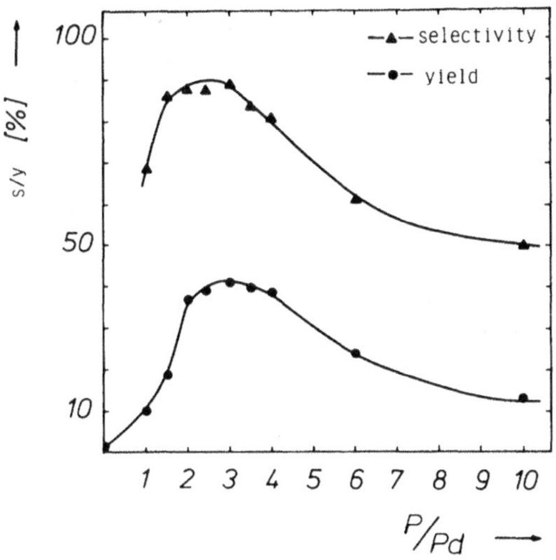

Fig. 7 Influence of the P/Pd-ratio on yield and selectivity of the δ-lactone

Also the influence of the catalyst concentration at the constant phosphorus/palladium ratio of 3:1 was investigated. Figure 8 shows the results which were obtained when 250 mmol of butadiene per experiment were reacted with CO_2. The yield of the δ-lactone passes a maximum at about 0.16 mmol of catalyst and decreases very fast at lower catalyst concentrations. Also the selectivity reaches at 0.16 mmol the favourable value of about 90 % and only few byproducts such as esters or oligomers of butadiene are formed.

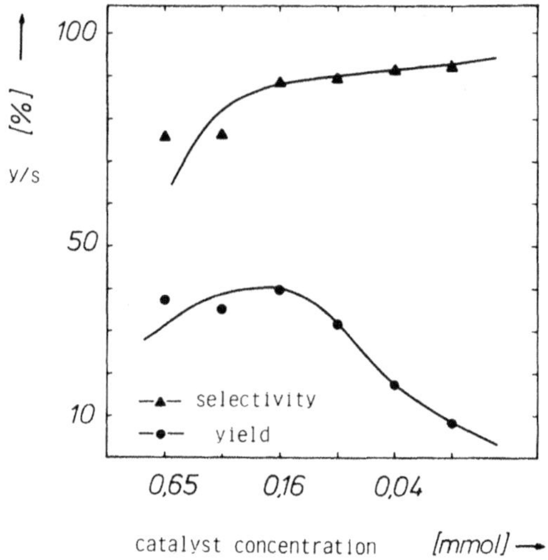

Fig. 8 Influence of the catalyst concentration

The influence of the reaction temperature is shown in Figure 9. At low temperatures of 60-70°C, the reaction works very selectively, however, the yields are lower than 30 %. At 80-90°C the yield of the δ-lactone reaches a distinct maximum of about 50 %. At temperatures higher than 100°C the formation of the esters increased, at temperatures above 130°C only oligomers of butadiene were observed.

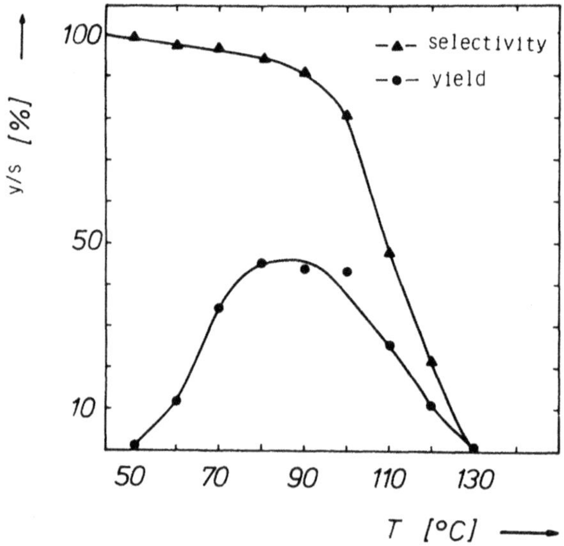

Fig. 9 Influence of the reaction temperature

Also the reaction time is of great importance (Figure 10): The yield of the δ-lactone increases very quickly during the first 15 hours; the selectivity reaches its maximum in the range of 8-15 h.

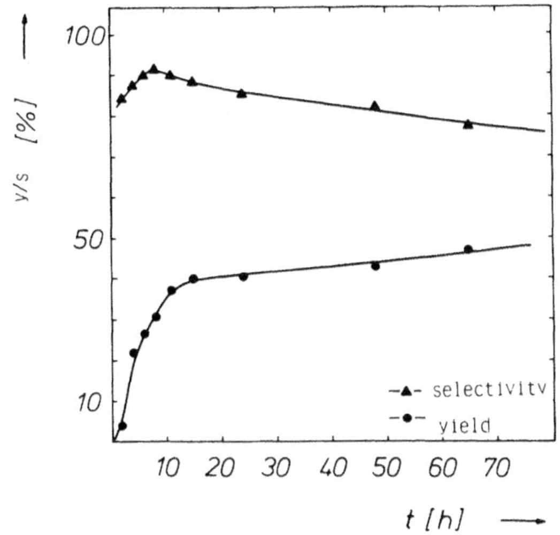

Fig. 10 Influence of the reaction time

The optimized reaction conditions cited above enabled the synthesis of the δ-lactone in a technical scale. Using a 1 l autoclave more than 100 g of the δ-lactone could be produced per experiment. The δ-lactone could be isolated by vacuum distillation or water steam distillation in purities higher than 99 %, because under the optimized conditions no interfering by-products of CO_2 and butadiene are formed.

The presumed mechanism of the palladium-catalyzed formation of the δ-lactone is shown in cycle A of Figure 11. Two molecules of butadiene combine to a C_8-chain and form a palladium-bis-η^3-allyl complex. Carbon dioxide is able to react with the bis(allyl)compound yielding a C_9-carboxylate complex. The oxygen of the carboxylate group then attacks the allyl group to form the δ-lactone by a cyclization step.

As shown in Figure 11, the δ-lactone is able to coordinate to the palladium yielding again the C_9-carboxylate complex. Under particular reaction conditions, especially at higher temperatures and high Pd-concentrations, this complex affords the two γ-lactones by an isomerization step.

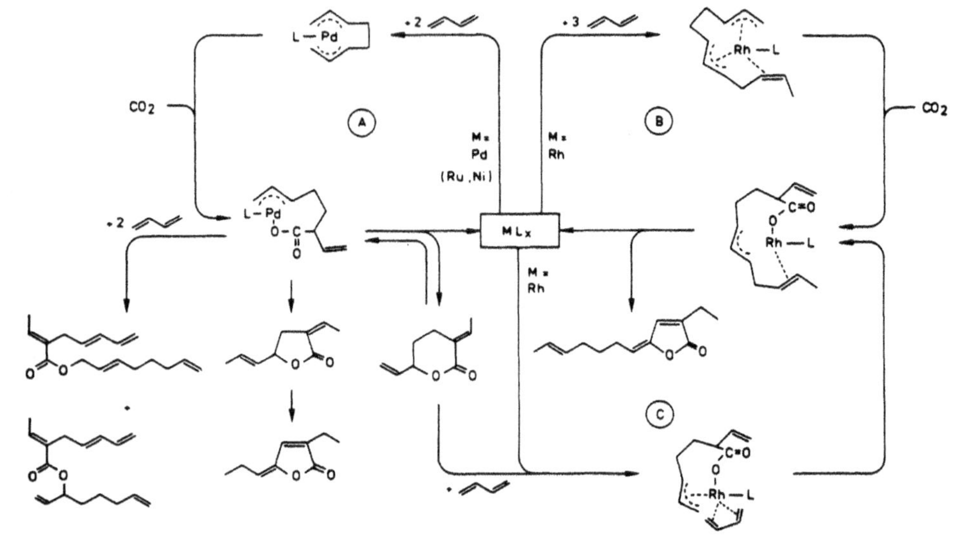

Fig. 11 Reactions of carbon dioxide with butadiene

Figure 12 shows the reaction sequence of this isomerization at a temperature of 90°C and at a ratio of lactone to metal as 40 to 1. After about two hours the nonconjugated γ-lactone is formed, after further 8 hours the conjugated γ-lactone is the single product.

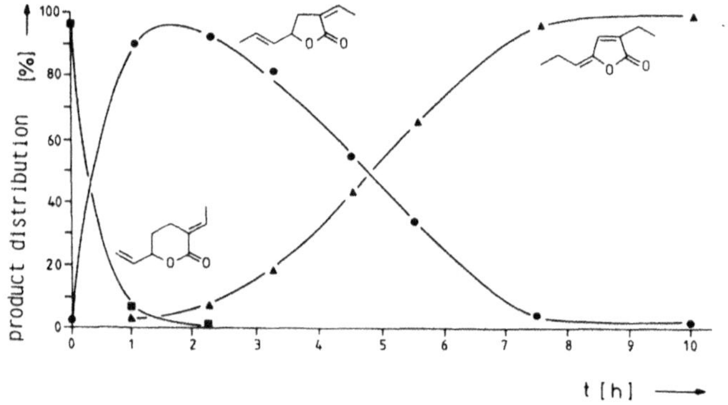

Fig. 12 Palladium-catalyzed isomerization of the δ-lactone

It can be assumed that the reaction of butadiene with carbon dioxide yields first the kinetically preferred δ-lactone and that - at higher temperatures and catalyst concentrations - the thermodynamically more stable γ-lactones are formed.

The key intermediate in cycle A in Figure 11 is the C_9-carboxylate complex which contains a phosphine and a η^3-allyl ligand in the coordination sphere of palladium. A similar complex could be isolated by the reaction of palladiumbis(acetylacetonate) with two moles of triisopropylphosphine and two moles of the δ-lactone (Equation 14). By ring opening of the lactone a palladium complex with two long-chained carboxylate substituents was formed. It is remarkable that the same ring opening reaction which is discussed in the catalytic cycle could be observed in a related stoichiometric experiment.

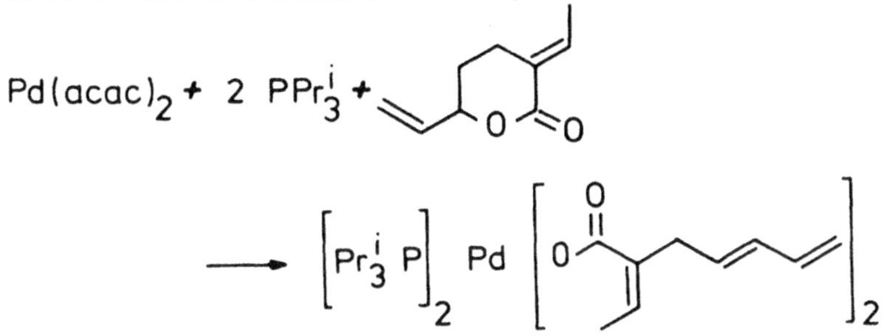

(14)

It is noteworthy that the isolated complex contains no allylic carboxylate ligands, but the ethylideneheptadienoat ligand with a 1,3-diene structure. As shown by x-ray analysis (Figure 13) the carboxylate ligands are in trans position to each other without any bonding between the diene-system andthe palladium.

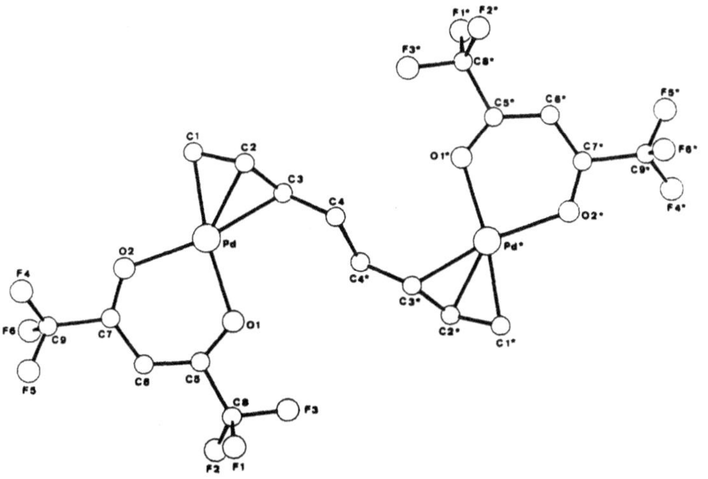

Fig. 13 x-ray of bis(triisopropylphosphine)-bis(2-ethylidene-4,6-heptadienoat)palladium

The novel isolated complex is itself a good catalyst for the synthesis of the δ-lactone. The further addition of a phosphine is not necessary; that means that this palladium complex is a real one-component-catalyst for the reaction of butadiene and carbon dioxide.

At this point it must be mentioned that in palladium-catalyzed conversions of butadiene also bimetallic complexes are discussed as mechanistic key-intermediates [28]. This was the motive to investigate in stoichiometric experiments whether the reaction of butadiene and carbon dioxide also proceeds via bimetallic palladium compounds. Therefore the model compound $Pd_2(F_6acac)_2(C_8H_{12})$ was synthesized, in which two palladium atoms are bound to a hexafluoroacetylacetonate ligand and linked between each other by a bridging octadienyl chain formed by the dimerization of two butadiene molecules (Figure 14). This complex can add one equivalent of triisopropylphosphine yielding selectively a complex in which only one palladium metal is attacked by the ligand. This complex is a very favourable model compound to study the mechanism of the reaction of butadiene and carbon dioxide, because the analogous "in situ" catalyst system formed by palladiumbis(hexafluoroacetylacetonate) and triisopropylphosphine afforded the δ-lactone in the good yield of 33 % (Compare Table 3).

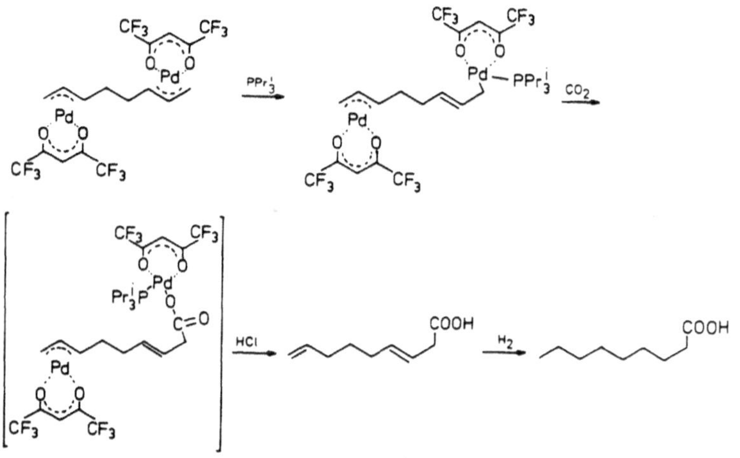

Fig. 14 Insertion of carbon dioxide into $Pd_2(F_6acac)_2(C_8H_{12})$

The reaction of the palladium/phosphine-complex with carbon dioxide yielded a carboxylate complex which was further characterized by treatment with hydrochlorid acid thus yielding a nonadienoic acid. After hydrogenation of this acid, pelargonic acid was obtained in a 44 % yield referred to the starting complex. The fact that the linear pelargonic acid is formed in the reaction of the bimetallic octadienyl-bridged palladium complex with CO_2, and not the branched δ-lactone invokes that the catalytic reaction of butadiene and carbon dioxide goes by a monometallic, and not by a bimetallic mechanism [29].

So far, only the palladium catalyzed reaction of butadiene with CO_2 has been considered. However, besides of palladium catalysts also other transition metal complexes are active in this reaction. Nickel(0)-compounds, such as nickelbis(cyclooctadiene) enabled the selective synthesis of the δ-lactone, however, in yields less than 2 %. In contrast to palladium-(II)-compounds, nickel-(II)-systems such as nickel-bis(acetylacetonate) are inactive.

It is also remarkable that ruthenium complexes, especially ruthenium(II)-hydrido-phosphine compounds yielded the δ-lactone in high selectivity, but only in low yields of about 7 %.

Quite different results have been obtained using neutral or ionic complexes of rhodium(I) and rhodium(III) [30]. A γ-lactone could be isolated, in which three molecules of butadiene are connected to one molecule of carbon dioxide (Cycle B in Figure 11). As shown by glc analysis the C_{13}-lactone exists in four isomeric configurations based on the two double bonds in its hydrocarbon chain. The best rhodium-catalyst proved to be bis(ethene)acetylacetonate-rhodium which was activated by basic trialkylphosphines such as triethylphosphine. Also ionic rhodium(I)-compounds and monomeric as well as dimeric rhodium(III)-complexes yielded the lactone selectively.

The optimum reaction temperature proved to be 120°C, whereas at higher values the decomposition of the catalyst prevailed. The most favourable solvent for the reaction was acetonitrile. In alcohols such as ethanol or isopropanol only few amounts of the new C_{13}-lactone were obtained and n-pentane or acetone desactivated the catalyst.

The interpretation of the formation of the C_{13}-lactone requires a sequence of mechanistical pathways which are unknown so far in rhodium-catalysis. A proposal for the mechanism is given in cycle B of Figure 11. Three molecules of butadiene react with the starting rhodium compound forming a C_{12}-chain, which is bound to the rhodium by two η^3-allyl systems and one olefinic double bond. Carbon dioxide inserts into one of the rhodium allyl bonds thus forming a C_{13}-carboxylate complex, which yields the new C_{13}-lactone. The highest yields obtained in cycle B were in the range of 5-8 %.

These modest results encouraged the search for an alternative route to the C_{13}-lactone which is shown in cycle C of Figure 11. The C_9-lactone is able to coordinate on rhodium and to react with a further molecule of butadiene. In the presence of bis(ethene)acetylacetonate-rhodium and triethyl phosphine the C_{13}-lactone was obtained in yields up to 44 %.

9. REACTIONS OF CARBON DIOXIDE WITH BUTADIENE AND FURTHER DIENES

Besides of butadiene also other 1,3-dienes are theoretically possible starting chemicals for the reaction with carbon dioxide. All our attempts to react isoprene with carbon dioxide failed. However, in our investigations concerning the co-reaction of butadiene, isoprene and carbon dioxide we found tow novel δ-lactones (Equation 15).

(15)

The selectivity towards these co-products reached values higher than 80 %, as outlined in Figure 15. The high selectivities of 80 % could only be reached at isoprene/butadiene ratios greater than 7:1. At a ratio of 1:1 the butadiene/CO_2-products were formed in an 85 % selectivity whereas the co-products were formed only in 15 % selectivity.

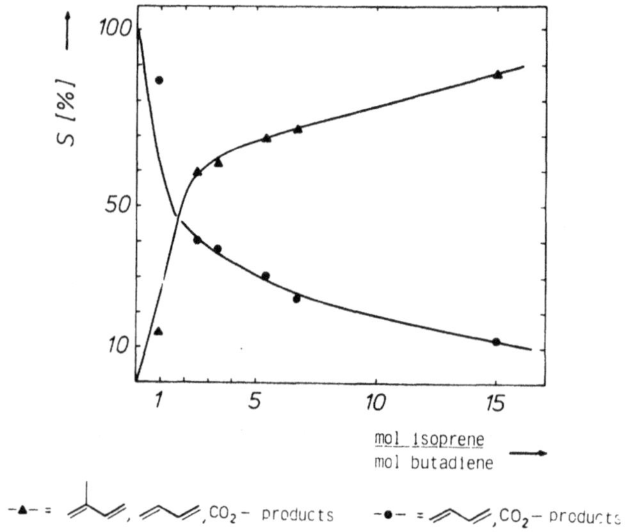

Fig. 15 Reaction of carbon dioxide with butadiene and isoprene

The proposed reaction mechanism for the formation of the two co-products is shown in Figure 16. Butadiene, isoprene and the palladium complex form two different bis-η^3-allyl complexes, in which the isoprene unit is both head- and tail-connected to butadiene. Carbon dioxide inserts into both complexes to yield a carboxylate species, but the insertions occur only on the η^3-allyl part steming from the butadiene unit perhaps by electronic reasons. Finally, the two novel lactones are formed by reductive elimination.

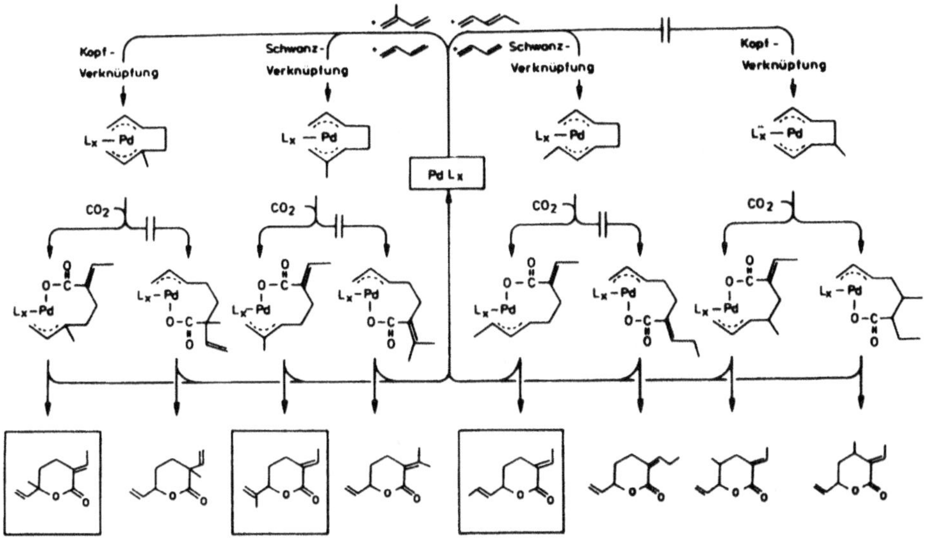

Fig. 16 Potential products in the reaction of carbon dioxide with butadiene and isoprene/piperylene

In the reaction of carbon dioxide with butadiene and piperylene only one tail-linked product could be isolated and no head-linked lactone was observed. The insertion of carbon dioxide occured - as already mentioned in the case of isoprene - exclusively at the non-substituted allyl group.

10. REACTIONS OF CARBON DIOXIDE WITH BUTADIENE AND OXIRANES

Both oxiranes [31-35] and butadiene (see chapter 8) proved to be highly active partners in reactions with carbon dioxide. The question arose which products occur, if both molecules were reacted simultaneously with CO_2. Both the oxidative addition of the oxirane and the addition of butadiene seemed to be possible first reaction steps leading to higher membered lactones or longer chained esters. Therefore, we extended our investigations of carbon dioxide to the three-component co-reaction of butadiene, oxiranes and CO_2 [36].

Butadiene, carbon dioxide and ethylene oxide react in the presence of palladium catalysts yielding a novel glycol ester of the 2-ethylidene-4,6-heptadienoic acid in a "one pot reaction". A related synthesis - starting with three components - could be carried out with propylene oxide. The additional methyl group of the propylene oxide caused the formation of two isomeric esters (Equation 16).

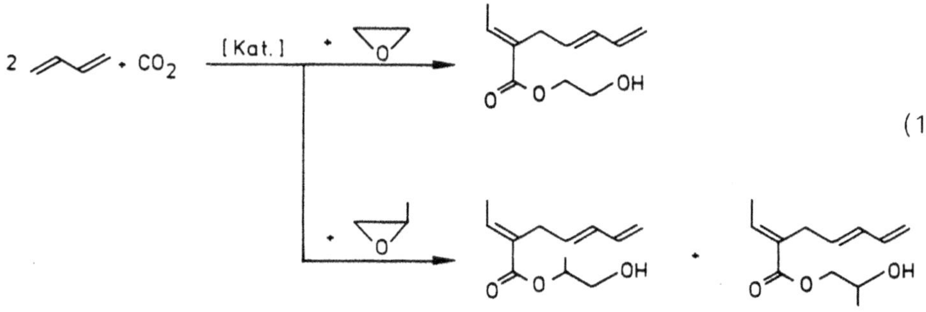

(16)

In side reactions also the products deriving from butadiene and CO_2, the lactones and the octadienyl esters, are obtained. It is remarkable, that the octadienyl esters and the novel glycol esters are derivatives of the same 2-ethylidene-4,6-heptadienoic acid, thus making it obvious that the mechanistic pathways are closely related to eachother. It must also be emphasized, that in the reaction of butadiene, CO_2 and oxiranes no carbonates, neither ethylene carbonate nor the corresponding propylene carbonate, could be identified as by-products.

In order to optimize the reaction and to obtain more details of the mechanism the influence of the reaction parameters on the formation of the glycol esters had been studied. A systematic variation of the palladium catalyst was carried out using palladiumbis(acetylacetonate) as catalyst precursor which was modified by different ligands, especially by mono- and bidentate phosphines. Table 4 shows the yields of the glycol esters and of the by-products.

Table 4 Reaction of CO_2 with butadiene and propylene oxide, variation of the ligands

Ligand	Yields (mol-%)	
	Glycol esters	By-products
PBu^n_3	2	3
bipy	3	3
PEt_3	3	1
PPh_3	5	1
dppe	6	1
dcpe	8	17
PCy_3	10	17
PPr^i_3	22	3

Trialkylphosphines with linear substituents, mono- and bidentate phosphines with phenyl groups and 2,2'-bipyridine yield only small amounts both of the products and of the by-products. Phosphines with cyclohexyl groups, 1,2-Bis(dicyclohexylphosphino)ethane and tricyclohexylphosphine, lead to a higher formation (8-10 %) of the glycol esters, however, also the yield of the by-products increases up to 17 %. The only exception in this comparison is triisopropylphosphine: The yield of the glycol esters achieves 22 % and only small amounts of the butadiene/CO_2-by-products are formed. In all experiments some dimers of butadiene are produced, too.

Not only the type of ligand, but also its quantities relative to the metal have a great influence on catalysis. Best yields of the glycol esters are obtained with phosphine/palladium ratios between 3:1 and 5:1; the - by far - best selectivity of 62 % is achieved exclusively at a ratio of 3:1.

The formation of the glycol esters is very sensitive towards the reaction temperature applied. As seen from Figure 17 only in the small range of 85°C-95°C reasonable yields and selectivities could be obtained.

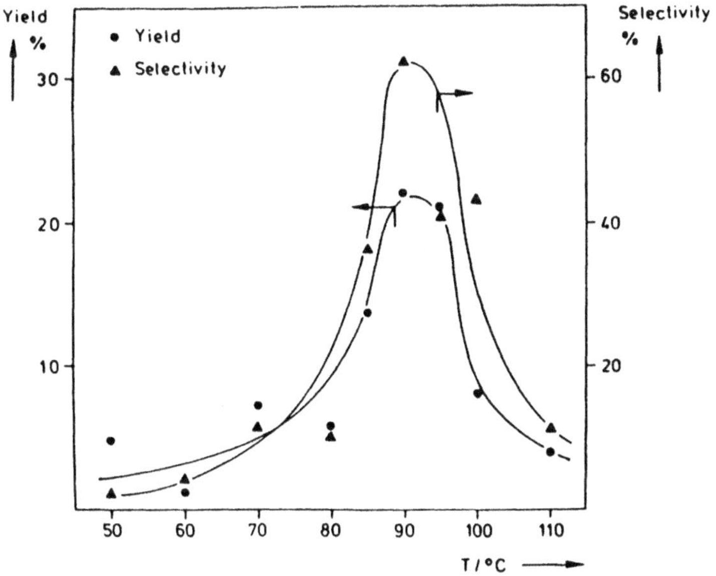

Fig. 17 Influence of the reaction temperature on yield and selectivity of the glycol esters

All investigations described so far were carried out with the catalyst precursor palladiumbis(acetylacetonate). As shown in Table 5, this compound is indeed one of the best catalysts only slightly exceeded by tris(palladiumbisacetate). Also cationic palladium compounds support the catalysis leading to the glycol esters in yields of 14-20 %. It is remarkable, that the esters are also formed using the rhodium compounds tris(rhodiumbisacetate) or (bisethylene)rhodiumacetylacetonate as catalysts, however, yields are very low.

Table 5 Variation of the transition metal catalyst precursor

Metal compound	Yields (mol-%) Glycol esters	By-products
$Pd(dba)_2$	9	22
$Pd(acac)_2$	22	3
$[Pd(OAc)_2]_3$	24	3
$[(cod)Pd(Cp)]BF_4$	14	1
$[\eta^3\text{-}C_3H_5\text{-}Pd(CH_3CN)_2]BF_4$	20	4
$[Rh(OAc)_2]_3$	3	1
$(C_2H_4)_2Rh(acac)$	1	5

The mechanism proposed for the reaction with ethylene oxide is shown in Figure 18. The first step is obviously the addition of butadiene and carbon dioxide yielding the already presented C_9-carboxylate complex. As seen from the optimization experiments only under very particular reaction conditions the oxirane may be coordinated by the carboxylate complex. By a synchroneous reaction the oxirane ring opens and inserts into the existing metal-oxygen bond. A carbon-oxygen and a new metal-oxygen bond are formed yielding an alkoxy allyl complex which then detaches the glycol ester by reformation of the catalyst ML_x. The release of the ester must be accompanied by a hydrogen-shift from the acid chain to the oxygen of the former oxirane thus forming the hydroxy group. By this migration of hydrogen in α-position to the allyl group the 1,3-diene structure of the acid part is formed.

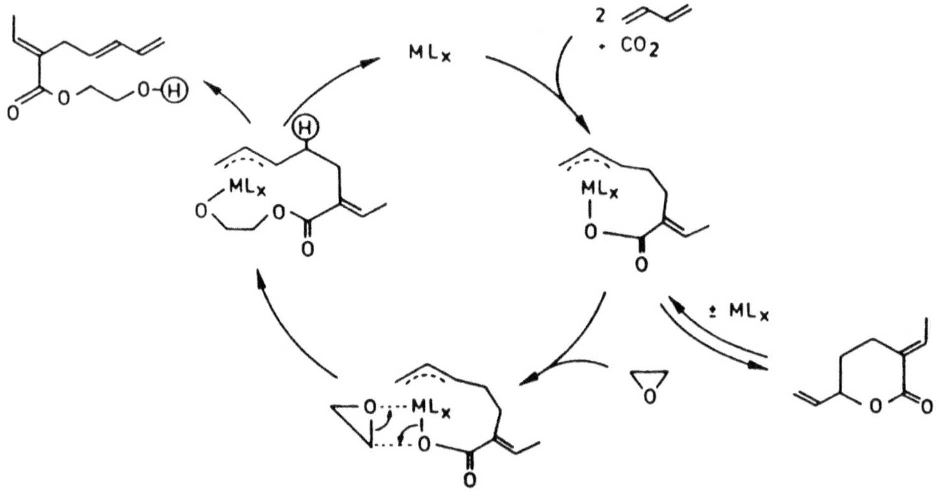

Fig. 18 Proposed mechanism of the reaction of carbon dioxide with butadiene and ethylene oxide

The formation of the δ-lactone by an elimination step starting from the carboxylate complex seems to compete with the synthesis of the glycol ester. Indeed, kinetic investigations prove that at reaction times till 10 h the δ-lactone prevails, whereas at longer reaction times the δ-lactone disappears almost quantitatively from the reaction mixture with an optimum of glycol esters after about 25 h. This observation suggests that the synthesis of the ester is a consecutive reaction of the lactone. In fact, a comparing test experiment with the optimum palladium catalyst $Pd(acac)_2/PPr^i_3$, propylene oxide and a pure sample of the isolated δ-lactone led to the formation of esters in the same yield (24 %) as obtained by starting from butadiene and carbon dioxide. Comparative investigations proved that this connection of δ-lactone and the oxirane occurs via transition metal catalyzed reaction steps, since no ester formation could be observed applying thermal or base (PPr^i_3) catalyzed reactions.

REFERENCES

[1] A. Behr, in: Catalysis in C Chemistry (Ed. W. Keim), D. Reidel Publ. Compl, Dordrecht 1983, p. 169.

[2] Ullmanns Encyclopädie der technischen Chemie, 4. Ed., Vol. 13, p. 163; Vol 20, p. 299, VCH Verlagsgesellschaft, Weinheim.

[3] I. Hirao, T. Kito, Bull. Chem. Soc. Jpn., 46 (1973), 3470.

[4] T. Yatsuka, A. Ito, O. Manabe, M. Dehara, H. Hiyama, Yuki Gosei Kagaku Kyokai Shi, 30 (1972), 1030.

[5] DOS 2 426 850, 2. Jan. 1976, Hoechst A.-G. (Inv. W. Bachmann, C. Gnabs, K. Janecka, E. Mundlos, T. Papenfuhs, G. Waese),

[6] Z. Weglinski, T. Talik, Rocz. Chem. 51 (1977), 2051, C.A. 89, 43036w.

[7] F. Mutterer, C.D. Weis, J. Heterocycl. Chem. 13 (1976), 1103.

[8] G. Bottaccio, G.P. Chiusoli, Chem. Commun. 1966, 618.

[9] U.S. Patent 4 016 185, 5. Apr. 1977, Chevron Research Co. (Inv. J.B. Wilkes).

[10] Y. Inoue, Y. Itoh, H. Hashimoto, Chem. Lett. 1977, (8), 855.

[11] a) H. Hoberg, D. Schaefer, G. Burkhart, J. Organomet. Chem. 228 (1982) C21.
b) H. Hoberg, D. Schaefer, J. Organomet. Chem. 238 (1982) 383.
c) H. Hoberg, D. Schaefer, G. Burkhart, C. Krüger, M.J. Ramão, J. Organomet. Chem. 266 (1984), 203.

[12] P. Albano, M. Aresta, J. Organomet. Chem. 190 (1980), 243.

[13] A.L. Lapidus, S.D. Pirozhkov, A.A. Koryakin, Bull. Akad. Sci. U.S.S.R., Div. Chem. Sci. 1978, 2513.

[14] S. Besecke, G. Schröder, Ger. Offen., DE 2.948.888 A1 (1981).

[15] A. Behr, U. Kanne, W. Keim, J. Mol. Catal. 35 (1986), 19.

[16] A. Behr, W. Keim, Arabian Journal for Science and Engineering, 10 (1985), 377.

[17] A. Behr, W. Keim, G. Thelen, J. Organomet. Chem. 249 (1983), C38.

[18] Y. Inoue, T. Hibi, M. Satake, H. Hashimoto, J.C.S. Chem. Comm. 1979, 982.

[19] P. Binger, H.-J. Weintz, Chem. Ber., 117 (1984), 654.

[20] A. Döhring, P.W. Jolly, Tetrahedron Lett., 21 (1980), 3021.

[21] H. Hoberg, B.W. Oster, J. Organomet. Chem., 266 (1984), 321.
[22] a) Y. Sasaki, Y. Inoue, H. Hashimoto, J.C.S. Chem. Comm. 1976, 605.
b) Y. Inoue, Y. Sasaki, H. Hashimoto, Bull. Chem. Soc. Jpn. 51 (1978), 2375.
[23] a) A. Musco, C. Perego, V. Tartiari, Inorg. Chim. Acta 28, (2) (1978), L 147.
b) A. Musco, R. Santi, G.P. Chiusoli, Ger. Offen. 2.838.610 (1979).
c) A. Musco, J. Chem. Soc. Perkin Trans. I, 1980 (3), 693.
[24] a) A. Behr, K.-D. Juszak, W. Keim, Synthesis, 1983 (7), 574.
b) A. Behr, K.-D. Juszak, J. Organomet. Chem. 255 (1983), 263.
c) A. Behr, K.-D. Juszak, R. He, Proc. 8th Int. Congr. Catalyis, Berlin (Germany), July 1984, Vol. V, 565.
d) A. Behr, R. He, K.-D. Juszak, C. Krüger, Y.-H. Tsay, Chem. Ber., 119 (1986), 991.
[25] C.A. Tolman, J. Am. Chem. Soc., 92 (1970), 2953.
[26] C.A. Tolman, J. Am. Chem. Soc., 92 (1970), 2956.
[27] C.A. Tolman, Chem. Rev., 77 (1977), 313.
[28] a) A. Behr, in: Aspects of Homogeneous Catalysis, (Ed.: R. Ugo), Reidel Publ. Co., Vol. 5, (1984), p. 3.
b) A. Behr, G. v. Ilsemann, W. Keim, C. Krüger, Y.-H. Tsay, Organometallics, 5 (1986), 514.
[29] A. Behr, G. v. Ilsemann, J. Organomet. Chem., 276 (1984), C77.
[30] A. Behr, R. He, J. Organomet. Chem. 276 (1984), C69.
[31] R.J. De Pasquale, J.C.S. Chem. Comm. 1973, 157.
[32] Y. Fujita, T. Morimoto, Japan. Kokai 72 15,951 (1972); C.A. 77, 33923w.
[33] Y. Fujita, T. Morimoto, T. Nakano, Japan. 72 26, 786 (1972); C.A. 77, 139451f.
[34] M. Ratzenhofer, H. Kisch, Angew. Chem. 92 (1980), 303.
[35] T. Fujinami, T. Suzuki, M. Kamiya, S. Fukuzawa, S. Sakai, Chem. Lett. 1985, 199.
[36] A. Behr, U. Kanne, J. Organomet. Chem., in print.

FIXATION AND ACTIVATION OF CARBON DIOXIDE WITH METALLOPORPHYRIN

Shohei Inoue
Department of Synthetic Chemistry
Faculty of Engineering
University of Tokyo
Hongo, Bunkyo-ku, Tokyo, Japan

ABSTRACT. Irradiation with visible light induces the reaction of CO_2 with aluminum porphyrin carrying an axial alkyl-aluminum group, (P)Al-R, in the presence of 1-methylimidazole. The reaction of zinc N-methylated porphyrin (Me-P)Zn-R with the mixture of CO_2 and a secondary amine to form a zinc carbamate is much accelerated by visible light. Aluminum porphyrin with an Al-OR group in the presence of 1-methylimidazole reacts reversibly with CO_2 to form a carbonate. CO_2 thus trapped is activated enough to react with epoxide under mild conditions to form cyclic carbonate catalytically. Aluminum porphyrin with an Al-O_2CR group reacts with the mixture of CO_2 and a secondary amine to form a novel hexa-coordinate porphyrin complex. Catalytic formation of a carbamic ester from CO_2, secondary amine, and epoxide proceeds with (P)Al-O_2CR under mild conditions.

1. INTRODUCTION

In the most common pathway of photosynthesis, carbon dioxide (CO_2) is not directly reduced, but is fixed by an activated ketose to form a carboxylic acid, which is subsequently reduced to aldehyde by NADPH. The formation of ATP required for the activation of the ketose and the formation of NADPH result from light reactions. In this respect, the fixation of CO_2 with a compound activated by light is of much interest as a simulation of photosynthesis, and also for the development of novel chemical utilization of CO_2. In this connection, we have studied the reactivity of metalloporphyrin having an axial ligand bound covalently to the metal, and the effect of visible light on the reactivity.

2. REACTION OF CARBON DIOXIDE AND METALLOPORPHYRIN ACCELERATED BY VISIBLE LIGHT

2.1. Reaction of carbon dioxide and aluminum porphyrin with axial aluminum-alkyl group

The effect of visible light on the reaction of carbon dioxide with

M. Aresta and G. Forti (eds.), Carbon Dioxide as a Source of Carbon, 321–329.

(tetraphenylporphinato)aluminum ethyl (1, X=Et: (TPP)AlEt) was investigated.

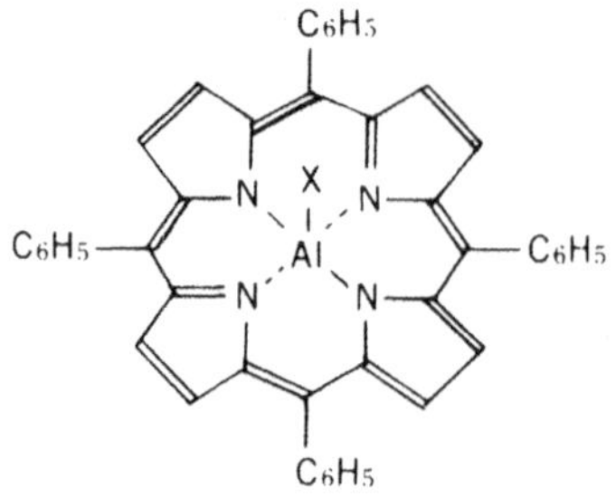

1

Aluminum-alkyl bond does not absorb visible light; triethylaluminum, for example, is a colorless liquid. The bond is not susceptible to homolysis by the effect of light, differently from transition metal-alkyls. However, we considered that the reactivity of the aluminum-alkyl bond in (TPP)AlEt would be affected by the excitation of the porphyrin ring with visible light. In the excitation of chlorophyll, a magnesium complex of a porphyrin analog, an electron is transferred to an acceptor. Thus, the increase in nucleophilic reactivity of the aluminum-alkyl group by visible light would be expected.

In fact, the insertion reaction of CO_2 to the aluminum-ethyl bond of (TPP)AlEt was found to be induced by visible light in the presence of 1-methylimidazole (MeIm)[1].

$$\text{(TPP)AlEt} + CO_2 \xrightarrow{\text{MeIm}} \text{(TPP)Al-}O_2\text{C-Et} \qquad (1)$$

The reaction does not proceed without irradiation and in the absence of 1-methylimidazole. The reactivity enhancement of the aluminum-ethyl bond of (TPP)AlEt is also observed in other nucleophilic addition and substitution reactions[2)-4)].

$$\text{(TPP)AlEt} + \text{tBu-}\underset{\underset{\text{O}}{\|}}{\text{C}}\text{-CH=CH}_2 \longrightarrow \text{(TPP)Al-O-}\overset{\overset{\text{tBu}}{|}}{\text{C}}\text{=C-CH}_2\text{-Et} \qquad (2)$$

$$\text{(TPP)AlEt} + \text{ArOH} \longrightarrow \text{(TPP)AlOAr} + \text{EtH} \qquad (3)$$

$$\text{(TPP)AlEt} + \text{RSH} \longrightarrow \text{(TPP)AlSR} + \text{EtH} \qquad (4)$$

This interesting effect of visible light is considered general for the reaction of axial ligand on aluminum porphyrin. For example, reversible exchange of aryloxyl group on (TPP)Al-OAr with another phenol is much accelerated upon irradiation[5)].

$$(TPP)AlOAr \quad + \quad Ar'OH \rightleftharpoons (TPP)AlOAr' \quad + \quad ArOH \qquad (5)$$

2.2 Reaction of carbon dioxide and zinc porphyrin with axial zinc-alkyl group

(N-methylporphinato)zinc ethyl (2: (NMTPP)ZnEt) is more susceptible to homolysis by visible light[6], but is enough stable in benzene solution.

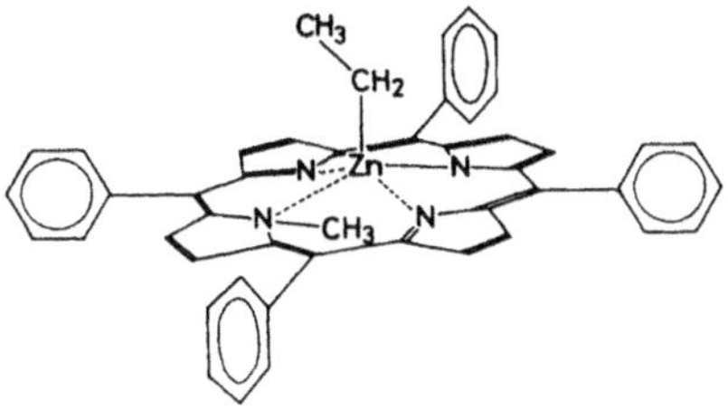

2

(NMTPP)ZnEt does not react with CO_2, but in the presence of a secondary amine the formation of a carbamate complex takes place[7].

$$(NMTPP)ZnEt + R_2NH + CO_2 \longrightarrow (NMTPP)Zn\text{-}O_2C\text{-}NR_2 + EtH \qquad (6)$$

Similarly, the reaction of (NMTPP)ZnEt, CO_2, and alcohol gives a carbonate complex.

$$(NMTPP)ZnEt + ROH + CO_2 \longrightarrow (NMTPP)Zn\text{-}O_2C\text{-}OR + EtH \qquad (7)$$

Since (NMTPP)ZnEt does not react with amine or alcohol under the conditions examined, these reactions are considered to proceed by the substitution of the zinc-ethyl group with dialkylcarbamic acid or monoalkyl carbonate, formed from the amine or alcohol, respectively, and CO_2. Of particular interest is the remarkable acceleration by visible light in these reactions (Figure 1).

3. CATALYTIC FORMATION OF CYCLIC CARBONATE FROM CARBON DIOXIDE AND EPOXIDE BY ALUMINUM PORPHYRIN

In order to develop further reaction of the fixed CO_2 and to realize a catalytic process, the reactivity of aluminum porphyrin with an axial metal-alkoxyl bond was examined.

(Tetraphenylporphinato)aluminum alkoxide (1, X=OR: (TPP)AlOR) does not react with CO_2, but in the presence of 1-methylimidazole rapid and reversible reaction takes place even in the dark, to form a (porphinato)aluminum carbonate[8][9].

$$(TPP)Al\text{-}OR \quad + \quad CO_2 \xrightleftharpoons{MeIm} (TPP)Al\text{-}O_2C\text{-}OR \qquad (8)$$

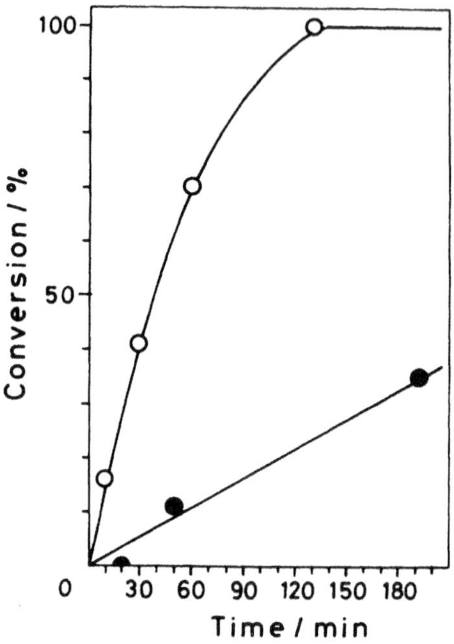

Figure 1. Reaction of (NMTPP)ZnEt, CO_2, and $Pr^i{}_2NH$ in benzene.
o: Irradiated. •: Dark.

Visible, ^{1}H-NMR, and ^{27}Al-NMR[10)] spectra of the (TPP)AlOR-MeIm system indicates the formation of a hexa-coordinate complex (Figure 2).

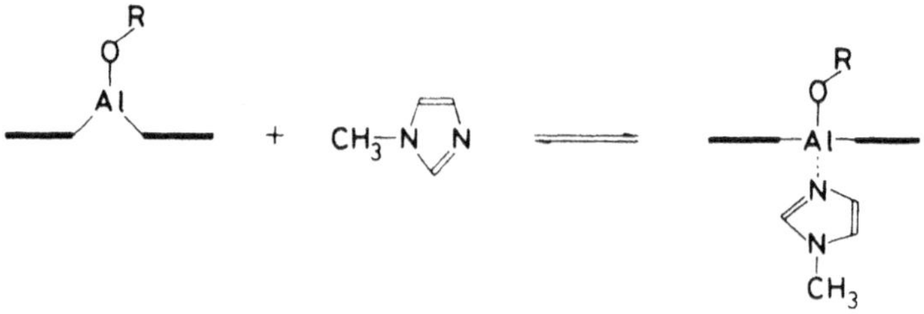

Figure 2. Side view of (TPP)AlOR and its MeIm complex.

Thus, coordinating MeIm as the axial ligand is responsible for the enhancement of the nucleophilic reactivity of the (porphinato)aluminum alkoxide toward CO_2.

CO_2 thus trapped by the aluminum porphyrin is activated enough to react with epoxide at room temperature to form cyclic carbonate and regenerate (porphinato)aluminum alkoxide.

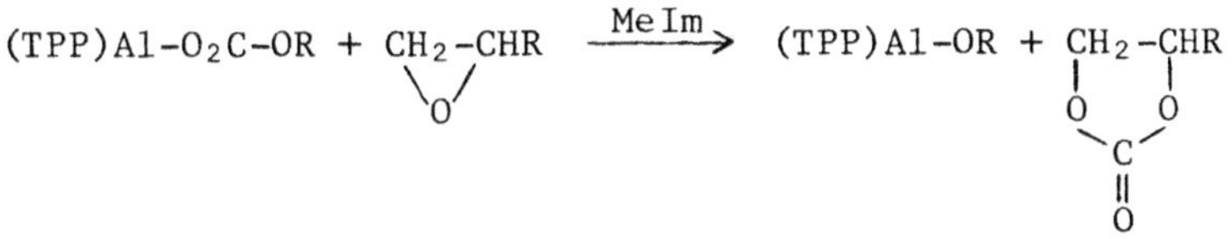

(9)

When epoxyethane is introduced in the (TPP)AlO$_2$COR-MeIm system in the absence of free CO_2, an infrared absorption at 1800 cm^{-1} assigned to ethylene carbonate appears at the expense of the band at 1680 cm^{-1} due to AlO$_2$COR. Addition of CO_2 regenerates the 1680 cm^{-1} band (Figure 3).

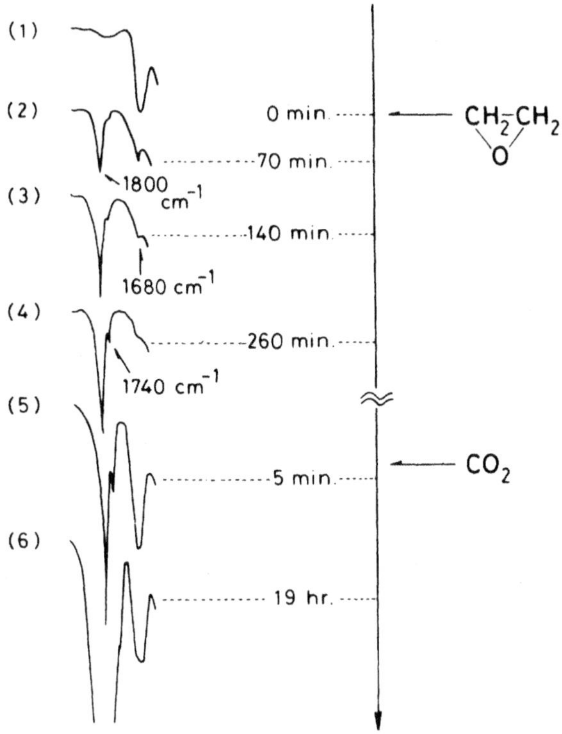

Figure 3. Reaction of CO_2 and epoxyethane with the (TPP)AlOR-MeIm system in CH_2Cl_2 at r.t. Infrared profile of the reaction at the carbonyl region. (1) (TPP)AlO$_2$COR-MeIm system in the absence of free CO_2.

In the presence of a large excess of CO_2 and epoxide to (TPP)AlOR-MeIm system, catalytic formation of cyclic carbonate proceeds under mild conditions[9] (Figure 4). Although various catalyst systems are known for the production of cyclic carbonate from CO_2 and epoxide, most of them require more rigorous conditions such as higher temperature. In this respect, aluminum porphyrin system is a good catalyst under mild conditions.

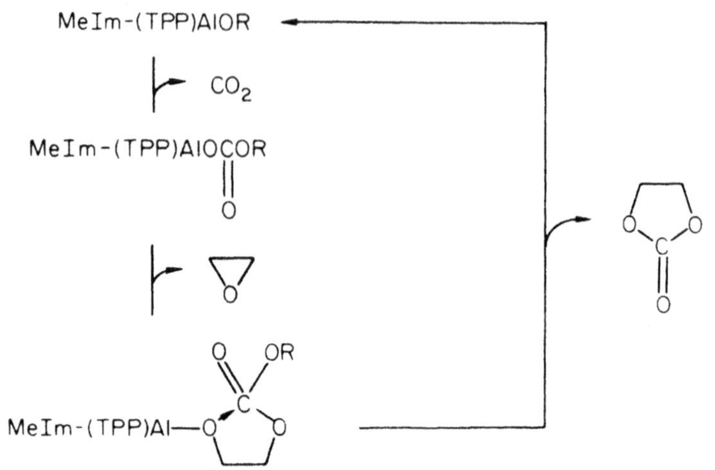

Figure 4. Catalytic cycle of the formation of cyclic carbonate from CO_2 and epoxide with (TPP)AlOR-MeIm system.

4. CATALYTIC FORMATION OF CARBAMIC ESTER FROM CARBON DIOXIDE, AMINE, AND EPOXIDE BY ALUMINUM PORPHYRIN

We have recently found that the aluminum porphyrin carrying an axial carboxylate group forms a novel hexa-coordinate complex with tetraalkylammonium carboxylate[11] (Figure 5).

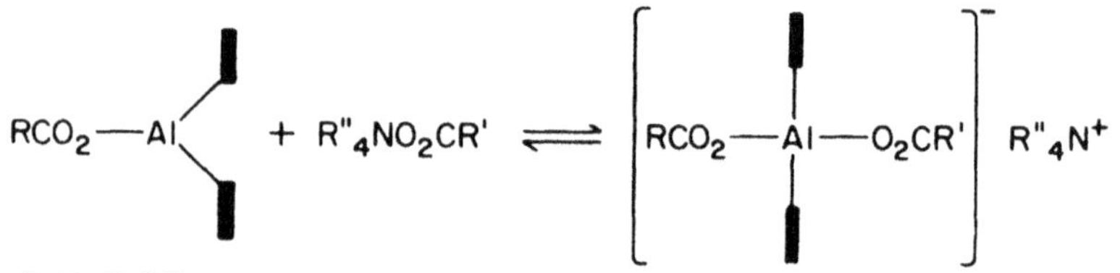

1, X = O_2CR

Figure 5. Formation of hexa-coordinate complexes from (TPP)AlO_2CR and tetraalkylammonium carboxylate.

Our studies on the reactivity of (porphinato)aluminum carboxylate have indicated that the nucleophicity of the carboxylate group on the hexa-coordinate complex is higher than that of the corresponding penta-coordinate complex.

These findings led us to examine the possible formation of a similar hexa-coordinate complex from (TPP)AlO_2CR, CO_2, and amine, since an ammonium salt of a carbamic acid readily forms from CO_2 and amine (Figure 6).

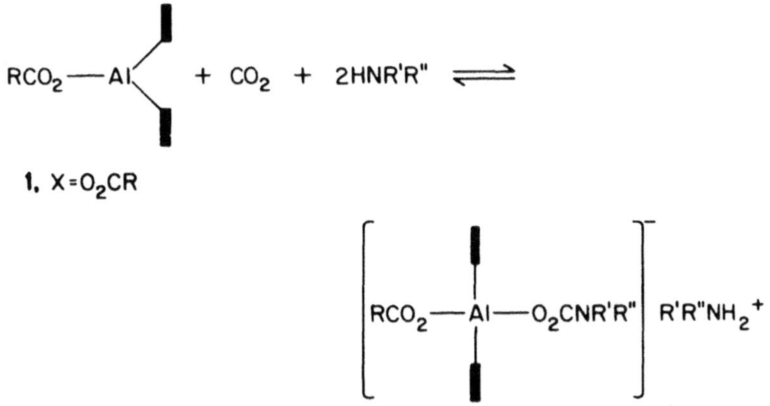

Figure 6. Formation of hexa-coordinate complex from (TPP)AlO_2CR, CO_2, and a secondary amine.

In fact, the formation of a hexa-coordinate aluminum porphyrin carrying a carboxylate group and a carbamate group was observed as evidenced by ^{1}H-NMR and ^{13}C-NMR spectra of the mixture of $TPPAlO_2CCH_3$, CO_2, and, for example, N-methylbenzylamine[12] (Figure 7). ^{1}H-NMR signals at high magnetic field (for example, b') indicates that the corresponding group is in the proximity of the porphyrin ring and is affected by the ring current. Reaction mixture using CO_2 enriched in ^{13}C shows a strong ^{13}C-NMR signal assigned to (porphinato)metal carbamate.

This carbamate group on the hexa-coordinate aluminum porphyrin is considered to exhibit high nucleophilic reactivity, and catalytic process by virtue of this is expected. In fact, (TPP)AlO_2CCH_3 was found to catalyze the reaction of CO_2, secondary amine, and epoxide to afford carbamic ester[12] (Figure 8). The reaction proceeds under mild conditions such as room temperature. The highest turnover observed was 47 for the reaction involving diethylamine and 1,2-epoxypropane, in pyridine with CO_2 pressure of 50 atm at 60°C for 6 h.

REFERENCES

1) S. Inoue, N. Takeda, Bull. Chem. Soc. Jpn., **50**, 984 (1977).
2) H. Murayama, S. Inoue, Chem. Lett., **1985**, 1377.
3) H. Murayama, S. Inoue, Chem. Lett., **1983**, 381.
4) H. Murayama, Y. Ohkatsu, S. Inoue, Nippon Kagaku Kaishi, **1984**, 219.
5) H. Murayama, S. Inoue, 52th Spring Meeting of Chem. Soc. Jpn. (April, 1986, Kyoto) Abstract, p. 576.
6) S. Inoue, H. Murayama, N. Takeda, Y. Ohkatsu, Chem. Lett., **1982**, 317.
7) S. Inoue, M. Nukui, F. Kojima, Chem. Lett., **1984**, 619.
8) N. Takeda, S. Inoue, Bull. Chem. Soc. Jpn., **51**, 3564 (1978).
9) T. Aida, S. Inoue, J. Am. Chem. Soc., **105**, 1304 (1983).

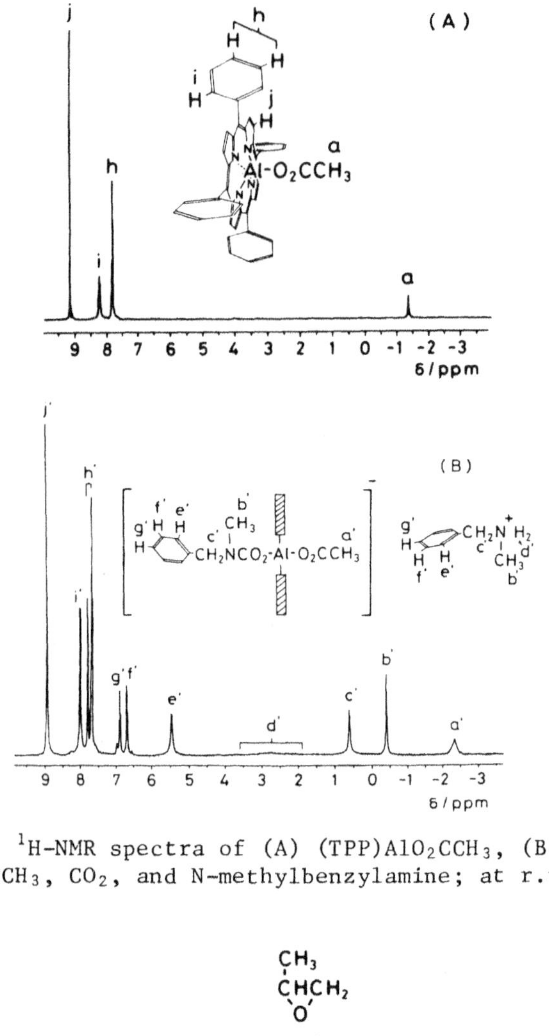

Figure 7. ^{1}H-NMR spectra of (A) (TPP)AlO_2CCH_3, (B) the mixture of (TPP)AlO_2CCH_3, CO_2, and N-methylbenzylamine; at r.t. in $CDCl_3$.

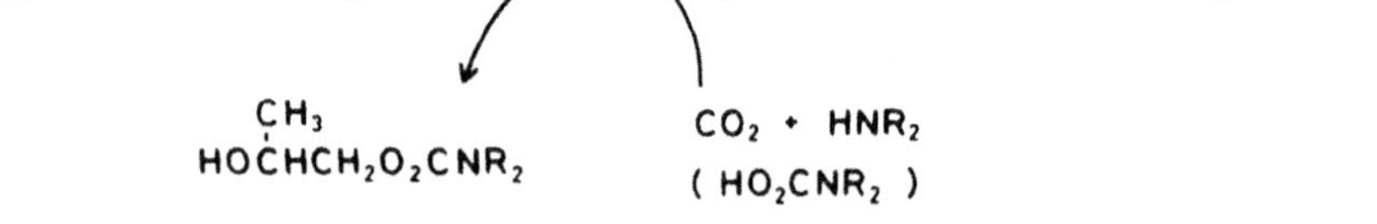

Figure 8. Catalytic cycle of the formation of carbamic ester from CO_2, secondary amine, and epoxide with (TPP)AlO_2CCH_3.

10) M. Kato, S. Tero, T. Ito, H. Murayama, S. Inoue, 52th Spring Meeting of Chem. Soc. Jpn. (April, 1986, Kyoto) Abstract, p. 657.
11) T. Aida, S. Inoue, J. Am. Chem. Soc., **107**, 1358 (1985).
12) F. Kojima, T. Aida, S. Inoue, J. Am. Chem. Soc., **108**, 391 (1986).

COPOLYMERIZATION OF CARBON DIOXIDE AND EPOXIDE

Shohei Inoue
Department of Synthetic Chemistry
Faculty of Engineering
University of Tokyo
Hongo, Bunkyo-ku, Tokyo Japan

ABSTRACT. Alternating copolymerization of CO_2 and epoxide to give aliphatic polycarbonate with high molecular weight proceeds under mild conditions by organometallic catalyst systems. The reaction product of diethylzinc with a polyhydric compound, such as water or polyhydric phenol, exhibits high catalytic activity. Elementary reactions for the copolymerization of CO_2 and epoxide with an organozinc catalyst system is considered to be the reaction of a zinc alkoxide with CO_2 to form a zinc alkyl carbonate, and the reaction of the zinc alkyl carbonate with epoxide to regenerate zinc alkoxide. The system of aluminum porphyrin such as (tetraphenylporphinato)aluminum chloride, coupled with a quaternary ammominum or phosphonium salt, such as ethyltriphenylphosphonium bromide, is a good catalyst for this copolymerization. Of much interest is the very narrow molecular weight distribution of the copolymer obtained by this system. Aluminum porphyrin and the quaternary salt forms a hexa-coordinate complex, and the copolymerization reaction is considered to proceed on both sides of the porphyrin ring. Some properties of the copolymer for possible application are described.

1. INTRODUCTION

Examples of catalytic fixation of CO_2 with metal complexes have been rather limited[1)2)]. In order to achieve catalytic processes there are considered two fundamental types. One is the oxidative addition of unsaturated compound or hydrogen to low valent transition metal complex, followed by the insertion of CO_2 to the metal-carbon or metal-hydrogen bond, and the reductive elimination of a derivative of carboxylic acid or formic acid, regenerating the low valent complex. The other type is the repetition of addition reactions, in which CO_2 reacts with the metal complex to form a metal carboxylate that can add to unsaturated or cyclic compound to regenerate the original metallic species. No change in the valency of the metal is involved, and the product is a linear macromolecule or, sometimes, a cyclic compound.

M. Aresta and G. Forti (eds.), Carbon Dioxide as a Source of Carbon, 331–337.

$$\text{M-X} \xrightarrow{CO_2} \text{M-}O_2\text{C-X} \xrightarrow{A} \text{M-A-}O_2\text{C-X}$$

$$\xrightarrow{CO_2} \text{M-}O_2\text{C-A-}O_2\text{C-X} \xrightarrow{A} \text{M-A-}O_2\text{C-A-}O_2\text{C-X} \longrightarrow \text{---} \qquad (1)$$

The first example of the latter type of catalytic fixation of CO_2 to form polymeric product was discovered by the author and the coworkers in 1968[3]. This is the alternating copolymerization of CO_2 and epoxide to give aliphatic polycarbonate.

$$\underset{\diagdown O \diagup}{CH_2\text{-}CHR} + CO_2 \longrightarrow \left(CH_2\text{-}CHR\text{-}O\text{-}\underset{\underset{O}{\|}}{C}\text{-}O \right)_x \qquad (2)$$

By using an appropriate catalyst system, the reaction proceeds under mild conditions such as at room temperature and under a moderate pressure of CO_2, to give the product with molecular weight of 100,000-150,000.

Catalyst is of primary importance for the copolymerization of CO_2 and epoxide. A representative catalyst is the equimolar reaction product of diethylzinc and water. Numerous catalyst systems have been investigated mainly based on organozinc compound in a number of places all over the world, in order to develop the system with high activity[4)-7)]. Catalyst systems for the copolymerization of CO_2 and epoxide, mainly propylene oxide, are listed in TABLE I[5)].

TABLE I. Catalyst systems for the copolymerization of CO_2 and epoxide

$ZnEt_2\text{-}H_2O$	$AlEt_3\text{-}PPh_3$
$ZnEt_2\text{-}H_2O\text{-}CO_2$ (aged)	$AlEt_2NEt_2$
$ZnEt_2\text{-}RNH_2$	Al(TPP)(OMe)[a)]
$ZnEt_2\text{-}C_6H_4(OH)_2$ ($\underline{m}$, $\underline{p}$)	$Al(TPP)Cl\text{-}PPh_3$
$ZnEt_2\text{-}C_6H_4(COOH)_2$ ($\underline{m}$, $\underline{p}$)	$Al(TPP)Cl\text{-}PPh_3EtBr$
$ZnEt_2\text{-}C_6H_4(OH)(COOH)$ ($\underline{m}$, $\underline{p}$)	
$ZnEt_2\text{-}C_6H_3(OH)_3$	$MgEt_2\text{-}H_2O$
$ZnEt_2\text{-}C_6H_4(NH_2)_2$ ($\underline{o}$)	$MgEt_2\text{-}RNH_2$
$ZnEt_2\text{-}C_6H_4(SH)_2$ ($\underline{m}$)	$MgEt_2\text{-}C_6H_4(OH)(COOH)$ ($\underline{m}$)
$ZnEt_2$-poly($\underline{p}$-OH-styrene)	$Mg(OAc)_2$
$ZnEt_2$-γ-Alumina	
$ZnEtNPh_2$	$Co(OAc)_2$
$Zn(CO_3)_2$	$Ni(OAc)_2$
Zn(OAc)	
$ZnO\text{-}C_6H_4(COOH)_2$ ($\underline{m}$)	
$Zn(OH)_2\text{-}C_6H_4(COOH)_2$ ($\underline{m}$)	

a) (TPP): Tetraphenylporphinato; see $\underline{4}$.

2. ORGANOZINC CATALYST FOR CO_2-EPOXIDE COPOLYMERIZATION

As shown in TABLE I, there have been reported a number of catalyst systems based on organozinc compound[8]. Elementary reactions for the copolymerization of CO_2 and epoxide with an organozinc catalyst system is considered to be the reaction of a zinc alkoxide with CO_2 to form a zinc alkyl carbonate (3), and the reaction of the zinc alkyl carbonate with epoxide to regenerate zinc alkoxide (4).

$$\text{-C-C-O-ZnX} + CO_2 \longrightarrow \text{-C-C-O-}\underset{\underset{\text{O}}{\|}}{\text{C}}\text{-O-ZnX} \qquad (3)$$

$$\text{-C-C-O-}\underset{\underset{\text{O}}{\|}}{\text{C}}\text{-O-ZnX} + \underset{\diagdown\text{O}\diagup}{\text{C-C}} \longrightarrow \text{-C-C-O-}\underset{\underset{\text{O}}{\|}}{\text{C}}\text{-O-C-C-O-ZnX} \qquad (4)$$

Of the two elementary steps, reaction (3) is a typical reaction of CO_2 involving the addition of a metal alkoxide. It should be noted that the alkoxide can also react with epoxide. In fact, epoxide is known to homopolymerize to a polyether with similar catalyst systems. Therefore, CO_2 is much more reactive with zinc alkoxide than epoxide. The key step by which the whole copolymerization reaction proceeds catalytically is reaction (4), in which zinc alkoxide is regenerated. Although epoxide appears more reactive than CO_2 in the latter step, the factors governing this apparent reversal of reactivity, thermodynamic or kinetic, are not yet clear.

As seen in TABLE I, many of the effective catalyst systems are the reaction product between diethylzinc and a compound with two or more protic hydrogens. The product of the reaction between polyfunctional compounds is not considered simple, but will give a mixture of oligomeric products. As for the reaction of diethylzinc with water or resorcinol, respectively:

$$n\ Et_2Zn + n\ H_2O \longrightarrow Et\text{-}Zn(\text{-O-Zn-})_{n-1}OH \qquad (5)$$

$$n\ Et_2Zn + n\ C_6H_4(OH)_2(m) \longrightarrow Et(\text{-Zn-O-}C_6H_4\text{-O(m)-})_nH \qquad (6)$$

With the diethylzinc-resorcinol system (6), the copolymerization reaction takes place at the zinc-resorcinolate linkage of the catalyst, but not at the zinc-ethyl bond, since the end group of the copolymer is always a resorcinol group[9].

The products of the reactions of diethylzinc and polyhydric compound such as (5) and (6) tend to undergo intermolecular association, and many of such systems are heterogenous. Some of the related systems, however, are homogenous, probably because intramolecular coordination prevents intermolecular association. Structures of the active species of some homogeneous systems have been suggested as 1-3.

In general, the structure of X in reactions (3) and (4) varies depending upon the combination of the catalyst components, and is considered to affect 1) the nucleophilicity of the zinc alkoxide or zinc alkyl carbonate, 2) the ability of the zinc atom to be coordinated by carbon dioxide or epoxide, and 3) the strength and the extent of the association of the catalyst system.

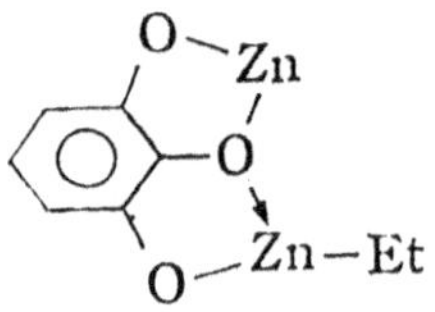

1

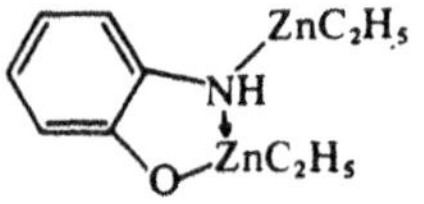

2

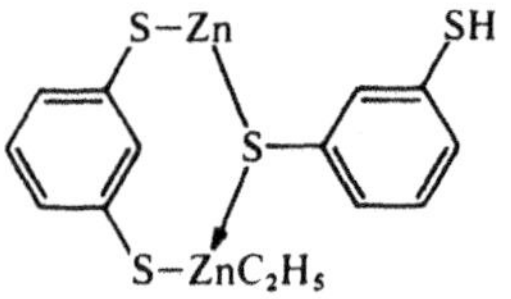

3

In this connection, the reactivities of organozinc and related compounds with simpler structures are of interest. Diethylzinc does not react with CO_2 under ordinary conditions. Complexation with 2,2'-bipyridyl or 1-methylimidazole much enhances the reactivity. This is in clear contrast with the reaction of triethylaluminum and CO_2, where tertiary amine suppresses the reaction. However, diethylzinc-2,2'-bipyridyl system, for example, does not show catalytic activity for the CO_2-epoxide copolymerization. Ethylzinc methoxide does not react with CO_2, but zinc dimethoxide, though insoluble, readily reacts with CO_2. Neither of these alkoxides, however, does not catalyze the copolymerization. Thus, the factors governing the activity of effective catalyst systems have not yet been fully understood.

3. ALUMINUM PORPHYRIN CATALYST FOR CO_2-EPOXIDE COPOLYMERIZATION

Although organoaluminum systems are known as effective catalysts for the homopolymerization of epoxide, triethylaluminum-water system, for example, is not very effective for the CO_2-epoxide copolymerization. The product is not alternating, but rich in the epoxide units or ether linkages. Recently, we have found that the system based on aluminum porphyrin is effective for the alternating copolymerization of CO_2 and epoxide[10]. Another very interesting behavior of this catalyst system is the fact that the copolymer has a very narrow molecular weight distribution. This is exemplified by the comparison of the gel permeation chromatography (GPC) curves of the copolymers obtained by aluminum porphyrin catalyst and by diethylzinc-water catalyst (Figure 1).

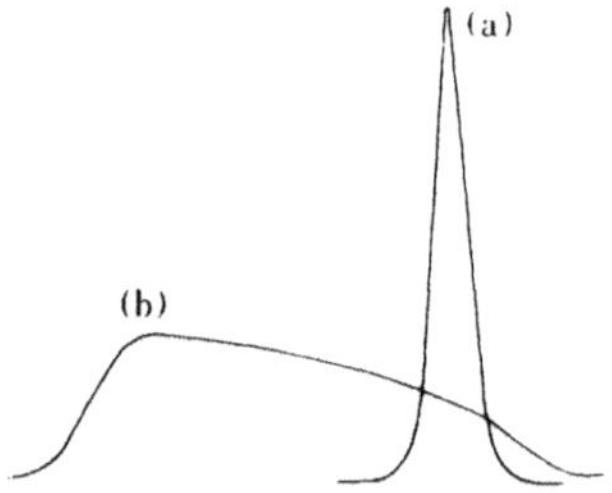

Figure 1. GPC curves for CO_2-propylene oxide alternating copolymer obtained by (a) aluminum porphyrin(4, X=Cl)-triphenylphosphine system, and (b) diethylzinc-water system.

The synthesis of macromolecules with controlled molecular weight is of importance for the molecular design of polymeric materials. The present finding is the first example of the synthesis of alternating copolymer with narrow molecular weight distribution.

Our recent finding that some aluminum porphyrins, such as (tetraphenylporphinato)aluminum chloride (4, X=Cl: TPPAlCl), is an excellent initiator or catalyst for the ring-opening polymerization of epoxide[11] and lactone[12], to give the corresponding polymers with narrow molecular weight distribution, prompted us to examine the possible development of the catalyst for the CO_2-epoxide copolymerization.

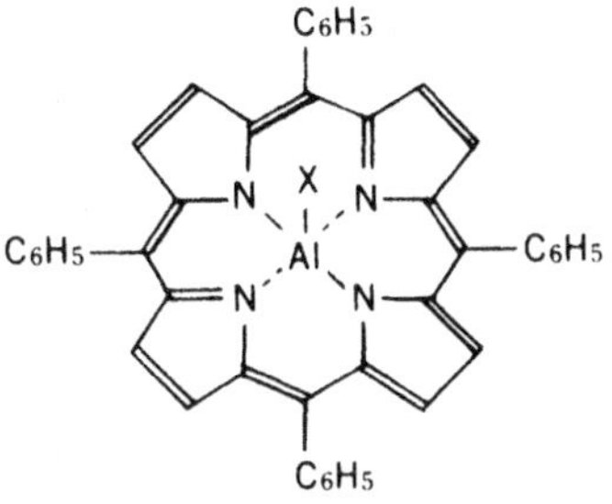

4

However, aluminum porphyrin was not very effective for the copolymerization. The reaction proceeded slowly to give the product rich in epoxide units or ether linkages, but the narrow molecular distribution of the product was noteworthy.

In order to enhance the reactivity of the catalyst to CO_2, the addition of the sixth coordinating ligand to the porphyrin was attempted. Although the addition of an aromatic tertiary amine such as 1-methylimidazole much accelerated the reaction, the exclusive product was not the copolymer but cyclic carbonate[13]. In contrast, the addition of triphenylphosphine brought about the formation of the alternating copolymer with narrow molecular weight distribution[10].

Since ^{31}P-NMR studies of the related reaction system indicated the quaternization of triphenylphosphine, the catalytic activity of aluminum porphyrin-quaternary phosphonium or ammonium salt system was examined. As the result, (TPP)AlCl-ethyltriphenylphosphonium bromide ($EtPh_3PBr$) system, for example, was found to be a good catalyst for the alternating copolymerization of CO_2 and epoxide to give the product with narrow molecular weight distribution[10]. As shown in Figure 2, the molecular weight of the copolymer can be controlled by the ratio of the epoxide to the catalyst in the presence of excess CO_2, retaining the narrow molecular weight distribution as indicated by the ratio of the weight average to number average molecular weights ($\bar{M}w/\bar{M}n$) ranging from 1.1-1.2. As described above, aluminum porphyrin alone is not a very good catalyst for the CO_2-epoxide copolymerization. Neither quaternary phosphonium nor ammonium salt alone brings about the reaction. Since (TPP)Al-halide and (TPP)Al-O_2CR form hexa-coordinate complex with

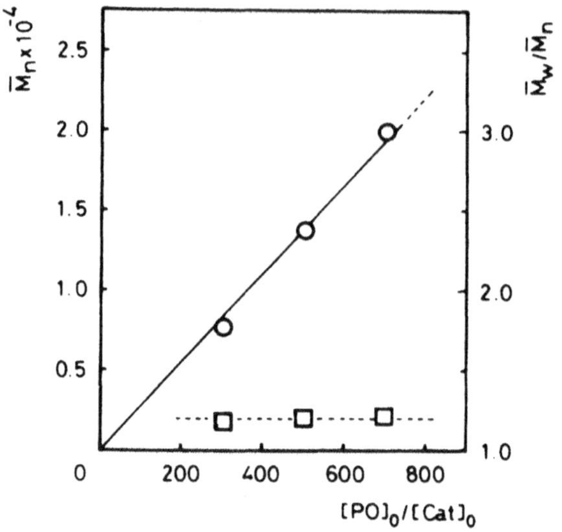

Figure 2. Copolymerization of CO_2 and propylene oxide (PO) with (TPP)AlOR-PPh_3 system. Relationship between $\bar{M}n$ (o) ($\bar{M}w/\bar{M}n$ (□)) of copolymer and $[PO]_o/[Cat]_o$; conversion, 100 %.

quaternary phosphonium or ammonium halide and carboxylate, respectively[15], the CO_2-epoxide copolymerization is considered to proceed on both sides of the ring of the aluminum porphyrin (Figure 3).

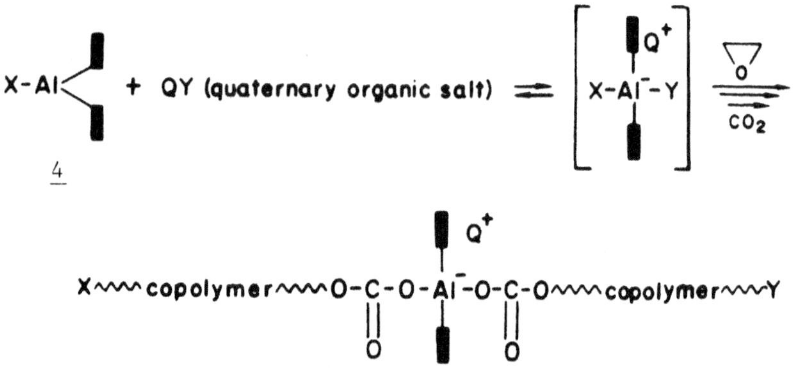

Figure 3. Mechanism of CO_2-epoxide copolymerization with the aluminum porphyrin-quaternary salt system.

In conformity with this mechanism, (TPP)AlCl-$EtPh_3PBr$ (1/1) system, for example, gives twice as many copolymer molecules as molecules of aluminum porphyrin.

4. SOME PROPERTIES OF CO_2-EPOXIDE COPOLYMER

High molecular weight alternating copolymer of CO_2 and propylene oxide or ethylene oxide may be molded to sheet or film by pressing it with heating at 140-150°C. Physical and other properties were reported[6]. Low permeability of oxgen is one of the characteristics. Low heat of combustion, because of the high oxygen content, and relatively low temperature of decomposition which takes place within a narrow temperature range are noteworthy. The product of the thermal decomposition of CO_2-propylene oxide copolymer, for example, at 180°C is exclusively propylene carbonate. Structure of aliphatic polycarbonate suggests the biodegradability. In fact, CO_2-ethylene oxide copolymer was confirmed to be biodegradable[14]. In this connection, fundamental studies as to the possible application to the controlled release drug system have been carried out[15].

REFERENCES

1) S. Inoue, N. Yamazaki, ed., 'Organic and Bio-organic Chemistry of Carbon Dioxide', Kodansha, Tokyo (1981).
2) S. Inoue, H. Koinuma, Rev. Inorg. Chem., 6, 291 (1984).
3) S. Inoue, H. Koinuma, T. Tsuruta, Polymer Letters, 7, 287 (1969).
4) S. Inoue, Progr. Polym. Sci. Jpn., 8, 1 (1975).
5) S. Inoue, in ref. 1), p. 167.
6) S. Inoue, Chemtech., 6, 588 (1976).
7) A. Rokicki, W. Kuran, J. Macromol. Sci., Rev. Macromol. Chem., C21, 135 (1981).
8) Recent examples: P. Gorecki, W. Kuran, J. Polym. Sci., Polym. Lett. Ed., 23, 299 (1985); P. W. Ledner, N. C. Rol, J. Chem. Soc., Chem. Commun., 1985, 598.
9) Y. Hino, Y. Yoshida, S. Inoue, Polym. J., 16, 159 (1984).
10) T. Aida, M. Ishikawa, S. Inoue, Macromolecules, 19, 8 (1986).
11) T. Aida, S. Inoue, Macromolecules, 14, 1162 (1981).
12) T. Yasuda, T. Aida, S. Inoue, Macromolecules, 17, 2217 (1984).
13) N. Takeda, S. Inoue, Bull. Chem. Soc. Jpn., 51, 3564 (1978); T. Aida, S. Inoue, J. Am. Chem. Soc., 105, 1304 (1983).
14) T. Kawaguchi, M. Nakano, K. Juni, S. Inoue, Y. Yoshida, Chem. Pharm. Bull., 31, 1400 (1983).
15) T. Kojima, M. Nakano, K. Juni, S. Inoue, Y. Yoshida, Chem. Pharm. Bull., 32, 2795 (1984).

ELECTROCHEMICAL SYNTHESES OF CARBOXYLIC ACIDS FROM CARBON DIOXIDE

Giuseppe Silvestri
Istituto di Ingegneria Chimica, Universita' di Palermo
Viale delle Scienze
90128 Palermo
Italy

1. INTRODUCTION

The first report on the reduction of carbon dioxide is due to the venerable name of Hermann Kolbe, who, in 1861, published with R. Schmitt the reduction of carbon dioxide to formate with metallic potassium (1). Seven years later he introduced to the Chemical Society in London his assistant E. Drechsel, who "after many unsuccessful experiments, made during several years" (2) at the end succeeded in reducing carbon dioxide to oxalate with metallic sodium or with potassium amalgam in anhydrous conditions. Soon afterwards Beketov (3) and Royer (4) opened the series of papers dealing with the direct involvement of carbon dioxide in electrolytic processes. Since then the reduction of carbon dioxide has been object of continuous investigations, and direct or indirect methodologies have been applied extensively.

To the thermodinamic and kinetic aspects of the reduction of carbon dioxide a separate chapter is devoted in this book. Here, the most important synthetic aspects of the electrochemistry of carbon dioxide are taken under consideration : in the first part the syntheses from carbon dioxide alone are shortly reviewed, and in the second part a series of processes in which carbon dioxide is used as a co-starting material for the reductive synthesis of carboxylic acids are examined .

2. SYNTHESES FROM CARBON DIOXIDE ALONE

The electrolytic reduction of carbon dioxide in water leads essentially to formic acid or to formates. In several of the papers dealing with this reaction at the beginning of this century (5 - 11), very high current efficiencies were reported for short-range electrolyses, but the wasteful reoxidation at the anode of the formate, even if glass frit diaphragms were used, hindered any possibility of the

M. Aresta and G. Forti (eds.), Carbon Dioxide as a Source of Carbon, 339–369.

use of this reaction for large scale productions. The problem was solved after a long time by the use of a cation exchange membrane to separate anolyte from catholyte (12). In this way the current efficiency remained very high even at concentrations of formate in the catholyte up to 20 %. An indirect process, making use of metal amalgams in water (13a) gives similar formate concentrations, but has the drawback of the formation of significant quantities of hydrogenocarbonates which precipitate in the reaction medium and have to be continuously filtered off. In alcoholic media (13b) very high rates of decomposition of the amalgams were reported, with electric yields arriving up to 90% . A further increase of the decomposition rate of the amalgam was reached by increasing the pressure of carbon dioxide up to 10 Atm. In fact, direct or indirect electrochemical routes are not competitive with the present technologies of production of formic acid, separated as by-product of the oxidation of hydrocarbons to acetic acid, or obtained from carbon monoxide.

The reduction of carbon dioxide to oxalate was performed, until the diffusion of the electrochemical processes in organic media, mainly by gas phase reaction with metals of the groups IA and IIA and their amalgams.

Patents (14,15) and scientific papers (16 - 21) reporting a progressively more refined technology of this reaction appeared until the sixties. The improvement in the analytical techniques led to the identification, depending on the reaction conditions and on the kind of reductant used, of products of further reduction of oxalic acid, like glycolates (20,21),or even of more complex structures,as tartronic acid (20). In non aqueous media, like hydrocarbons, DMF, dioxane or even halogenated solvents or acetone, reduction of carbon dioxide with amalgams gave rise only to carbon monoxide, with the exception of hexane, in which the reduction gave a mixture of products consisting of 40% of oxalate and 60% of carbonate (22).

The electrochemical reduction of carbon dioxide to oxalate was patented and published by several research groups in the seventies (23 - 29). Most of the procedures proposed made use of divided cells. Aluminum (25) or zinc (29) sacrificial anodes were proposed in undivided cells, being the corresponding oxalates the primary products of the syntheses.

As already seen for the reaction with metals or amalgams, even by electrolysis in media of low proton availability the reduction can proceed further, leading to glycolic acid (26,30) and to glyoxylic, malic, tartaric and succinic acids (26). From the published results it does not seem that these reactions are selective enough to be used for synthetic purposes.

The synthesis of oxalic acid from carbon dioxide does not seem to be competitive if compared with the existing technologies, among which the oxidation of naturally occurring carbohydrates with nitric acid is the most diffused and particularly advantageous (31). Nevertheless it is interesting to note that a cost estimate on an electrochemical integrated system with sacrificial zinc anodes (32), gave figures of the production costs near to the selling price of oxalic acid (W. Germany in 1978).

In this system zinc is anodically dissolved in the carboxylation cell, then recovered in a separate aqueous electrolysis unit, from which the zinc bars can be recycled to the synthesis unit.The reactions taking place in the subsequent steps of such a system are listed in scheme 1.

SCHEME 1

electrolytic synthesis of zinc oxalate

cath. $2\ CO_2 + 2\ e^- \longrightarrow C_2O_4^{2-}$

an. $Zn - 2\ e^- \longrightarrow Zn^{2+}$

overall $2\ CO_2 + Zn \longrightarrow ZnC_2O_4$

zinc oxalate hydrolysis $ZnC_2O_4 + H_2SO_4 \longrightarrow ZnSO_4 + H_2C_2O_4$

zinc electrolytic recovery $ZnSO_4 + H_2O \longrightarrow Zn + H_2SO_4 + 1/2\ O_2$

SUM OF REACTIONS $2\ CO_2 + H_2O \longrightarrow H_2C_2O_4 + 1/2\ O_2$

The system had Cr.Ni-steel cathode, solvent acetonitrile, with tetrabutylammonuim perchlorate as supporting electrolyte. Current and product yields were respectively 85 % and 90 % .
This system, which in principle does not pose problems of waste disposal, could become attractive in the case of an increase of the cost of conventional methodologies due to more severe restrictions in the environmental control.

If divided cells are used (33) with an aprotic solvent in the catholyte and an aqueous solution in the anolyte, current yields are much lower (50-60 %) and the process looks as completely unattractive.

Direct or indirect electrochemical processes have been proposed for the removal of carbon dioxide in atmosphere regeneration units in space systems (34a) and in submarines (34b,c).

The use of carbon dioxide as a source of chemicals, in the context of the electrochemistry of biomass, has been recently reviewed (35).

3. ELECTROCARBOXYLATIONS.

The term "electrocarboxylation", was proposed several years ago (23) to classify the rather substantial group of electrochemical processes by which one or more carboxylic groups are introduced into suitable organic substrates.

In the large majority of examples carbon dioxide is used as the source of the carboxylic groups, but some reactions are quoted here, in which the carboxylic groups arise from the anodic oxidation of oxalates, which, in some extent, can be considered as CO_2 derivatives.

It is well known, in the study of reaction mechanisms, that carbon dioxide is a useful diagnostic reactive for the trapping of intermediate anions, whose structure can be determined on the basis of that of the carboxylic acid isolated. For many years carbon dioxide entered the electrochemical studies for this purpose, even if the synthetic implications of these processes were underlined in several cases.

Synthetic aspects have assumed in the last years predominant importance, in such a way that now it is possible to define the electrocarboxylations as an effective synthetic method competitive, in several cases, with the existing conventional processes.In fact, it must be stressed that the interest of these processes should go far beyond the scale of the laboratory synthesis, as many of the carboxylic acids which could be prepared by electrocarboxylations, such as tetra- or hexa- carboxylic acids, α-hydroxyacids, α-aminoacids, α-arylpropionic acids find present industrial application.

The applicative interest of these reactions has to be considered in view of the fact that different electrocarboxylation processes can be performed in the same reactor, and, in many cases, a proper choice of the experimental conditions allows to obtain different products from the same starting material.

Electrocarboxylations by insertion of carbon dioxide may be grouped in two main sub-classes :

a) e.c. by addition, taking place with starting materials bearing different kinds of unsaturations, such as olefins, ketones, aldehydes, aromatics, imines, acetylene derivatives, heterocycles and so forth;

b) e.c. by substitution, taking place with starting materials, bearing good leaving groups, which are transformed into the corresponding carboxylic acids, or their esters.

These two groups shall be examined separately, following, as far as possible, the reaction mechanism as the classification guide inside each sub-class.

3.1. ELECTROCARBOXYLATIONS BY ADDITION.

3.1.1 Electrocarboxylation of mono-olefins.

The electrocarboxylation of olefinic compounds may lead to different products, depending on the molecular structure of the starting material, on the composition of the electrolytic medium and on the cathodic potential maintained during the electrolysis.

Scheme 2 lists the four fundamental reaction pathways which may be realized by electrocarboxylation of a mono-olefin.

Dicarboxylations (i) and monocarboxylations (ii) have been realized in several cases with good yields, although ii are less selective. Carboxylative dimerizations (iii) may be applied only to some particular starting materials and telomerizations (iv) have been quoted for lower olefins only. All these reactions are performed in organic, generally aprotic solvents. Residual water, or protic substances added on purpose in low concentrations, address the syntheses towards the monocarboxylation.

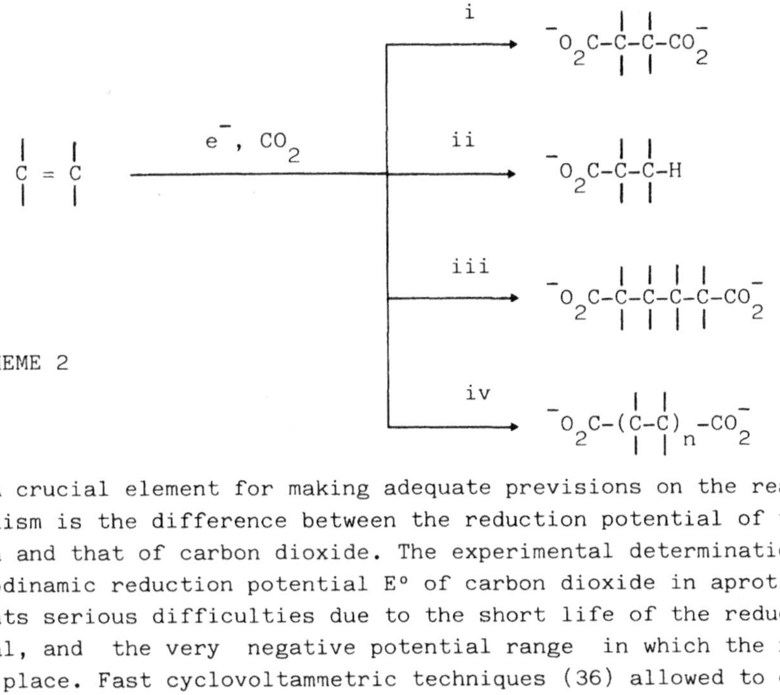

SCHEME 2

A crucial element for making adequate previsions on the reaction mechanism is the difference between the reduction potential of the olefin and that of carbon dioxide. The experimental determination of the thermodinamic reduction potential E° of carbon dioxide in aprotic media presents serious difficulties due to the short life of the reduced anion radical, and the very negative potential range in which the reduction takes place. Fast cyclovoltammetric techniques (36) allowed to observe a rather pronounced reversibility of the reduction wave , and a value of E° = – 2.21 V vs. SCE was derived. In anhydrous DMF an irreversible cathodic peack, attributable to a slow one-electron transfer to carbon dioxide, for a sweep rate of 0.1 V/s , is observed at -2.8 V on Pt microelectrode, i. e. at potentials far more negative than E°.

If the cathodic peack of the olefin I is more positive than that of carbon dioxide in the same conditions, the first cathodic step involves the reduction of I and the formation of the corresponding anion radical II. As the following step, II may react with carbon dioxide not only as a nucleophile, forming the corresponding carboxylate anion radical III, but even as a reducing agent, by an electron transfer to carbon dioxide,

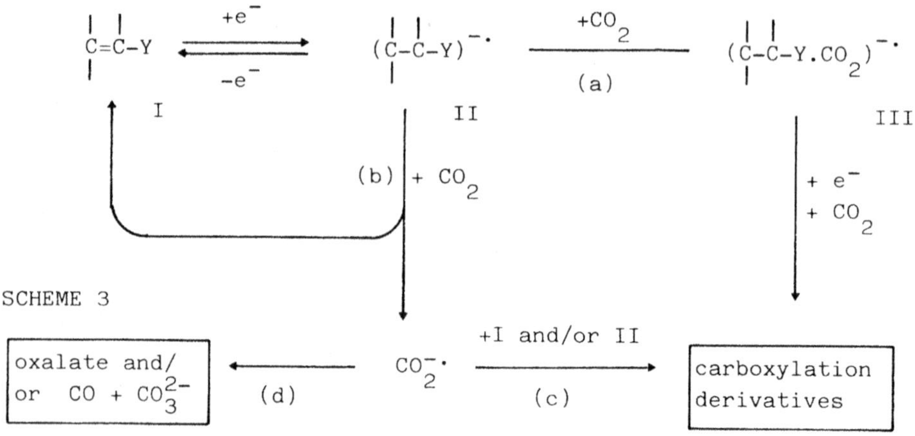

SCHEME 3

giving rise to the $CO_2^{-\cdot}$ anion radical (see scheme 3; Y = electron withdrawing group). Since, according to the Marcus theory (37), the rate of the outer sphere charge transfer between II and carbon dioxide is expected to increase with the standard Gibbs energy of the reaction, from the standard potential difference

$$\Delta E° = E°_{CO_2/CO_2^{-\cdot}} - E°_{I/II}$$

it is possible to have useful information on the competition between the two possible pathways. It has been deduced (37) on grounds of kinetic evidence, that when $\Delta E°$ is lower than about - 400 mV the mechanism involving nucleophilic attack as the initial step should be predominant.

Let us consider separately the following three possible cases:

a) $\Delta E° < -400$ mV. In this case the cathodic depolarizer is almost certainly the olefin, and the carboxylations (i , ii or iii depending on the electrolysis conditions) are performed with good yields for many starting materials .

b) $\Delta E° > -400$ mV but the peack potential of the olefin is more positive than that of carbon dioxide .In this case the reaction mechanism involves the reduction of the olefin, and processes of electron transfer in solution (path b in scheme 3) are likely to occur . Side products , due to the high reactivity of the intermediate species originated in the electrolytic medium, are to be expected .

c) $\Delta E° > -400$ mV but the peack potential of the olefin is more negative than that of carbon dioxide : CO_2 is the cathodic depolarizer, and the electrocarboxylation goes through a radical attack of $CO_2^{-\cdot}$on the unsaturation. The formation of oxalates, carbon monoxide and carbonates is to be expected .

As far as case a) is considered, the mechanism accepted for the reduction of substituted olefins is recalled in scheme 4 : the olefin I is reduced to its corresponding anion radical II . A higher spin density in ß and a negative charge in α with respect to the activating group have been hypothized for II on the basis of the results of the hydrodimerization of some olefins (38) or deduced from structural investigations (39). II may dimerize to V or may be furtherly

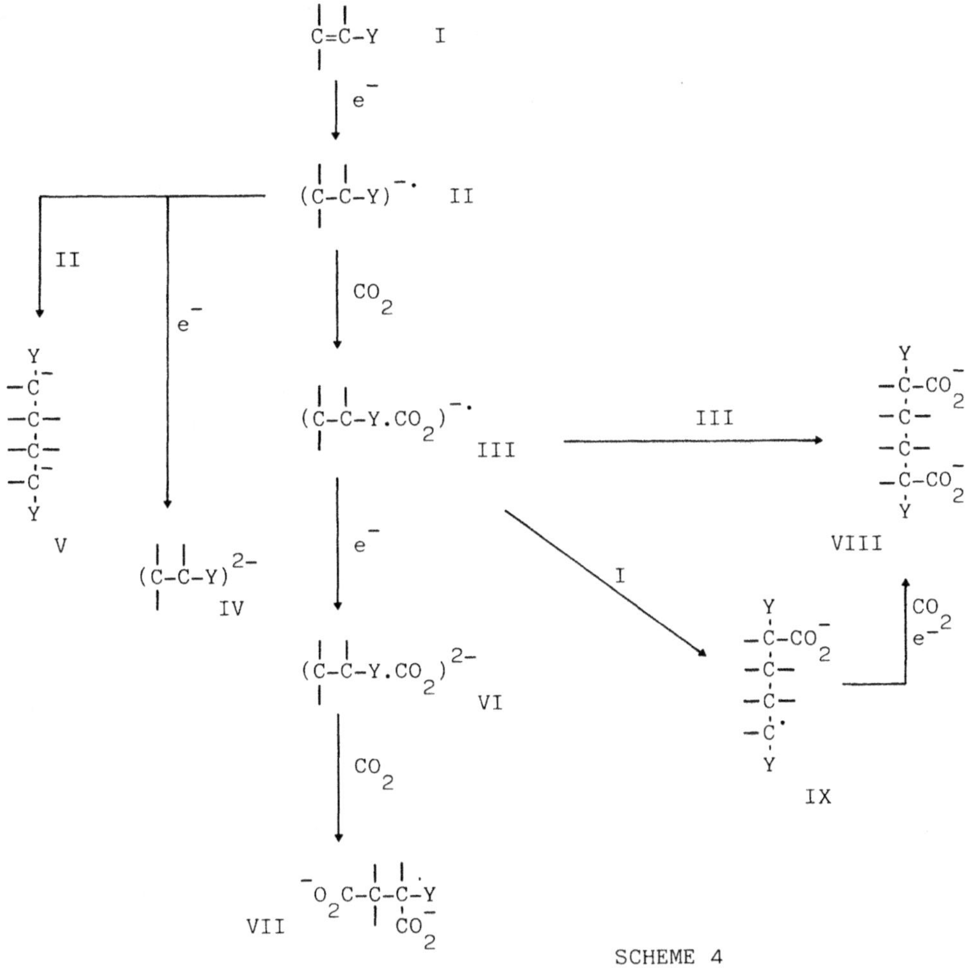

SCHEME 4

reduced to the dianion IV . If carbon dioxide is present in the reaction medium, in aprotic conditions, II in most cases reacts faster with it, and the corresponding carboxylate anion radical III is obtained. The further reduction of III to the corresponding dianion VI may derive from a further cathodic reaction, or from an electron exchange in solution with II . VI, by reaction with a second carbon dioxide molecule leads to the derivative of dicarboxylation VII. Alternatively III might undergo a radical-radical coupling to the dimer VIII.

In reaction scheme 4 , different pathways are possible for the carboxylation of different starting materials. From the point of view of the reaction mechanism, the following possibilities shall be examined separately :

1) III → VI takes place at potentials more positive or equal to I → II. This most common case is evidentiated by the polarographic or voltammetric paths : the first reduction wave, which is monoelectronic without carbon dioxide , shows with carbon dioxide a current increase, approaching the double of its initial value at higher concentrations of CO_2. In this case the sequence of steps of the dicarboxylation in aprotic conditions, is I → II → III → VI → VII. The dicarboxylation of acenaphtylene belongs to this case and is a simple and straightforward process (40,41).

2) III → VI takes place at potentials more negative than I → II. The polarographic paths show in this case a single monoelectronic reduction wave when I is reduced alone in aprotic conditions. A second distinct reduction wave at more negative potentials appears when carbon dioxide is added to the electrolytic medium, each wave still corresponding to the uptake of a single electron . Bulk electrolyses at the cathodic potential of the first reduction step give dimeric derivatives of dicarboxylation VIII, whereas at the cathodic potential corresponding to the uptake of the second electron the monomeric dicarboxylated product VII is obtained . This behavior was observed for the electrocarboxylation of dialkyl maleates and fumarates, thoroughly explored by industrial laboratories, in view of the production of polydentate ligands to be used as builders for detergents (23,42 - 46).

The reductive sequence I → II → III → VIII was demonstrated for dialkylfumarates and maleates by Yeh and Bard (47). Other alternate paths as the addition of III onto I followed by reduction and carboxylation were discarded on the basis of the results of bulk electrolyses in which the concentration of maleate did not exert any influence on the yields of the dimer VIII (43). The carboxylation of maleates at the potential corresponding to the second reduction step leads to the monomeric dialkyl tetracarboxylate dianion VII, and the reductive sequence is summarized as I → II → III → VI → VII (42).

Olefins belonging to case b are very interesting starting materials for the production of useful derivatives both by dicarboxylation and by hydrocarboxylation.As already pointed out, these olefins could give rise, when reduced in the presence of carbon dioxide, to homogeneous charge transfer processes by which the $CO_2^{-\cdot}$ anion radical is formed in the reaction medium.A typical example is that of ethyl acrylate : the monoelectronic reduction wave displayed in the absence of carbon dioxide shows a current increase corresponding to the uptake of far more than two electrons when CO_2 is added to the electrolytic medium (48). An enhancement factor of the reduction wave higher than two is indicative of the occurrence of the sequence of steps already shown in scheme 3, in which paths b and d have to take part in the evolution of $CO_2^{-\cdot}$ anion radical.

The mechanism of the electrocarboxylation should involve therefore a radical attack of $CO_2^{-\cdot}$ on the olefin, thus leading to the intermediate III of scheme 4. III should evolve in the way shown in that scheme, and/or with radical-radical couplings involving $CO_2^{-\cdot}$ and II, III or IX.

E° values have been reported in the literature only for very few olefins. Nevertheless on the basis of the polarographic E 1/2, at a first approximation it is possible to attribute to case b the olefins listed below.

Acrylonitrile (42,50,52), methyl acrylate (23,42,43), methyl methacrylate (23,42), methylcrotonate (23,42) were dicarboxylated with yields between 40 and 60 % .The electrocarboxylative dimerization of methyl acrylate to tetramethyl 1,1,4,4-tetracarboxybutane takes place, at high concentration of acrylate, with yields up to 47 % (43). The reaction mechanism should involve steps II → III → IX → VIII (see scheme 4). It is worth to quote here that many of the carboxylated products quoted in this review are recovered as alkyl esters, obtained by nucleophilic substitution of the carboxylate anions on the corresponding alkyl halides, catalyzed by the tetraalkylammonium salts used as supporting electrolytes (74).

The dicarboxylation of stilbenes leads to 2,3-diphenylsuccinic acid (49 - 51). This synthesis was studied in the course of an investigation on the stereochemical effects in the electrochemistry of hindered molecules (51). DMF proved to be the best solvent for the carboxylation of both stilbenes to 2,3-diphenylsuccinic acid . The cis isomer gave a ratio of dl to meso isomers of 1.4, whereas the trans isomer gave a ratio of 2.7 , with overall yields up to 86 % .

Styrene can be carboxylated to 2-phenylsuccinic acid (85 % yield) or to 3-phenylpropionic acid (65 % yield)(53 - 55). This olefin has a standard reduction potential E° = - 2.58 V (S.C.E.), i.e. more negative than E° of carbon dioxide. The mechanism of these electrocarboxylations has been recently investigated (54). A comparison of the cyclovoltammograms of styrene alone and of carbon dioxide alone, on Pt in the same conditions, shows that Ep_{Sty} is more positive than Ep_{CO_2}. On grounds of electroanalytical and preparative results, the existence of a homogeneous electron transfer mechanism was hypothized. The use of an inert charge transfer agent, as benzonitrile, which has a standard reduction potential E° = - 2.28 V, allows to perform the electrocarboxylation of styrene at this last cathodic potential, instead of that of styrene. The resulting reductive processes are summarized in scheme 5. From the synthetic point of view routes i) and ii) gave very similar results, and this could indirectly support the hypothesis that without benzonitrile (route i) or with benzonitrile (route ii) the carboxylation of styrene could follow the steps 1 → 2 → 4 → 5 → 6, being the radical-radical coupling 3 not involved.

It is interesting to note that, without styrene, benzonitrile acts as a charge transfer agent towards carbon dioxide, leading to good yields of oxalate (54).

	i) without benzonitrile		ii) with benzonitrile
1	$S + e^- \rightarrow S^{-\cdot}$		$B + e^- \rightarrow B^{-\cdot}$
2	$S^{-\cdot} + CO_2 \rightarrow S + CO_2^{-\cdot}$		$B^{-\cdot} + CO_2 \rightarrow B + CO_2^{-\cdot}$
3	$S^{-\cdot} + CO_2^{-\cdot} \rightarrow SCO_2^{2-}$		
4		$S + CO_2^{-\cdot} \rightarrow SCO_2^{-\cdot}$	
5		$SCO_2^{-\cdot} + e^- \rightarrow SCO_2^{2-}$	
6		$SCO_2^{2-} + CO_2 \rightarrow S(CO_2)_2^{2-}$	SCHEME 5

B = benzonitrile

S = styrene

An intriguing patent filed at the very beginning of the applied studies on the electrocarboxylation of olefins (56) claims the production of styrene carbonate in low yield by electrolysis of styrene in DMF at potentials more negative than those required for the carboxylation to phenylsuccinic acid.

Very few informations are available for mono-olefins belonging to case c. The carboxylation takes place through the reduction of carbon dioxide to its anion radical, and some further sequences of radical-olefin and radical-radical reactions can be imagined, but no studies on the reaction mechanisms were performed. Among the substrates for which synthetic results are reported, only trans-ß-methoxy-acrylate has a reduction potential in a range for which reliable electroanalytical data were recorded (42) . For norbornadiene and dihydropyran (23) only the formation of carboxylated products was quoted; ethylene was carboxylated to succinic acid (25,58), but variable amounts of derivatives of telomerization were always present as by products (59).

The influence of protic substances has been investigated for the electrocarboxylation of olefins falling into cases a and b . Looking at case a , from the results of a synthetic study on the hydrocarboxylation of alkylcynnamates and cynnamonitrile (37), the possible reaction pathway was deduced. The total absence of hydrocarboxylated products in which the carboxyl group is linked to the position α with respect to the activating group, together with the already quoted structural data concerning the charge distribution in the anion radical II, in which the higher charge density should be in the α position, allowed the Authors to propose the sequence summarized in scheme 6.The first chemical step after the cathodic reduction is the α-protonation of II to the neutral radical X , which is reduced to the anion XI. The two possible evolutions of XI are the carboxylation to XII or the complete hydrogenation to XIII. Yields between 55 and 90 % were reported (37,41).

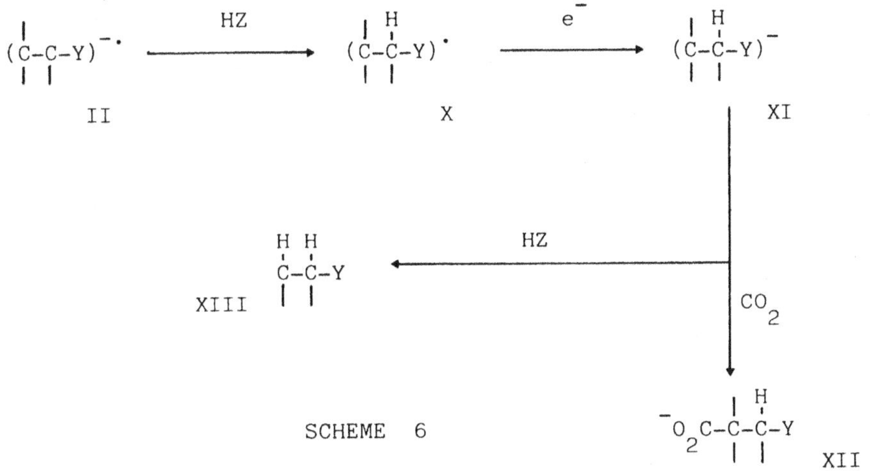

SCHEME 6

It is still not clear the reason for the inversion of the reactivity of the two intermediates II and XI , as the first is singularly selective towards the protonation, and the second, in the same reaction medium, rather selective towards the carboxylation . This reaction scheme can be applied to the hydrocarboxylation of other activated olefins falling into case a as acenaphtylene, 67% yield (40,41) and benzalacetone, 73% yield (41).

Things are of course different if hydrocarboxylation derivatives are obtained in anhydrous conditions . In this case the formation of the hydrocarboxylate (with the carboxyl group invariably in the ß position with respect to the activating group) may arise from a chemical step in the ECEC sequence, in which the solvent or the supporting electrolyte act as proton donors, or from the work-up of the effluents from the electrolysis.The last is the case of benzalacetophenone XIV (60), whose hydrocarboxylation goes through the electrochemical production of the dicarboxylated derivative XV, which decarboxylates upon acid hydrolysis leading to the final product XVI:

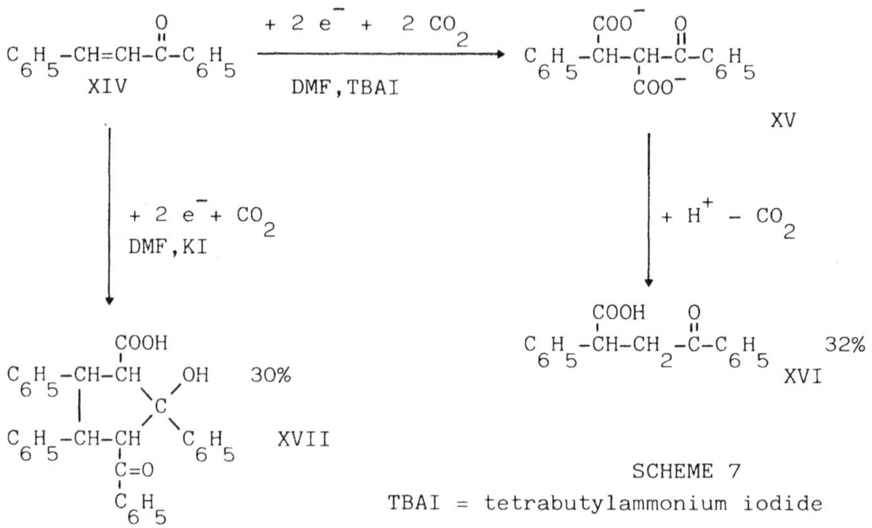

SCHEME 7

TBAI = tetrabutylammonium iodide

The electrolysis medium has a strong influence on the reactivity of the intermediates : from benzalacetophenone the dimeric cyclo--monocarboxylated XVII was obtained using potassium iodide instead of TBAI (60). The formation of XVII arises most probably from a radical or anionic addition of a monocarboxylate intermediate on XIV .

No informations are available on the carboxylation of maleates and fumarates in the presence of protic substances.

As far as case b is considered, it is interesting to observe that these olefins, which undergo very easily the hydrodimerization in aqueous media, require an aprotic solvent, to which only limited amounts of water are to be added, in order to achieve the hydrocarboxylation. In the case of acrylonitrile a continuous electrolytic system was proposed (52) in which propionitrile saturated with water was the electrolytic medium. The product, 3-cyano-propionic acid, was continuously extracted with water from the organic phase in order to avoid the cathodic hydrogen evolution due to the conversion of the weak carbonic acid into the stronger carboxylic one. A yield of 48 % was claimed.

The electrocarboxylation of a series of substituted ketene-S,S-acetals of general formula XVIII shows the unusual behavior summarized in scheme 8 (61,62). In this case the anion XX arising from the ECE first carboxylation step is unstable and decomposes loosing a S-Me⁻ anion, thus leading to a substitution reaction. The hydrocarboxylation can take place only if the thiogroups are linked as in XXI. Good yields are reported for starting materials having general formulae XVIII and XXI, and bearing various substituents.

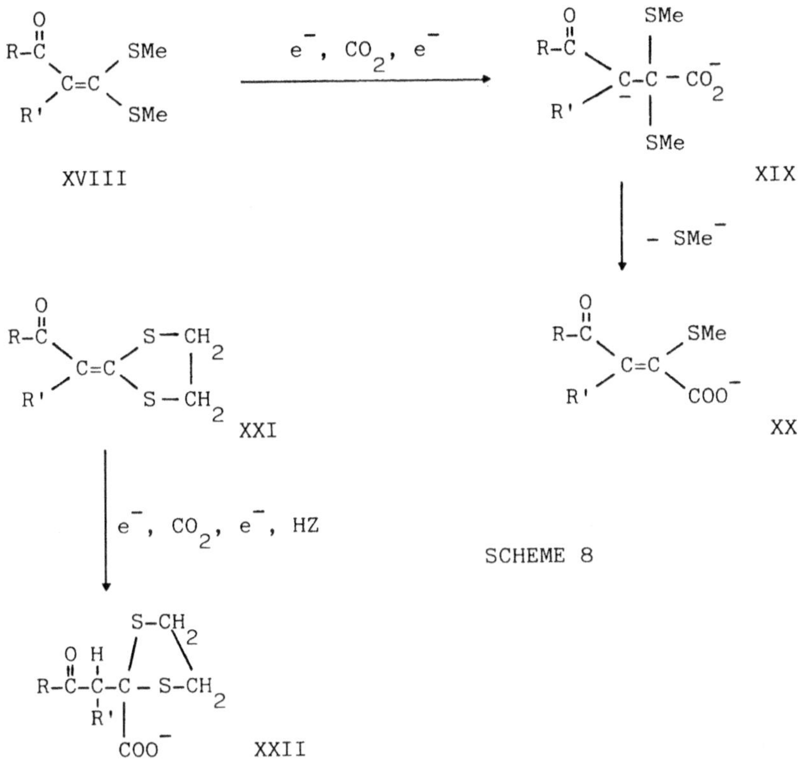

SCHEME 8

3.1.2. Electrocarboxylation of poly-olefins

The electrocarboxylation of butadiene and of other conjugated olefins is reported in several patents and papers, mainly arising from laboratories of applied research (46,50,56,63 - 66). No specific studies on the mechanism of these electrocarboxylations have been published so far, but the rather negative reduction potentials of conjugated diolefins, very near to that of carbon dioxide, should authorize the hypothesis that the anion radical of carbon dioxide , either arising from the direct reduction of CO_2 , or from an electron transfer from the butadiene anion radical, is involved in the process. Whatever mechanism is supposed for the evolution of the butadiene monocarboxylate anion radical or dianion, the presence of two reactive sites is demonstrated by the formation of a series of products, i.e. dicarboxylates , monocarboxylates, dimers and oligomers , resulting from the additions in the 1-4 and 1-2 positions. Therefore the process shows a moderate selectivity which frustrates any applicative project. The enhancement of the selectivity in the 1,4-dicarboxylation has been claimed recently in a patent (67): performing the reaction in tetrahydrofurane, in the presence of catalytic amounts of the complex $Cp_2Fe_2(CO)_4$, the carboxylation is achieved with an overall current yield of 76% and a selectivity in the 1,4-dicarboxylated product (mixture of cis + trans 3-hexenedioic acid) up to 75% . The catalyst seems to be more active towards the conjugated system; the carboxylations of mono olefins, such as methylacrylate, styrene or ethylene, do not show relevant improvements if compared with the uncatalyzed processes. This system constitutes, to our knowledge, the only example of the use of transition metal catalysts in the carboxylation of olefins: the very good results obtained with butadiene should encourage further efforts in this stimulating field.

If the unsaturations are isolated and the steric configuration of the molecule is favourable, the carboxylation may lead to high yield intramolecular cyclizations. Thus a series of bis-activated diolefins of general formula :

$$H_3COOC{-}CH{=}CH{-}(CH_2)_n{-}CH{=}CH{-}COOCH_3 \qquad n = 2,3,4.$$

has been investigated (43), and the dicarboxylative cyclizations summarized in scheme 9 were described .

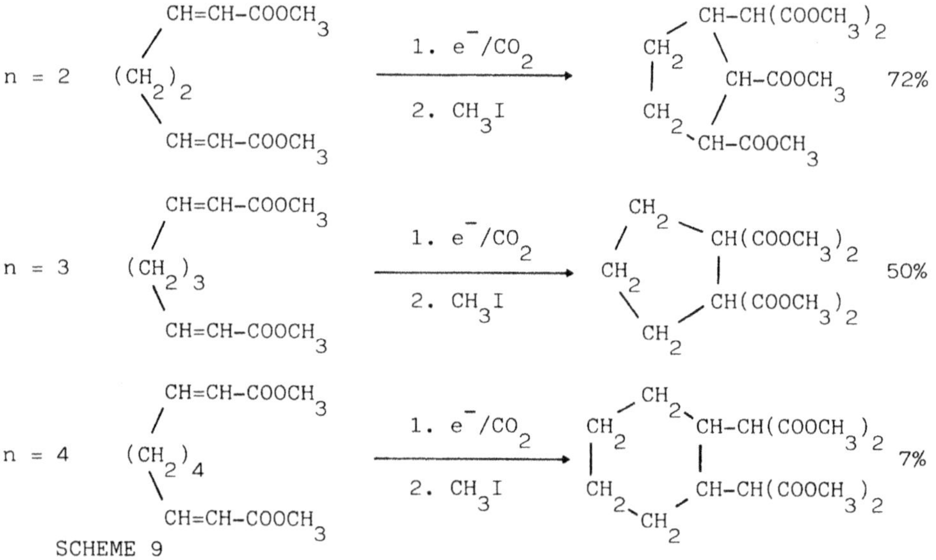

SCHEME 9

For n = 2,4 the Authors suggest a reaction sequence in which a) the diolefin is reduced at one of the unsaturations; b) the resulting anion radical undergoes a first carboxylation, c) the radical mojety, or, after further reduction, the corresponding anion give rise to an intramolecular addition on the second unsaturation , followed by the last carboxylation step.

For n = 3 a concerted ring closure is proposed, analogous to the electrohydrocyclization of the same starting material. In fact, further investigations ought to be made on these reaction mechanisms, as at the very negative reduction potentials of these starting materials, it is possible that homogeneous charge transfer processes are involved, with formation of $CO_2^{-\cdot}$ anion radical.

3.1.3. Electrocarboxylation of acetylenes and aromatics.

The electrocarboxylation of diphenylacetylene leads to a mixture of fumaric acid, maleic anhydride and meso diphenylsuccinic acid, with an overall yield of 40% (68).

The electrocarboxylation of condensed polynuclear aromatic rings has been object of considerable attention, and quite a large number of examples was reported in the literature. As a general rule, the carboxylation results in the reductive addition of two molecules of carbon dioxide, or of a proton and carbon dioxide, on one of the rings of the aromatic starting material, thus breaking the aromaticity of that part of the molecule.

Despite the applicative interest of the transformation of naphtalene and other polycyclic aromatic compounds into valuable difunctional derivatives, disappointingly low yields are reported in the literature (53,68) . Thus naphtalene is carboxylated to 1,4-dihydro-1,4-dicarboxy naphtalene with 43% yield and phenanthrene to trans-9,10-dicarboxy-9,10-dihydro-phenanthrene with 20 % yield (68).

Among the aromatic heterocycles, 85% yield is reported for the hydrocarboxylation of 2-carboxymethylthiophene to 4,5-dihydro-5-carboxy--2-carboxymethylthiophene (69).

Nitrogen heterocycles are starting materials of great interest for the production of pharmaceuticals, and appreciable yields have been quoted in the literature for the carboxylation of some of them.

The electrocarboxylation of a series of six membered heterocycles, containing almost one nitrogen atom in the cycle, to the corresponding 1,4-dihydro-1,4-dicarboxy derivatives is described in a patent from Rhone-Poulenc (70) . The best results were claimed for pyrazine 75.9%, and for pyridine 90% ; among the other starting materials poorer yields were reported for quinoxaline, 25%, and quinoline, 32% .

Quinoline (71) and substituted quinolines, isoquinolines, 9,10-phenanthroline and 5,6-benzoquinoline (72) were electrocarboxylated to the corresponding dihydrocarboxyderivatives. The best yield, 55%, was reported for benzoquinoline. Catalytic currents were observed for several heterocycles at the voltammetric sweep by addition of carbon dioxide.

An interesting study on the electroanalytical and synthetic aspects of the electrocarboxylation of thirty N-heteroaromatic compounds, mainly quinoline derivatives, was published in 1981 (73). Several halogenates compounds were among the substrates, in order to investigate the possibility, for such compounds, of a carboxylation without loss of halogen. The Authors describe the cyclovoltammetric behavior of the different halogenated compounds, and propose a diagnostic criterion for the preliminary prevision of the prevalence of the addition or of the cleavage when preparative electrolyses are to be performed. From the synthetic point of view, these carboxylations gave for almost all the starting materials a complex mixture of products. In some cases an increase of selectivity, performing the electrolysis at low temperature, was observed.

3.1.4. Electrocarboxylation of aldehydes and ketones.

Ketones are good starting materials for chemical (78,85) and electrochemical syntheses of α-hydroxyacids.

As many other electrocarboxylations described in this review, even the addition of carbon dioxide on the carbonyl group was pioneered by Wawzonek (75). In that first paper a yield of 49 % was quoted for the carboxylation of benzophenone to benzilic acid. More recently very good results were achieved for a series of substituted acetophenones, acetonaphtones and benzophenones (76,79,80 - 83). The reduction potentials of these ketones are more positive than that of carbon dioxide, and no homogeneous charge transfer processes have been so far described. The reaction paths of these additions depend on the reactivity of the radical anion of the ketone after the first electrochemical reduction : competitive with the carboxylations are the dimerization to pinacols, and the hydrogenation to alcohols.

A detailed study on the synthetic aspects of the electrocarboxylation of benzophenone and of several substituted benzophenones has been published very recently (76). A molar ratio 10 : 1 of potassium iodide, used as supporting electrolyte, to the ketone, provided the best yields for the electrocarboxylation. The influence of the substituents on the yields of the carboxylation was also examined: electron withdrawing groups may enhance the density of the negative charge on the carbon atom of XXIV, thus enhancing the the rate of the carboxylation (route a, scheme 10) as the first chemical step following the formation of the anion radical of the ketone XXIV. On the contrary electron donor substituents may enhance the electron spin density on the same carbon, thus leading to dimers or tars by radical-radical coupling or polymerizations. 4-dimethylamino and 4-hydroxy-benzophenone gave mainly tarry products. The carboxylation of 4-nitro-benzophenone led to 4,4'-dibenzoylazobenzene with 60% yield.The reductive coupling of two aromatic nitro groups to the corresponding azobenzene is well known in the presence of water. The formation of azobenzene in anhydrous conditions in the presence of carbon dioxide was also observed in the attempt of carboxylating 4-nitro-bromobenzene (77).

In strictly anhydrous conditions XXVI should react with molecular carbon dioxide to give a final dianion XXVIII , from which the hydroxyacid is recovered by hydrolysis. In fact, in several cases the evolution of carbon dioxide by hydrolysis of the effluents from the electrolysis, is observed.

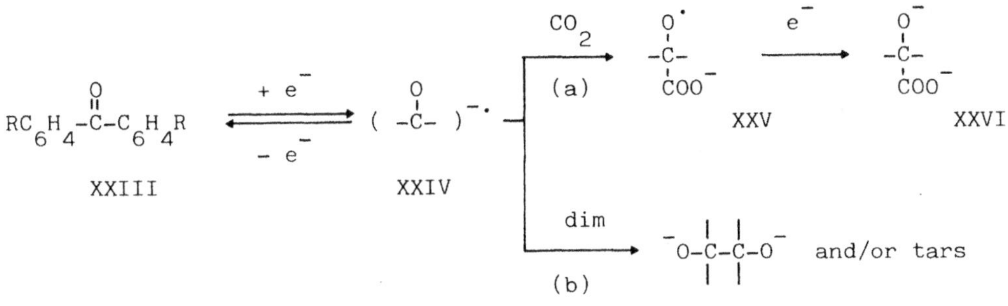

SCHEME 10

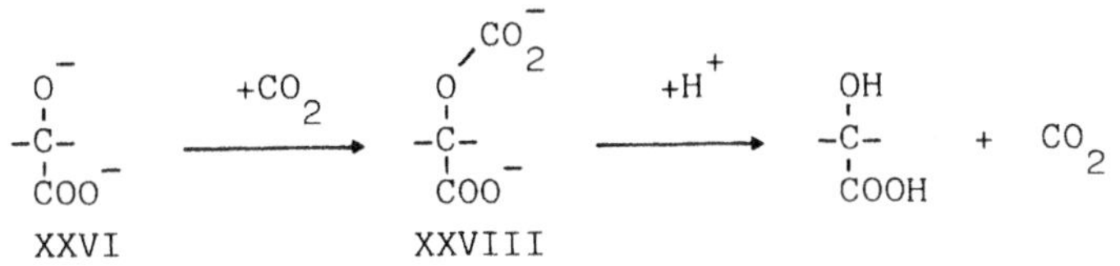

The use of conventional electrochemical techniques, in which divided cell are employed, does not seem to allow the carboxylation of aldehydes (75), which, even if reduced in the presence of carbon dioxide, give rise to dimers or to tars. It has been recently observed that in undivided cells, with sacrificial aluminum anodes (80) feeding the reaction medium with Al(III) species, it is possible to achieve the carboxylation of some aromatic aldehydes, and, with poor yields, of acetaldehyde. Although the way in which Al(III) modifies the reaction pattern has not been investigated, it is possible to imagine that the metallic cation should act as a coordination centre for the unreacted aldehyde or for some charged intermediates. In this context it can be recalled that other examples of changes of the reactivity of aromatic aldehydes induced by metallic species present in the reaction medium have been published (84).

3.1.5. Electrocarboxylation of carbon-nitrogen groups.

The chemical addition to an imine of hydrogen cyanide, to give the nitrile of the corresponding aminoacid (85), finds its electrochemical equivalent, milder and safer, in the electrocarboxylation of the same class of substrates . This synthesis, affording dl-N-substituted

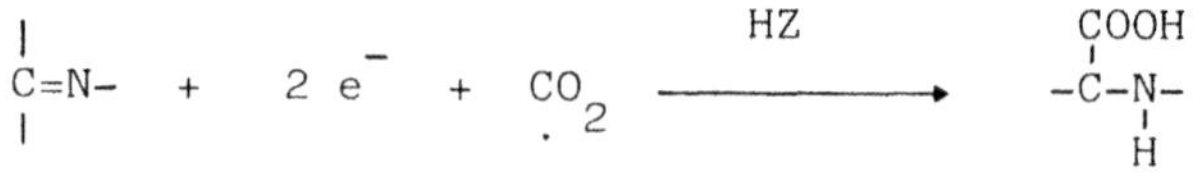

XXIX

SCHEME 11

aminoacids, leads to interesting results for several imines. Hess and coworkers devoted large attention to the electroanalytical and synthetic aspects of these reactions (87 - 89), and evidentiated the possible existence , depending on the reduction potential of the imine, of two mechanisms : a) direct reductive carboxylation via the formation of the corresponding anion radical, and its nucleophilic addition on carbon dioxide, analogous to that summarized in scheme 8 for ketones, or b) radical-radical coupling between the anion radical of the imine and the anion radical of carbon dioxide, if the reduction potential of the imine approaches that of carbon dioxide. Side reactions, involving the hydrogenation of the >C=N- double bond, couplings between two imine radical anions, the monocarboxylation at the nitrogen of the imine, or even a dicarboxylation on both carbon and nitrogen, are often present and, depending on the molecular structure of the imine, may become the prevalent processes. Yields in the range 40 - 60 % for the monocarboxylation to XXIX were reported for a substantial number of imines.

Aminoacids unsubstituted at the nitrogen atom can be obtained by carboxylation of azines XXX or of benzonitrile.

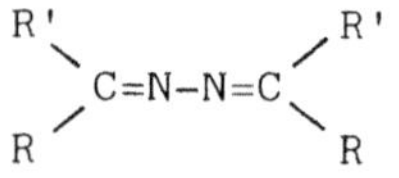

R'= aromatic group

R = R' or H or aliphatic

XXX

The electrocarboxylation of azines (91) at Hg cathodes in acetonitrile with tetra-alkylammonium salts leads to N-benzilideneamino-2-phenylglicines, derivatives of monocarboxylation of XXX, at a cathodic potential from -1.7 to -2.0 V vs. SCE ; if the potential is kept at more negative values (-2.1 to -2.2 V) the N-N bond is broken, and the corresponding phenylglycine is formed.

The electrocarboxylation of benzonitrile is claimed in a recent patent (92) : the process is performed in benzonitrile, with tetraethylammonium p-toluensulfonate as supporting electrolyte, and in the presence of the strongly acid mixture HCl + $AlCl_3$. The current efficiency is 11% if carbon dioxide is fed at atmospheric pressure, and raises to 19% at 10 Atm. It is interesting to note that benzonitrile, in aprotic conditions, proved to be totally unreactive towards the electrocarboxylation, in such a way that it was used as inert charge transfer agent for the carboxylation of styrene (54).

The formation of N,N'-dicarboxyhydrazobenzene from azobenzene (126) shall be discussed in the context of the electrocarboxylations via electrogenerated bases.

3.1.6. Anodic carboxylations : decomposition of oxalates.

The anodic oxidation of oxalates generates carbon dioxide and the anion radical $CO_2^{-\bullet}$. The influence of the solvent on this reaction has been studied recently by Vasil'ev and coworkers (93), in connection with the synthesis of dicarboxylic acids by anodic decomposition of oxalates in the presence of lower olefins. In fact, the anion radical of carbon dioxide can add on the unsaturation with the same reaction scheme of the electrocarboxylation of olefins belonging to case c in section 3.1.1., and the only difference between the two processes lies in the different adsorption phenomena taking place respectively on the anode and on the cathode. The first report dealing with this interesting synthetic route was published in 1965 (94). By decomposition of oxalic acid in the presence of butadiene, in ethanol on Pt mesh anode, 2,6-octadiene -1,10-dioic acid was obtained with 40% yield with respect to oxalic acid. The anodic electrocarboxylation of lower olefins has been studied in detail, and the production of the diesters of the dicarboxylic acids of general formula

$$ROOC-(\underset{R'}{CH}-\underset{R''}{CH})_n-COOR$$

was reported with overall product yields around 40% (95,96).At low current density the yield of higher oligomers increases, and this is coherent with an interpretation of the reaction mechanism analogous to the polymerization processes : at higher current density the local concentration of initiators is higher and the molecular weight is lower. Anodes of porous titanium promoted with platinum were employed to increase the effective current intensity of the cells maintaining the same yield in higher oligomers. No ethylene oxidation was observed at the anodic potential at which the carboxylation was performed (97). Significant improuvements of the yields of carboxylated products were reported performing the electrolyses under ca. 60 Atm of ethylene

(98,99). A successful anodic carboxylation of dimethylmaleate has been recently proposed (100) . In this last case the formation of a mixture of dicarboxylate monomer (62%) and dicarboxylated dimer (34%) was reported.

In this connection the cathodic electrocarboxylation of various starting materials in undivided cells has to be quoted. In fact, to avoid the anodic oxidation of the carboxylated anions produced at the cathode, it is proposed the use, as supporting electrolytes, of easily oxidizable salts, as oxalates, azides and formates (45,46). In the case of oxalates, it is very probable that the carboxylation could take place both at the cathode and at the anode.

3.2. ELECTROCARBOXYLATION BY SUBSTITUTION

3.2.1. Electrocarboxylation of halides

The electroreduction of halides is one of the most studied topics in organic electrochemistry. The very complex reaction sequences which follow the first monoelectronic reduction of RX to its corresponding anion radical have been recently reviewed by Feoktistov (101) and examined in detail in a series of monographies by Hawley (102). Both the Authors stress the very complex reactivity of halides, strongly dependent on the nature of the radical bearing the halogen and on the nature of the halogen itself. Furthermore a large spectrum of products is to be espected, depending on the nature of the halide and on the experimental conditions .

As many other electrocarboxylation processes described in this review, even those involving organic halides were thoroughly explored by the electrochemical research group active at the Monsanto Laboratories in St. Louis. Those researchers greatly improved the yield of the electrocarboxylation of benzyl halides reported by Wawzonek in 1964 (103), and transformed it into a valuable synthetic procedure. An important finding was that alkyl, allyl and benzyl halides, once carboxylated according to the stoichiometry of scheme 12, if in the presence of tetra-alkylammonium salts as supporting electrolytes, give a chemical follow up reaction leading to the corresponding ester XXXI (23,104 - 106).

$$\mathrm{RX} \xrightarrow[-\,X^-]{+\,2e^-\,+\,CO_2} \mathrm{R{-}COO^-} \xrightarrow[R_4N^+]{+\,RX,\,-X^-} \underset{\text{XXXI}}{\mathrm{R{-}COOR}}$$

SCHEME 12

In the case of alkyl halides the nature of the cathode has a strong influence on the reaction mechanism and on the product distribution (105). In the presence of carbon dioxide, on mercury at - 2.3 V, butyl bromide is reduced to dibutyl mercury and no carboxylation is observed. To get the electrocarboxylation the cathodic potential has to be set at a more negative value, at which even carbon dioxide is reduced. In these conditions a mixture of products is obtained, among which dibutylmercury XXXII, dibutyloxalate XXXIII and butylvalerate XXXIV in approximately equal amounts were detected. The formation of XXXII by reduction of BuBr on mercury, and of XXXIII by reduction of carbon dioxide to oxalate followed by esterification with BuBr are easily understood. XXXIV, occurring only at the cathodic potential at which carbon dioxide is reduced, may arise from a charge transfer process between CO_2 and BuBr, followed by a radical-radical coupling :

$$CO_2 \xrightarrow{e^-} CO_2^{-\bullet}$$

$$CO_2^{-\bullet} + BuBr \longrightarrow CO_2 + Bu^{\bullet} + Br^-$$

$$Bu^{\bullet} \xrightarrow{+\ CO_2^{-\bullet}} BuCO_2^- \xrightarrow[-Br^-]{+BuBr} BuCO_2Bu \quad \text{XXXIV}$$

SCHEME 13

The reduction on mercury is apparently monoelectronic, whereas on

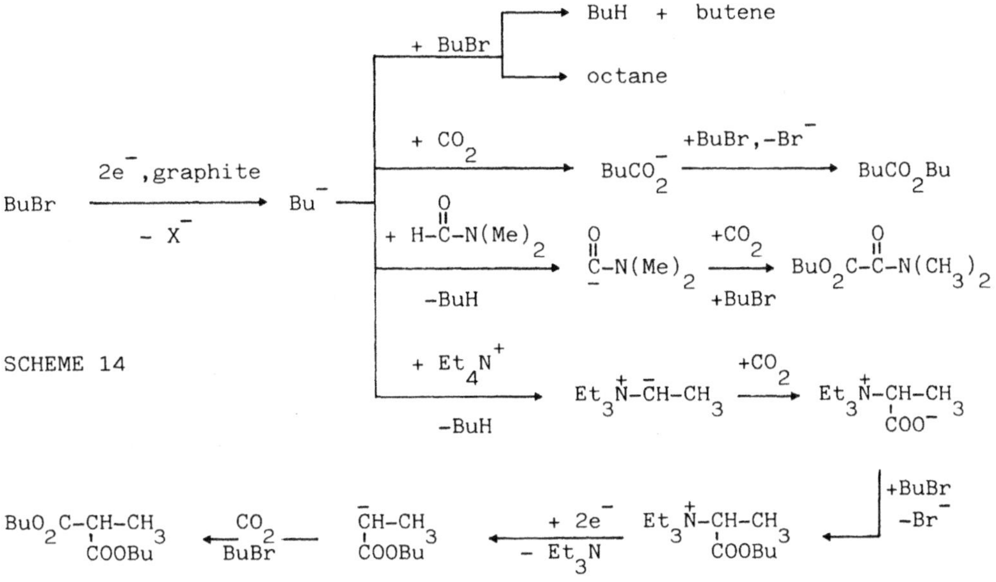

SCHEME 14

graphite appears to be bielectronic, leading to the butyl anion (105,106). In fact on graphite completely different products were obtained, such as butane, butene, octane, butylvalerate, butyl-N,N-dimethyloxamate and dibutyl-2-methylmalonate (30% current efficiency). The proposed pathway is summarized in scheme 14.

Some of the stages of this apparently complex scheme were demonstrated arriving to the same products from intermediates prepared elsewhere. So the α-ammonium ester $(CH_3)_3\overset{+}{N}-CH(CH_3)-COOEt$ was electrochemically reduced in DMF saturated with carbon dioxide and subsequently treated with ethyl bromide, giving a 76% yield of diethyl-2-ethylmalonate. This last reaction represents, up to now, the only example of a preparative electrocarboxylation by displacement of the nitrogen of a quaternary ammonium group.

A very high yield (95%) was reported for the electrocarboxylation of 1-bromo-2-methylpentane with a system using as anodic process the anodic decomposition of oxalates in undivided cells (46).

The electrocarboxylation of a perfluorurated iodide, $C_6F_{13}I$ on mercury cathodes can be realized only if the formation of diperfluoroalkylmercury is avoided by a proper choice of the supporting electrolyte (107). In fact, lithium chloride addresses the reduction towards the formation of the anion R_F^- , which is trapped by carbon dioxide thus giving high yields of the corresponding carboxylate anion. On the contrary lithium perchlorate favors the formation of the organomercury derivative, and no carboxylation at all was observed.

Benzyl and allyl chlorides and bromides are reduced to the corresponding anions at cathodic potentials more positive than the alkyl ones. Very good yields of the esters formed according to scheme 11 are reported (104,106) for benzyl and allyl chloride. The one step synthesis of polyesters from bis-halomethyl aromatics, shortly mentioned some years ago (23) should be explored in better detail.

The electrocarboxylation of several benzyl halides has been proposed for the synthesis of a series of α-aryl propionic acids of different structure, currently employed as pharmaceuticals. In this case the follow up reaction of esterification, wasting at least half of the expensive halide used as starting material, has to be avoided. This result was achieved by the use of undivided cells, in which the anodic reaction consists of the anodic dissolution of suitable metals, such as aluminum (108,109) or magnesium (110,111). With these systems very high yields of carboxylated products were obtained, being the primary products the aluminum or magnesium salts of the corresponding acids. A reasonable explanation of this behavior is that the formation of complex salts like:

$$[(R\text{-}COO)_nM]^{m+} \qquad M = Al,\ Mg$$

inhibits the nucleophilicity of the carboxylate anion, thus hindering the esterification.

The indirect carboxylation of benzyl halides with alkali and alkaline-earth metals or amalgams was patented in 1971 (112).

Some attempts of using some complexes of various transition metals, able to interact with the halide and possibly with carbon dioxide, have been recently published. Low valent nickel and palladium halides, stabilized by phosphine ligands, have shown acceptable catalytic activity towards the electrocarboxylation of benzyl or allyl halides (113). Better results were reported (114) particularly for the electrocarboxylation of allyl halides in the presence of Co Salen complexes. The sequence of steps proposed by the Authors is shown in scheme 15.

a) $$\text{Co(II)Salen} + e^- \longrightarrow \text{Co(I)Salen}^-$$

b) $$\text{Co(I)Salen}^- + RX \longrightarrow \text{R Co(III)Salen} + X^-$$

c) $$\text{R Co(III)Salen} + CO_2 + 2\,e^- \longrightarrow RCO_2^- + \text{Co(I)Salen}^-$$

SCHEME 15

No indications are at present available on the way in which the carboxylation in step c takes place .

The electrocarboxylation of aryl halides has been investigated in some detail . Very negative cathodic potentials are needed to reduce the organic substrate, and, in the case of chlorobenzene, the reduction potential is in the same range of that of carbon dioxide. Iodobenzene and bromobenzene were carboxylated on mercury cathodes with yields approaching 40 % (115,116), whereas for the carboxylation of chlorobenzene a yield of 35% was quoted, and a platinum cathode in DMF with Bu_4NClO_4 was needed (115). The formation of benzoic acid, even if with these moderate yields, was used to demonstrate that the mechanism of reduction involves the anionic intermediate $C_6H_5^-$ (115).

Of higher synthetic interest are the results reported in the case of electrocarboxylations of aromatic halides catalytically assisted by transition metal compounds. Up to now, good results have been obtained with low valent transition metal complexes able to catalyze the electroreductive dimerization of halobenzenes to biaryls, in which nickel (117) or palladium (118) are involved. The electrocarboxylations are achieved performing, in the presence of carbon dioxide, the electrochemical reduction of the complex deriving from the reaction of the aryl halide with the low valent metal complex. Both metals are stabilized in their low valent state with triphenylphosphine, used in large excess in the case of nickel, or 1,2-bis(diphenylphosphino)ethane. The catalytic cycle is summarized in scheme 16 .

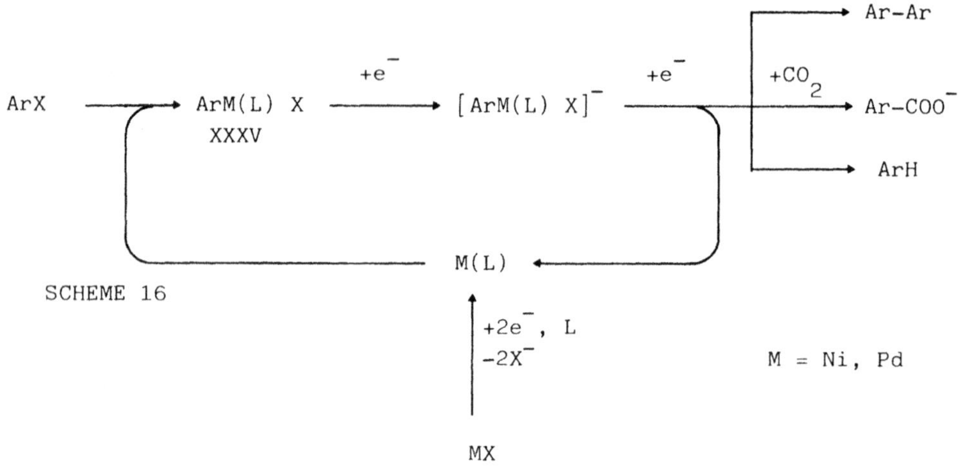

SCHEME 16

The electrocarboxylations are performed at cathodic potentials near to those corresponding to the reduction of the complex XXXV, i. e. at values more positive than those of the reduction of the halide alone. In the presence of carbon dioxide the formation of the dimer Ar-Ar is almost totally inhibited, whereas in some cases small amounts of the hydrogenated derivative are detected in the effluents.

Yields between 70 and 90 % were reported for the carboxylation of chloro and bromobenzene with nickel complexes (119). Chloro and bromo benzenes parasubstituted with F-, CF_3-, and C_6H_5O- gave, with the same catalytic system, yields in the range 40 - 94 % , depending on the experimental conditions (120). The nickel based catalytic system is deactivated irreversibly at low concentrations of the organic halide. This serious drawback, which could limit the application of this reaction to batch systems, can be overcome by the addition to the electrolytic medium of a second halide, such as butylbromide, which contributes to the stabilization of the catalyst, and, in the conditions of the synthesis does not undergo the carboxylation, but is transformed into volatile products which do not interfere with the isolation of the acids.

In a very recent preliminary communication the use of palladium(0) complexes is reported (121). Very high yields are quoted for the electrocarboxylation of iodobenzene and substituted iodobenzenes. Low conversions are reported for bromides whereas chlorides are scarcely reactive.

The electrocarboxylation of benzal chloride failed with conventional divided cells (122). The intermediate α-chlorophenylacetate anion was supposed to give an intramolecular nucleophilic substitution leading to an unstable lactone then decomposing to mandelic acid, isolated in low yields. If the same synthesis is performed with sacrificial aluminum anodes in undivided systems, the Al(III) species act in the same way as already seen for monohalides, i.e. inhibiting the nucleophilicity of the carboxylate anion, and the dicarboxylation to phenylmalonic acid can be realized, even if with yields not exceeding 30% (123). Among the effluents α-chlorophenylacetic acid and phenylacetic acid are also isolated each with ca. 30 % yield. The synthesis can be maximized towards the production of α-chlorophenylacetic acid (55 % yield).

3.2.2.Electrocarboxylations via electrogenerated bases.

Not many examples of electrocarboxylations have been so far quoted in the literature concerning this interesting and versatile synthetic route. The essential features of the syntheses via electrogenerated bases have been recently reassumed by Baizer (124) who, with his coworkers at Monsanto, pioneered this field. The synthetic sequence needs some organic compound of suitable structure (probase PB) which is reduced, in two or more steps, to a mono or a dianion, acting, in the reaction medium, as a base (electrogenerated base, EGB). The EGB can thus deprotonate relatively acidic carbon-hydrogen, nitrogen-hydrogen, sulfur-hydrogen or even other element-hydrogen bonds, thus forming the corresponding anion, able to react with electrophiles eventually present in the reaction medium, and the protonated EGB (EGBH). EGBH can be reoxidized to PB with chemical oxidants (oxygen, halides and so on), or electrochemically. If carbon dioxide is used as electrophile, a carboxylic group will replace the hydrogen removed by the EGB.

As an example, the electrocarboxylation of 9-phenyl-fluorene, RH (125,126) is examined in detail . The probase chosen for this synthesis was tetraethyl ethenetetracarboxylate, which was reduced to its corresponding dianion at - 1.2 V (vs. S.C.E.). In a divided cell the catholyte was a solution of the PB and RH in acetonitrile, containing TEABr, NaBr and 18-crown-6. The anolyte was a solution of the corresponding EGBH, i.e. tetraethyl ethanetetracarboxylate, in acetonitrile containing the same electrolytes as in the catholyte, with the addition of sodium carbonate. The electrochemical and chemical steps are summarized in scheme 17.

a) $$PB + 2\,e^{-} \longrightarrow EGB$$

b) $$EGB + 2\ RH \longrightarrow EGBH + 2\ R^-$$

c) $$2\ R^- + 2\ CO_2 \longrightarrow 2\ RCOO^-$$

SCHEME 17

d) $$2\ Br^- - 2\ e^- \longrightarrow Br_2$$

e) $$Br_2 + EGBH \longrightarrow 2\ Br^- + PB + 2\ H^+$$

Steps a) - c) take place in the catholyte, d) and e) in the anolyte. The EGBH obtained from an electrolytic run, is fed to the anolyte in the following one. Making use of this system, 9-phenyl-9-fluorenecarboxylate was obtained with 81% yield. A further example of a carboxylation via EGB is that of ethyl phenylacetate which is converted to ethyl phenylmalonate with a product yield of 78 %, making use of 2,2'-di-t-butylazobenzene as PB. If unsubstituted azobenzene is used as PB, the carboxylation does not involve ethyl phenylacetate but is stopped by the reaction of the EGB with carbon dioxide. Thus the dicarboxylation derivative N,N'-hydrazobenzenedicarboxylate XXXIV is obtained with 64 % yield.

$$C_6H_5\text{-N=N-}C_6H_5 + 2\ e^- + 2\ CO_2 \longrightarrow C_6H_5\text{-}\underset{COO^-}{N}\text{-}\overset{COO^-}{N}\text{-}C_6H_5$$

SCHEME 18

XXXVI

3.2.3. Other electrocarboxylations by substitution.

Some examples of electrocarboxylations by substitution have been already quoted in this review, but not properly classified, like that of ketene-S,S-acetals (61,62), resulting in the substitution of one of the -S-R groups, or that of the α-ammonium esters to malonates with displacement of the nitrogen of a quaternary ammonium group (105). A further substitution process has been quoted in the literature consisting in the electrocarboxylation, with poor yields and at very negative cathodic potentials, of acetonitrile and α-substituted acetonitriles to the corresponding α-cyanocarboxylic derivatives (127).

REFERENCES

1. H. Kolbe, R. Schmitt, Ann. Chem. **119**, 251 (1861).
2. E. Drechsel, J. Chem. Soc. **21**, 121 (1868).
3. N. N. Beketov, J. Russ. Phys. Chem. Soc. **1**, 33 (1869).
4. M. E. Royer, C. R. Seances Acad. Sci. Fr. (Sect. Chim.) **70**, 731 (1870).
5. J. Walther, Chem. Z. **102**, 1151 (1901).
6. A. Cohen, S. Jahn, Ber. Deutsch. Chem. Gesell. **37**, 2836 (1904).
7. R. Ehrenfeld, Ber. Deutsch. Chem. Gesell. **38**, 4138 (1905).

8. F. Fisher, O. Prziza, Ber. Deutsch. Chem. Gesell. 47, 256 (1914).
9. A. Jaeger, Abhdlg. z. Kenntnis d. Kohle 7, 201 (1925).
10. M. Rabinowitsch, A. Maschowetz, Z. Elektrochem. 36, 846 (1930)
11. T. E. Teeter, P. Van Rysselberghe, J. Chem. Phys. 22, 759 (1954).
12. K. S. Udupa, G. S. Subramanian, H. V. K. Udupa, Electrochimica Acta 16, 1593 (1971).
13. a) P. S. Chechel, L. I. Antropov, Zh. Prikl. Khim. 31, 1856 (1958); C. A. 53:6832g .
b) V. A. Smirnov, V. D. Semchenko, D. P. Semchenko, L. I. Antropov, Zh. Prikl. Khim. 40, 2345 (1967); C. A. 68:63070x.
14. K. Haupt, Ger. Pat. 286,461 (1913).
15. H. Hohn, E. Fitzer, O. Schlager, H. Nedved, Austrian Pat. 175,237 (1953); C. A. 47:8978f.
16. F. A. Englein, H. Sontheimer, Z. Anorg. Allgem. Chem. 267, 181 (1952).
17. C. Porlezza, G. Ginori Conti, Ann. Chim. Appl. 13, 53 (1923).
18. H. Hohn, E. Fitzer, H. Nedwed, Z. Anorg. Allgem. Chem. 274, 297 (1953).
19. R. Setton, Bull. Soc. Chim. Fr. 1958, 1400.
20. R. M. Miller, H. V. Knorr, H. J. Eichel, C. M. Meyer, H. A. Tanner, J. Am. Chem. Soc. 27, 2646 (1962).
21. K. Clusius, H. Mollet, Helv. Chim. Acta 39, 363 (1956).
22. V. A. Tushnalobova, V. A. Smirnov, D. P. Semchenko, Zh. Prikl. Khim. 41, 275 (1968).
23. D. A. Tyssee, J. H. Wagenknecht, M. M. Baizer, J. L. Chruma, Tetrahedron Lett. 47, 4809 (1972).
24. T. Hatayama, G. Inoue (to Asahi Chem. Ind.), Jap. Pat. 73 48409 (1973); C. A. 79:125861u
25. S. Gambino, G. Silvestri, Tetrahedron Lett. 32, 3025 (1973).
26. U. Kaiser, E. Heitz, Ber. Bunsenges. Phys. Chem. 77, 818 (1973).
27. L. Skarlos (to Texaco Inc.), U. S. Pat. 3,720,591 (1973); C. A. 78:158943e.
28. E. Heitz, U. Kaiser, Ger. Pat. 2,391,032 (1974); C. A. 81:130200y.
29. J. Fisher, Th. Lehmann, E. Heitz, 30th ISE Meeting, Trondheim (Norway) abs. 345 (1979).
30. J. C. Gressin, D. Michelet, L. Nadjo, J. M. Saveant, Nouv. J. Chim. 3, 545 (1979).
31. Kirk Othmer,'Concise Encyclopedia of Chemical Technology', p. 825. Whiley Interscience (1985).
32. J. Fisher, Th. Lehmann, E. Heitz, J. Appl. Electrochem. 11, 743 (1981).
33. F. Goodridge, G. Presland, J. Appl. Electrochem. 14, 791 (1984).
34. a) R. W. Treharne, C. M. Cox, U. S. Clearinghouse Fed. Sci. Tech. Inf. A. D. 679597 (1968); C. A. 70:99413s.

b) F. B. Leitz, U. S. Clearinghouse Fed. Sci. Tech. Inf. A. D. 654146 (1967); C. A. 68:26306h.

c) F. H. Meller, U. S. Clearinghouse Fed. Sci. Tech. Inf. A. D. 678427 (1968); C. A. 70:83579q

35. H. L. Chum, M. M. Baizer, 'The Electrochemistry of Biomass and Derived Materials', ACS Monograph 183, American Chemical Society, Washington DC (1985).
36. E. Lamy, L. Nadjo, J. M. Saveant, J. Electroanal. Chem. 78, 403 (1977).
37. E. Lamy, L. Nadjo, J. M. Saveant, Nouv. J. Chim. 3, 21 (1979).
38. C. P. Andrieux, D. J. Brown, J. M. Saveant, Nouv. J. Chim. 2, 157 (1977) .
39. a) H. D. Roth, M. L. Manion Shilling, J. Am. Chem. Soc. 101, 1898 (1979).

    b) H. D. Roth, M. L. Manion Shilling, J. Am. Chem. Soc. 102, 4303 (1980).
40. S. Gambino, G. Silvestri, G. Filardo, J. Appl. Electrochem. 12, 549 (1982)
41. C. Degrand, R. Mora, H. Lund, Acta Chem. Scand. B 37, 429 (1983).
42. D. A. Tyssee, M. M. Baizer, J. Org. Chem. 39, 2819 (1974).
43. D. A. Tyssee, M. M. Baizer, J. Org. Chem. 39, 2823 (1974).
44. D. A. Tyssee (to Monsanto Co.) U. S. Pat. 4.013.524 (1977) . C.A. 86:162812u
45. W. J. Van Tilborg, C. J. Smit, R. Engels (to Shell Co.) Brit. Pat. Appl. 79/37,859 (1979) . C.A. 95:69965x .
46. R. Engels, C. J. Smit, W. J. M. Van Tilborg, Angew. Chem. Int. Ed. Engl. 22, 492 (1983).
47. L. R. Yeh, A. J. Bard, J. Electrochem. Soc. 124, 355 (1977).
48. H. Lund, J. Simonet, J. Electroanal. Chem. 65, 205 (1975).
49. S. Wawzonek, E. W. Blaha, R. Berkey, M. E. Runner, J. Electrochem. Soc. 102, 235 (1955).
50. S. Tsutsumi, T. Murakawa, Amer. Chem. Soc. Div. Petrol. Chem. Prepr. 15, B-5 (1970).
51. R. Dietz, M. E. Peover, Discuss. Far. Soc. 45,154 (1968).
52. D. A. Tyssee, Ger. Pat. 2,356,657 (1974). C. A. 81:44737x.
53. W. C. Neikam (to Sun Oil Co.) U. S. Pat. 3,344,045 (1967). C. A. 68:8809y
54. G. Filardo, S. Gambino, G. Silvestri, A. Gennaro, E. Vianello, J. Electroanal. Chem. 177, 303 (1984).
55. G. Filardo, S. Gambino, A. Gennaro, G. Silvestri, E. Vianello, 36th ISE Meeting, Salamanca, Spain abs. 13110 (1985).
56. W. C. Neikam (to Sun Oil Co.) U. S. Pat. 3,344,046 (1967). C. A.68: 121859d.
57. D. A. Tyssee (to Monsanto Co.) U. S. Pat. 3,864,225 (1975). C. A. 82:161871t.

58. R. Ercoli, M. Guainazzi, G. Silvestri, S. Gambino, G. Filardo, B. Giannici, M. Galluzzo, Chim. Ind. (Milan) 55, 156 (1973).
59. G. Silvestri, S. Gambino, G. Filardo, Electrochemical Soc. Spring Meeting Washington, D. C. abs. 278 (1976).
60. S. Wawzonek, A. Gundersen, J. Electrochem. Soc. 111, 324 (1964).
61. H. Matschiner, H. H. Ruettinger, S. Austen, J. Prakt. Chem. 327, 45 (1985).
62. H. H. Ruettinger, W. D. Rudorf, H. Matschiner, Electrochim.Acta 30, 155 (1985).
63. J. W. Loveland (to Sun Oil Co.) U. S. Pat. 3,032,489 (1962). C. A. 57:4470i.
64. W. J. M. Van Tilborg, C. J. Smit, Recl. J. R. Neth. Chem. Soc. 100, 437 (1981). C. A. 96:51608u
65. W. J. M. Van Tilborg, C. J. Smit (to Shell Int. Res. Co.) Brit. Pat. 2,069,533 (1981). C. A. 96:199098y.
66. H. T. Chen, H. J. Tien, T. T. Lai, T. C. Chou, J. Chin. Inst. Chem. Eng. 16, 25 (1985). C. A. 103:29152g
67. I. B. M. Tkatchenko, D. A. Ballivet-Tkatchenko, Fr. Pat. 2,542,764. (1984). C. A. 102:69341d
68. S. Wawzonek, D. Wearring, J. Am. Chem. Soc. 81, 2067 (1959).
69. L. M. Korotaeva, A. M. Moiseenkov, V. S. Bogdanov, V. P. Gul'tyai, Izv. Akad. Nauk SSSR, Ser. Khim. 1983, 2098. C. A. 100:51386w
70. D. Michelet (to Rhone-Poulenc Co.) Fr. Pat. 2,444,030 (1978). C. A. 94:164821k.
71. H. Lund, Denki Kagaku, 45, 2 (1977).
72. U. Hess, P. Fuchs, E. Jakob, H. Lund, Z. Chem. 20, 64 (1980).
73. P. Fuchs, U. Hess, H. H. Holst, H. Lund, Acta Chem. Scand. B 35, 185 (1981).
74. J. H. Wagenknecht, M. M. Baizer, J. L. Chruma, Synth. Comm. 2, 215 (1972).
75. S. Wawzonek, A. Gundersen, J. Electrochem. Soc. 107, 537 (1960).
76. Y. Ikeda, E. Manda , Bull. Chem. Soc. Jpn. 58, 1723 (1985)
77. J. H. Wagenknecht, G. Silvestri, S. Gambino, G. Filardo, unpublished results
78. Y. Ikeda, E. Manda, J. Chem. Soc. Jpn. 1982, 830.
79. Y. Ikeda, E. Manda, Chem. Lett. 1984, 453.
80. G. Silvestri, S. Gambino, G. Filardo, Tetrahedron Lett. in course of publication.
81. Agency of Industrial Sciences and Technology Jpn. Pat. 85 103,193 (1985). C. A. 103:131247d
82. Agency of Industrial Sciences and Technology Jpn. Pat. 85 24,386 (1985). C. A. 103:13560c
83. Agency of Industrial Sciences and Technology Jpn. Pat. 85 100,536 (1985). C. A. 103:214989y

84. D. W. Sopher, J. H. P. Utley, J. Chem. Soc. Perkin Trans. II 1984, 1361.

85. J. March,'Advanced Organic Chemistry'2nd Ed. p. 875. McGraw Hill Intnl.(1977).

86. N. L. Weinberg, A. Kentaro-Hoffmann, T. B. Reddy, Tetrahedron Lett. 25, 2271 (1971).

87. U. Hess, Z. Chem. 20, 148 (1980).

88. U. Hess, M. Ziebig, Pharmazie 37, 107 (1982).

89. U. Hess, R. Thiele, J. Prakt. Chem. 324, 385 (1982)

90. S. Kato (to Tokuyama Soda Co.), Jpn. Pat. 80 06,432 (1980) . C. A. 93:46197t.

91. S. Kato (to Tokuyama Soda Co.)Jpn. Pat. 80 14,841 (1980). C. A. 93:176065j.

92. Ube Industries, Jpn. Pat. 80 119,185 (1980). C. A. 93:212419p

93. S. A. Sargisyan, Yu. B. Vasil'ev, Elektrokhimiya, 18, 961 (1982).

94. M. J. Fioshin, L. A. Mirkind, L. A. Salmin, A. G. Kornienko, Zh. Vses. Khim. Obshcestva Im. D. I. Mendeleeva 10, 238 (1965). C. A. 63:15858c.

95. S. A. Sargisyan, V. A. Grinberg, L. S. German, V. P. Polishchuk, Yu. B. Vasil'ev, Izv. Akad. Nauk. SSSR Ser. Khim. 1983, 579. C. A. 99:5163j.

96. Yu. B. Vasil'ev, V. S. Bagotzky, E. P. Kovsman, V. A. Grinberg, L. S. Kanevsky, V. R. Polishchuk, Electrochim. Acta 27, 919 (1982).

97. Yu. B. Vasil'ev, V. S. Bagotzky, E. P. Kovsman, V. A. Grinberg, L. S. Kanevsky, V. R. Polishchuk, Electrochim. Acta 27, 929 (1982).

98. V. A. Grinberg, L. S. German, L. S. Kanevsky, V. R. Polishchuk, A. M. Skundin, Yu. B. Vasil'ev, Izv. Akad. Nauk. SSSR Ser. Khim. 1980, 344. C. A. 92:155057x.

99. Yu. B. Vasil'ev, L. S. German, V. A. Grinberg, L. S. Kanevsky, V. P. Polishchuk, A. M. Skundin, Elektrokhimiya 16, 1449 (1980).

100. V. P. Nechiporenko, K. G. Bogoslovskii, N. A. Novikov, N. I. Guigina, L. A. Mirkind, Zh. Org. Khim. 20, 1799 (1984). C. A. 102:71763x.

101. L. C. Feoktistov, in'Organic Electrochemistry'2nd Ed., M. M. Baizer, H. Lund Editors, p. 259 . M. Dekker , New York (1980).

102. M. D. Hawley, in'Encyclopedia of Electrochemistry of the Elements', A. J. Bard Executive Editor, H. Lund Associate Editor, Vol XIV. M. Dekker, New York U.S.A.1980.

103. S. Wawzonek R. C. Duty, J. H. Wagenknecht, J. Electrochem. Soc. 111, 74 (1964)

104. M. M. Baizer, J. L. Chruma, J. Org. Chem. 37, 1951 (1972).

105. J. H. Wagenknecht, J. Electroanal. Chem. 52, 489 (1974).

106. M. M. Baizer, J. H. Wagencknecht, U. S. Pat. 3,764,492 (1973). C. A. 79:142364b.

107. P. Calas, A. Commeyras, J. Electroanal. Chem. 89, 363 (1978).
108. G. Silvestri, S. Gambino, G. Filardo, It. Pat. Appl. 21789A/82 (1982).
109. G. Silvestri, S. Gambino, G. Filardo, A. Gulotta, Angew. Chem. 96, 978 (1984).
110. O. Sock, M. Troupel, J. Perichon (to S.N.P.E.), Fr. Pat. Appl. 8409787 (1984).
111. O. Sock, M. Troupel, J. Perichon, Tetrahedron Lett. 26, 1509 (1985)
112. L. Moreschini, A. Del Vesco, G. Carraro, It. Pat. 894,274 (1971). C. A. 83:27924e.
113. J. F. Fauvarque, A. Jutand, C. Chevrot, F. Pfluger, M. Troupel, Fr. Pat. 2,530,266 (1984). C. A. 101:14051h.
114. J. C. Folest, J. M. Duprilot, J. Perichon, Y. Robin, J. Devynck, Tetrahedron Lett. 26, 2633 (1985).
115. F. Barba, A. Guirado, A. Zapata, Electrochim. Acta27, 1335 (1982).
116. T. Matsue, S. Kitahara, T. Osa, Denki Kagaku 50, 732 (1982).
117. M. Troupel, Y. Rollin, S. Sibille, J. Perichon, J. F. Fauvarque, J. Organomet. Chem. 202, 435 (1980).
118. S. Torii, H. Tanaka, K. Morisaki, Tetrahedron Lett., 51, 1655 (1985).
119. M. Troupel, Y. Rollin, J. Perichon, J.F. Fauvarque, Nouv. J. Chim. 5, 621 (1981).
120. J.F. Fauvarque, C. Chevrot, A. Jutand, M. Francois, J. Perichon, J. Organomet. Chem. 264, 273 (1984).
121. S. Torii, H. Tanaka, T. Hamatani, K. Morisaki, A. Jutand, F. Pfluger, J.F. Fauvarque, Chem. Lett. 1986, 169.
122. S. Wawzonek, J. M. Shradel, J. Electrochem. Soc. 126, 401 (1979).
123. G. Silvestri, S. Gambino, G. Filardo, G. Greco, A. Gulotta, Tetrahedron Lett. 25, 4307 (1984).
124. M. M. Baizer in'Organic Electrochemistry', M. M. Baizer, H. Lund Editors, p. 1038 . M. Dekker, New York (1983).
125. R. C. Hallcher, M. M. Baizer, D. A. White (to Monsanto Co), U. S. Pat. 4,072,583 (1978).
126. R. C. Hallcher, M. M. Baizer, J. Liebig Ann. der Chem. 737 (1977).
127. D. A. Tyssee (to Monsanto Co.) U. S. Pat. 3,945,896 (1976). C. A. 84:171509m.

SUPERCRITICAL CARBON DIOXIDE AS AN EXTRACTION AGENT

Wilhelm O. Eisenbach
Max-Planck-Institute für Kohlenforschung
P.O. Box 10 13 53
D 4330 Mülheim-Ruhr
West Germany

ABSTRACT. A new method will be presented in which a gas in its supercritical state is used as an extraction agent. An ideal example is carbon dioxide, because it is readily available, is cheap and has no toxicological or enviromental problems. The critical data of carbon dioxide (31°C, 72.9 bar and 468 g/l) make it suitable in the extraction of valuable components from natural products under mild thermal conditions. In most cases the extraction with supercritical carbon dioxide is more selectiv than the extraction using an organic solvent such as chlorinated hydrocarbons. The behavior of supercritical carbon dioxide and the features of the equipment used in the Max-Planck-Institut für Kohlenforschung and the range of applications will be described.

1. Introduction

The subject of the meeting was "Carbon Dioxide: Chemical and Biochemical Uses as a Source of Carbon". For this purpose the under normal conditions non reactive carbon dioxide must be transfered into reactive compounds or complexes, in most cases a very difficult process. But just this inactivity is the property which makes carbon dioxide suitable for the new method in which carbon dioxide is not a reactant but a solvent, in particular carbon dioxide in its supercritical state as an extraction agent for high-boiling and high-molecular weight substances. This procedure, the supercritical fluid extraction (SCF-extraction) [1], is very suitable to extract valuable components from natural products under mild thermal conditions.

1.1 Discovery

This process was discovered by Dr. Kurt Zosel at the Max-Planck-Institute für Kohlenforschung at Mülheim-Ruhr about 25 years ago and is now investigated besides our institute at several institutes all over the world very intensively. The phenomenon of the excellent solvent power of a gas in its supercritical state was discovered during the investigations on the field of organo-metallic reactions, in special using organo-alumnium compounds. In the earlier 50th years K. Ziegler and H.G. Gellert [2] found out the so called "Aufbau"-reaction of triethyl-aluminium with ethylene under pressure at about 100°C. Today this

M. Aresta and G. Forti (eds.), Carbon Dioxide as a Source of Carbon, 371–388.

reaction forms the basis of the large-scale technical production of long-chain primary alcohols which are used as a raw material for the production of biodegradable detergens [3].

"Aufbau"-reaction

$$\text{al-}C_2H_5 \;+\; x\,C_2H_4 \;\Rightarrow\; \text{al-}(C_2H_4)_x\text{-}C_2H_5 \;\Rightarrow$$

$$+\; 0.5\,O_2 \;\Rightarrow\; \text{al-O-}(C_2H_4)_x\text{-}C_2H_5 \;\Rightarrow$$

$$+\; H_2O \;\Rightarrow\; \text{alOH} \;+\; \underline{\text{HO-}(C_2H_4)_x\text{-}C_2H_5}$$

Cleavage-reaction

$$\text{al-}CH_2\text{-}CH_2\text{-R} \;\Rightarrow\; \text{al-H} \;+\; \underline{CH_2{=}CH\text{-R}}$$

$$(\text{al} = 1/3\ \text{Al})$$

In this reaction ethylene is inserted into the aluminium carbon bond so that in a stepwise manner long-chain alkyl groups bonded to the aluminium are formed. After oxidation of these alkyl-aluminium compounds and the following hydrolysis the desired alcohols are formed. Under the conditions of this reactions, specially at some higher temperatures (100 to 120°C) partly the trialkylaluminium undergoes cleavage to give dialkylaluminium hydride and α-olefines.

This reaction was carried out on a laboratory scale as follows: a 200 ml autoclave was charged with about 10 ml of triethylalumnium (b.p. 194°C at 760 torr) and filled with ethylene at room temperature. Under these conditions the ethylene is 10 - 15°C above its critical temperature of 9°C. In this temperature range the compressibility of ethylene is particulary high, so the pressure cannot be used as a guide for the amount in the autoclave: this was determined by weighing. In one case it was found that the autoclav was overfull, and heating at 100°C would have resulted in too high a pressure. It seemed inadvisable to simply vent out the autoclav since it was feared that traces of the spontaneously inflammable triethylalumnium might be carried over and ignite the gas. The gas was therefore vented through a cold trap, where condensation of triethylalumnium was indeed observed. It was assumed that this had been carried over either in form of droplets or that it had crept up the sides of the autoclav. But this phenomenon did not receive further attention at this time.

A systematic investigation of the ability of pressurized ethylene to transport high-boiling material was first undertaken in 1962 as the result of a difference of opinion with a licensee concerning the quality of the product of the "Aufbau"-reaction. The product prepared by the licensee contained large quantities of α-olefines which were virtually absent from our product obtained under the same conditions. The controversial results arose from differences in the experimental procedures. In our experiments the α-olefins formed were removed from the reactor by the supercritical ethylene and hence separated from the "Aufbau"-product. This effect could not be associated with the partial pressure of the α-olefin since this was too low. It was very quickly realized that a general principle had been discovered for the separation of substances, specially of high-boiling substances.

2. General Basic Principles

The physicochemical basis will not be discussed in detail here. It has been studied and described in detail, with reference to phase diagrams and equilibrium measurments, by various groups, in West Germany especially by Prof. Schneider and his associates in Bochum [4,5]. Only a few important points will be considered to understand the nature of the supercritical fluid extraction.

2.1 Supercritical Fluid

Normally three states of matter are recognised for a chemical compound: solid, liquid and gas. If the pressure/temperature diagram of carbon dioxide in Figure 1 is considered, the three regions »solid«, »liquid« and »gas« may be observed, defined by the sublimation, melting and vapor pressure curves.

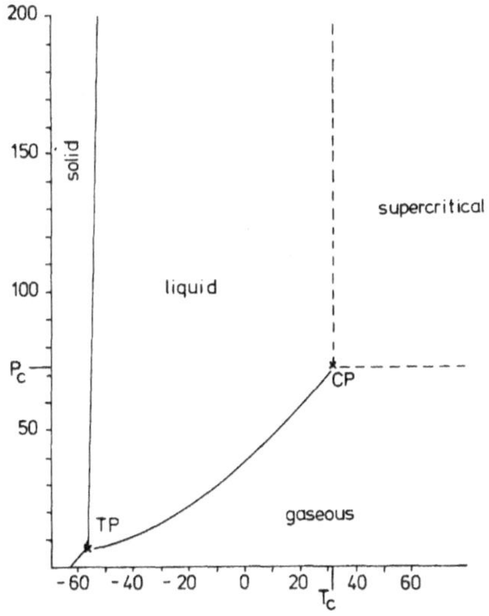

Figure 1 Pressure/Temperature Diagram of Carbon Dioxide
TP = triple point -56.6°C 5.2 bar, CP = critical point 31.0°C, 72.9 bar

If the vapor pressure curve is followed, the critical point (charaterised by the critical temperature and the corresponding critical pressure) is reached. Above the critical temperature of a compound it is impossible to liquify a gas irrespectiveof pressure applied. At the critacal point the densities of a liquid and a gas are equal. The region limited by the critical temperature and the critical pressure is described as a supercritical fluid, practically a »fourth state of matter«.

Such supercritical fluids show interesting solvent characteristics in relation to low volatility and high molecular weight compounds. Table 1 gives a comparison of some physical properties of liquids, gases and supercritical fluids.

Table 1 Physical Porperties of Gases, Liquids and Supercritical Fluids

	gas	liquid	supercritical
density (g/l)	0.001	1.0	0.7 *)
viscosity (cSt)	0.01	0.5 - 1.0	0.05 - 0.1
diffusion (cm^2/s)	0.1	10^{-5}	10^{-3} - 10^{-4}

This compilation of data shows that some characteristics of the supercritical fluid, e.g. density and diffusion coefficient, resemble those of a liquid, whereas the viscosity is similar to that of a gas. High denstity with low viscosity give supercritical fluids especially favourable solvent properties.

In contrast to an incompressible liquid the density of a supercritical fluid can be varied in a wide range by temperature and pressure as shown in Figure 2.

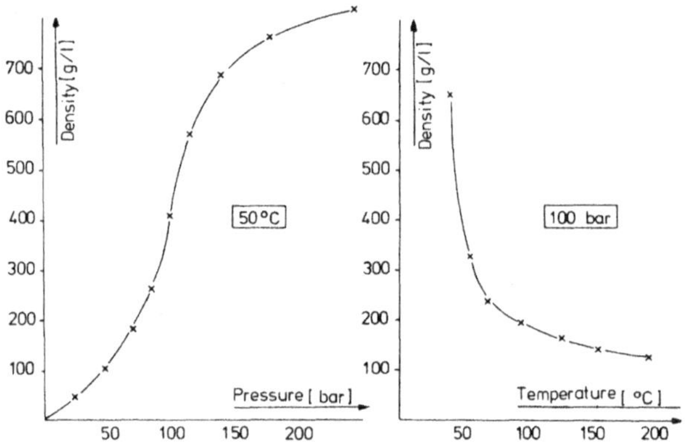

Figure 2 The Effect of Temperature and Pressure upon the Density of CO_2

The left-hand diagram shows the effect of varying the pressure upon the density at a constant temperature of 50°C that is a temperature of 19° above the critical temperature of 31°C for carbon dioxide and as can be seen the density of the gaseous carbon dioxide increases dramatically on passing the critical pressure of 72.9 bar, a particular marked effect is shown between 70 and 120 bar. Higher pressures are accompanied by an approximately linear increase in density.

In the right-hand diagram the density is plotted against the temperature at a constant pressure of 100 bar, i.e. a pressure above the critical pressure: it shows that the density decreases with incresaing the temperature, especially over the range 40 - 100°C.

As the density of a solvent is an important criterion of its dissolving capacity, the concentration of a compound dissolved in a supercritical fluid shows, in a first approximation, similar dependences on temperature and pressure. An experimental example of this behaviour is the solubility of paraffin-oil in ethylene. At room temperature, i.e. about 10°C above the critical temperature for ethylene, and atmospheric pressure 1 l of ethylene takes up only

1 mg of paraffin-oil. If the pressure is increased above the critical pressure of 50.6 bar for ethylene then the amount of oil dissolved increases dramatically so that at 100 bar 1 l of ethylene takes up 12 g of paraffin-oil. If on the other hand the temperature is increased then the amount of oil transported decreases. For example the solubility in 100 g of ethylene at a constant pressure of 200 bar is 25 g at 23°C and only 7 g at 70°C.

2.2 Separation Procedures

Based on these phenomena it is obvious that in principle two methods can be used to isolate the product dissolved in the supercritical phase, both compared with lowering the density of the loaded phase.

1. By increasing the temperature at constant pressure, the so-called »isobaric method« and
2. By decreasing the pressure at constant temperature, the so-called »isothermal method«.

2.2.1 The Isobaric Method. The equipment for the supercritical fluid extraction under isobaric conditions is shown by the flow scheme in Figure 3. The supercritical gas enters the extraction vessel (1) at the bottom. The temperature of the vessel is T_1. The extraction agent passes through the material to be extracted and the soluble components are taken up. The loaded supercritical phase leaves the autoclave at the top and passes the heat exchanger (2) in which the temperature is raised up to the temperature T_2. The density of the supercritical phase decreases and the dissolved material precipitates and can be collected in the separation vessel (3). The heated supercritical fluid is circulated by a metering pump (4) and cooled to extractor temperature T_1 by the second heat exchanger (5).

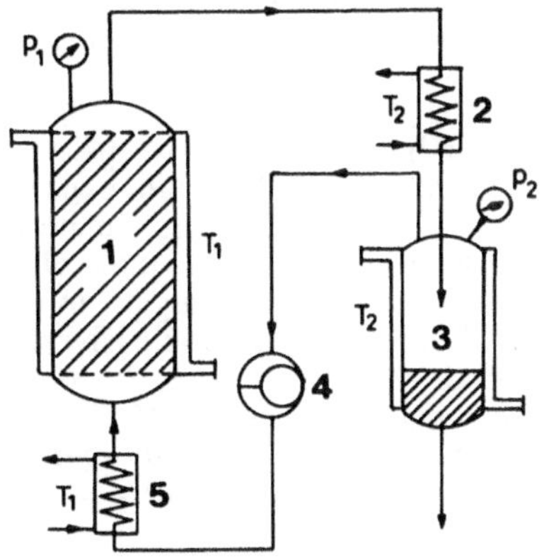

Figure 3 Flow Scheme of th Isobaric Method
1 Extractor, 2 Heat exchanger T_2, 3 Separator, 4 Circulating Pump, 5 Heat exchanger T_1

2.2.2 The isothermal Method. The disadvantage of the isobaric method is that a part of the dissolved material may remain in the supercritical phase even after raising the temperature while the phase is still supercritical and while more volatile components because of their vapor pressure at the high temperature are difficult to remove from the system. On the other hand several products, mainly natural products, do not agree with such high temperatures which are needed for this process. The temperature T_2 in most cases is 100 or more °C above the extraction temperature.

In this case the better method to isolate the product from the loaded supercritical phase is the isothermal approach which flow scheme is shown Figure 4.

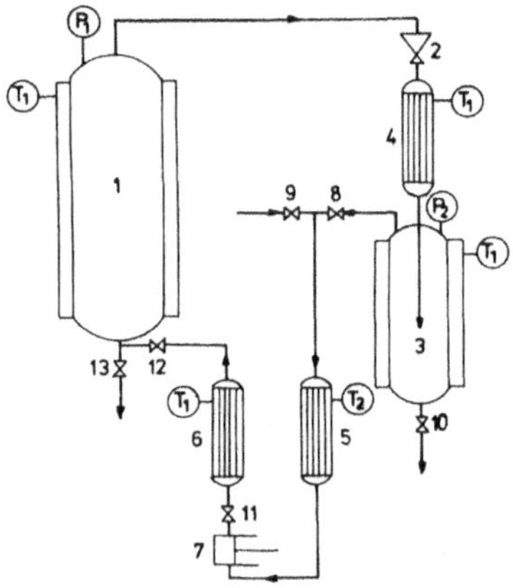

Figure 4 Flow Scheme of the Isothermal Method
1 Extractor, 2 Expansion valve, 3 Separator, 4-6 Heat exchangers, 7 Pump, 3-13 valves

The supercritical gas enters the extraction vessel (1) and is loaded by passing the material to be extracted as before. But now the loaded supercritical phase leaving the extractor at the top is expanded by an expansion valve (2) to pressures below the critical pressure. The dissolved material is deposited quantitatively and is collected in the separator vessel (3). A disadvantage of this approach is that the resulting subcritical extraction agent must be recycled to supercritical conditions. Here two posibilities present themselves:

1. Gases which can be liquified at normal pressure or slightly above by water cooling are condensed in the heat exchanger (5). The liquid then is pumped by the metering pump (7) and recycled to the extraction vessel after passing the second heat exchanger (6), where the compressed liquid is heated to the temperature of the extraction vessel.

2. Gases which are less easily to be liquified are compressed by a compressor at the extraction temperature before being recycled. The compressor is placed instead of the two heat exchangers (5) and (6) and the pump (7). In the case of carbon dioxide as the extraction agent both procedures are possible, but normally the expanded gas will be compressed and recycled.

2.3 Fractionation

One advantage of the use of supercritical fluids as extraction agents is that less volatile or high molecular weight material which would be pyrolysed under the conditions of a conventional distillation can be fractionated under mild thermal conditions. A number of variations is possible in which temperature and pressure are adjusted in the manner described above.

2.3.1 Variation of Pressure. The first possibility is the variation of the pressure. Discontinuous variation of the pressure in either the extraction vessel or in the separator can lead to fractionation. In the first case the pressure and hence the density of the extraction agent is increased in a step-wise manner with the result that material is taken up into the supercritical phase in order of increasing boiling point or molecular weight. Then the dissolved material can be isolated in either an isobaric or an isothermal manner. In practice the pressure is generally raised after the extraction at the lower pressure is complete, i.e. the extraction will be carried out until no extract can be collected in the separator.

Alternatively the extraction can be carried out at a constant pressure in the extraction vessel and the loaded supercritical phase will be expanded in a step-wise manner whereby the least volatile components are deposited first, while the better volatile components remain dissolved in the supercritical phase. A flow scheme of such an apparatus is shown in Figure 5. Here the pressure P_3 is minor than P_2, P_2 minor than P_1 and P_1 minor than the pressure P in the extraction vessel.

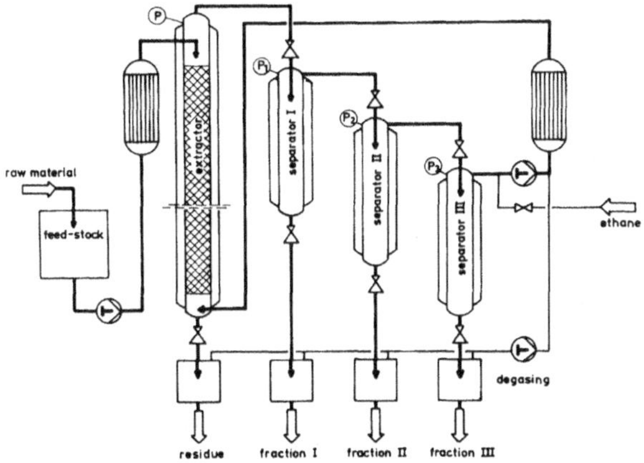

Figure 5 Fractionation by Pressure Reducing

The advantage of this procedure is that the extraction also can be carried out continuously. The raw material is pumped onto the top of a column, a stainless steel pressure tube, which is filled with stainless steel Raschig rings. The extraction agent enters the column at the bottom and the soluble components are taken up into the supercritical fluid while passing the column. The loaded supercritical phase leaves the column at the top and is now expanded to the pressure P_1 and the least volatile components are deposited first and collected in separator I. The supercritical phase leaving this separator at the top is now expanded to the pressure P_2 and a second portion is deposited and collected in separator II. This procedure is continued until the gas is subcritical and no product is still dissolved. By a suitable choice of the pressures it is possible to get very clean fractions.

2.3.2 The Hot Finger. The other application of fractionation is characterized by the phenomenon that the density of the supercritical phase decreases at the same pressure at higher temperatures. The equipment of such an apparatus is shown in Figure 6.

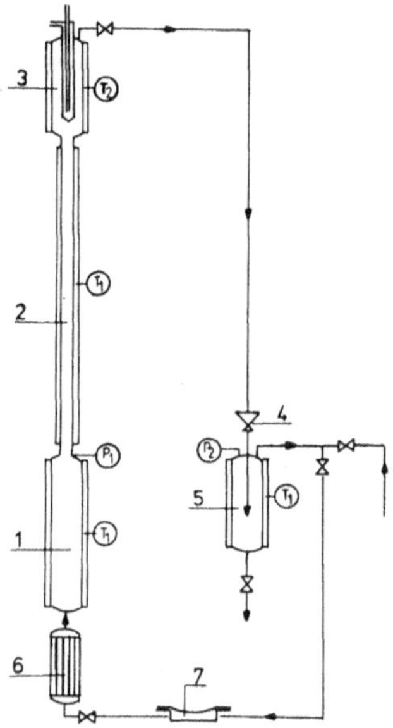

Figure 6 Hot Finger Apparatus

1 Extraction vessel, 2 Fractionation column, 3 Hot finger, 4 Expansion valve, 5 Separator, 6 Heat exchanger, 7 Compressor

The extraction vessel (1) has been combined with a pressure tube packed with stainless steel Raschig rings (2) like a rectification column in the conventional distillation process. The separately heatable head of the column, the so-called Hot Finger (3), is fitted onto the top of the column. The temperature of this Hot Finger is held at a higher temperature than the rest of the apparatus. On contact with this Hot Finger the density of the loaded supercritical phase decreases and the less volatile components of the mixture condense, drop back into the column and are subjected to rectification in the same manner as in conventional distillation process. The product still dissolved in the supercritical phase after passing the Hot Finger is isolated by lowering the pressure to subcritical conditions by an expansion valve (4) as described before. The dissolved product precipitates and is collected in the separator (5). The expanded gas is compressed by the compressor (7) and recycled to the extractor.

2.4 Adsorption

A wholly different method for separation of dissolved material from the loaded supercritical phase is the adsorption on a suitable adsorption agent, in most cases activ charcoal will be used. This method will be used if the concentration of the dissolved product is very low. The supercritical fluid is circulated at constant temperature and pressure; a column filled with the the adsorption agent is fitted in place of the separator in the flow scheme of the isobaric method (Figure 3). The disadvantage of this method is that the adsorption agent must be regenerated periodically and the adsorbed material must be extracted.

3. Applications

After the above general considerations, this separation method will be further explained by means of practical examples.

Table 2 gives a summary of practical trials on the two fields in which this method for isolation of low-volatility and high molecular weight compounds has been applied.

1. The isolation of valuable components or the removal of undesirable constituents from natural products and
2. Such processes to clear off residues and waste materials from the chemical and petrochemical industry which are difficult to process by conventional methods.

In this paper only the first group of applications is described with carbon dioxide as the extraction agent. In isolation of valuable or other constituents from natural materials, two main applications may be differentiated. Undesirable constituents may be eliminated, e.g. decaffination, debittering or deodorization, in which case the extraction residue comprises the desired refined product. Alternatively, the aim may be to isolate valuable constituents of the raw material, e.g. flavour compounds or oil. In this case the extract is the desired end product.

3.1 Decaffeination of Green Coffee Beans

The first example is the patented process discovered by K.Zosel for decaffeination of green coffee beans [6], which has been operated under lincence on a large scale by the company HAG in Bremen for more than five years. Previously, caffeine was extracted from raw coffee with methylene chloride. As methylene chloride is an unselective solvent, part of the fat and flavour compounds were also extracted, so that the flavour of the roasted decaffeinated coffee is less pronounced than that of un-decaffeinated coffee. It is also difficult to remove the last traces of the solvent from the coffee beans.

The treatment of the green coffee beans with supercritical carbon dioxide may have two results: inserting dry coffee beans and using dry carbon dioxide only fat and flavour compounds and no caffeine is extracted, with moist carbon dioxide and moist coffee beans caffeine is extracted selectively, fat and flavour compounds remaining in the beans. This leads to a considerable improvment in flavour quality; in contrast to methylene chloride decaffeination, there is practically no difference in flavour between treated and untreated coffee. By selection of suitable pressure, temperature and moisture content, extraction conditions may be established which permit the reduction of the caffeine content in the decaffeinated coffee to below the legally specified value.

Table 2 Extraction with Supercritical Fluids »Applications«

Natural Materials	
decaffeination	green coffee beans [6] tea [7]
deodorization	vegetable oils [8] animal oils and fats [8] spent brewers yeast [12]
isolation of flavours	spices (pepper, paprika, nutmeg) [9] hops [10,11] tobacco, tobacco dust
separation of oils [8]	soybeans, palmbeans, jojobabeans peanuts, coconuts cotton seed, sunflower seed indian corn yolk
fractionation [8]	cod-liver oil esters from fish oil
Petrochemical and Coal industry	
heavy oil recovery	residues from hydrogenation by the Bergius-Pier process of [15,16] * brown coal * bituminous coal * petroleum vacuum residue oil sands [8,17] oil shale [8,17]
regeneration	used motor oil [8,17] millscale
de-asphalting [1,8]	residues from petroleum distillation oil sand bitumen
fractionation	tall oil [8] brown coal high-temperature tar [8,14]

In the Max-Planck-Institut three process variants operating at the same temperature, pressure and moisture content were evaluated in batchwise experiments. In all the different procedures before the experiment commercial raw coffee with 1 - 2.5 % caffeine and approximatly 10 % water was moistened at

70 - 80°C to 40 % water content. The extraction was performed isobarically at 90°C and 200 bar and the caffeine was removed from the extraction agent, that is supercritical carbon dioxide, by adsorption or absorption.

Two variants of adsorption on activated charcoal are presented schematically in Figure 7.

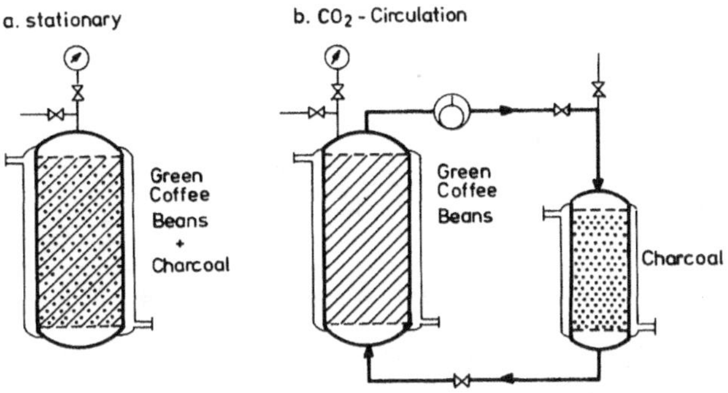

Figure 7 Decaffeination of Coffee: Adsorption on Charcoal

The left-hand diagram shows the technically most simple system: the raw green coffee beans are mixed with activated charcoal, the particle size of which is selected to occupy the interstices between the beans thereby optimising the use of the extractor volume. For a particle diameter of the charcoal of 3 mm the charcoal / coffee ratio is 1 : 3. This mixture is then held for 10 - 14 hours under carbon dioxide at a temperature of 90°C and a pressure of 200 bar. During this period caffeine diffuses out of the coffee beans into the activated charcoal. Finally the charcoal is separated from the coffee beans by sieving and the caffeine therein is isolated by extraction with alcohol or chloroform.

This process is more rapid when the equipment shown on the right-hand diagram is used. The green coffee beans and the charcoal are filled into two separate pressure vessels and the supercritical carbon dioxide is circulated by means of a pump.

If the activated charcoal column in this apparatus is replaced by a packed column through which water flows as schematically is shown in Figure 8, the optimum procedure may be achieved.

The disintigrated coffee beans are placed in the extractor, the temperature of which is held at 90°C. Carbon dioxide is compressed and the supercritical fluid is circulated at a pressure of 200 bar by means of a pump. The carbon dioxide enters the extractor from below. The supercritical carbon dioxide takes up caffeine from the beans and the loaded phase leaves the extractor at the top. This phase is fed into the bottom of the packed washing column through which the water, which is pumped into the column on the head, flows downwards. The countercurrent flow of water extracts the caffeine from the supercritical carbon dioxide which is recycled to the extractor. The aqueous solution of caffeine is collected below the washing column, the pressure is released and the solution is distilled for recovery of the caffeine. The water can be recycled to the process too.

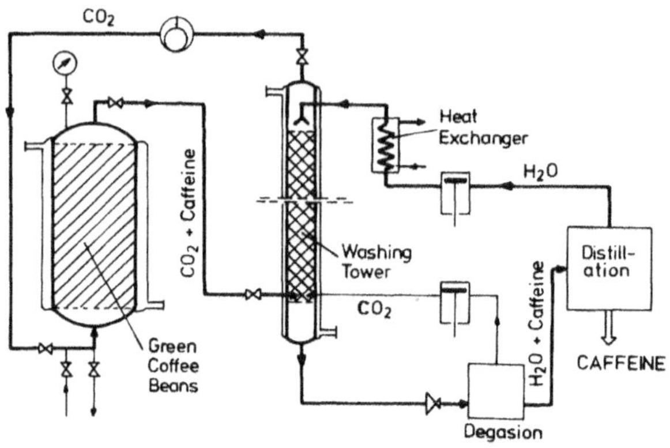

Figure 8 Decaffeination of Coffee: Water Washing Procedure

A major advantage of this procedure is the possibility of control of the decaffination rate by UV-spectroscopic analysis of the resulting caffeine solution. If the logarithm of the caffeine concentration in the aqueous solution is plotted against the duration of the extraction, a linear relationship is found. The diagram in Figure 9 shows the result of three extractions, in which raw coffee beans with different concentrations of caffeine are inserted.

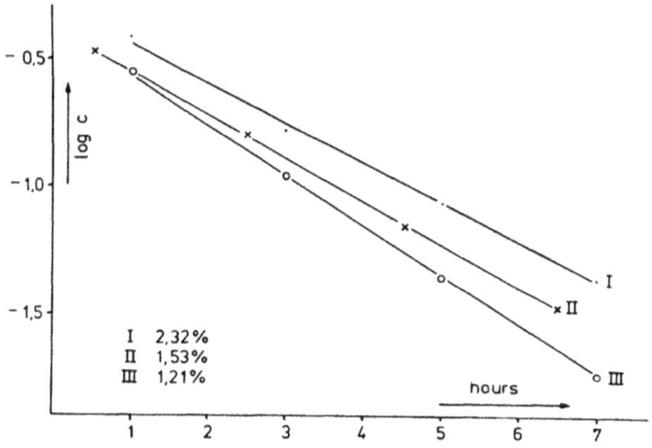

Figure 9 Dependence on Caffeine Concentration

The lines being almost parallel for the different initial caffeine concentrations of the raw coffee. These results show that under the selected extraction conditions no saturation of the supercritical carbon dioxide with caffeine occurs and that the extraction time increases only with increasing the caffeine concentration in the raw coffee beans. Therefore it seams that the diffusion of the caffeine out of the beans limits the rate of decaffeination.

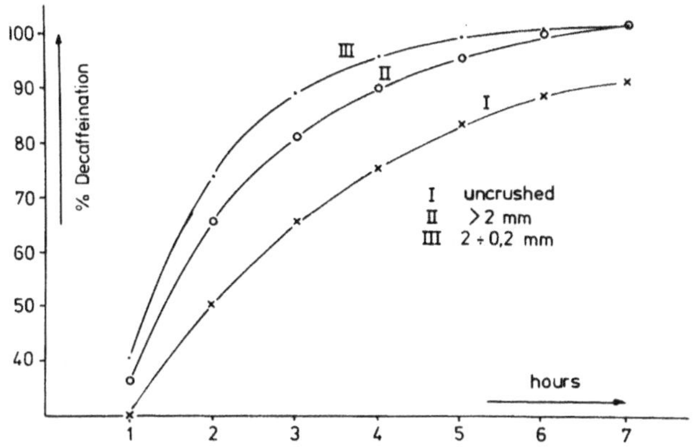

Figure 10 Dependence on Particle Size

This is confirmed by the following experiments. The curves I - III in Figure 10 illustrate the dependence of the degree of decaffeination on the extraction time and particle size. Whole and comminuted coffee beans of the same variety were decaffeinated under identical conditions, i.e. 90°C, 200 bar and 40 % moisture. 98 % decaffeination was achieved after 5 hours for finely comminuted (curve III) and 6 hours for the coarsely comminuted coffee beans (curve II). For the whole coffee beans the degree of decaffeination was only 90 % after an extraction time of 7 hours (curve I).

3.2 Deodorization of Vegetable and Animal Oils and Fats [8]

A second important procedure in the food industry is the deodorization of vegetable and animal oils and fats. Whereas in the decaffeination process of coffee it is economically adventageous to isolate the caffeine as a marketable product, the deodorization of vegetable and animal fats and oils deals with compounds which reduce only the product quality and generally are of no economic significance. The free fatty acids comprise a large proportion of the undesirable material in the crude oils; these, especially the unsaturated acids, may cause the rancity of the oils. Therefore natural oils must be refined, this is frequently a difficult oparation which may results in considerable losses.

The extraction with supercritical fluids, especially with supercritical carbon dioxide is a possible alternative technique as conditions are suitable for elimination of the free fatty acids.

1. the fatty acids are more polar than the triglycerids therefore they have a higher solubility in supercritical carbon dioxide which can be seen as a polar solvent.

2. the considerable lower molecular weight of the free fatty acids as compared to those of the triglycerids of the oils also results in a higher solubility.

This possibility was first tested in an isobaric process, in which the component dissolved in the supercritical carbon dioxide are adsorbed on activated charcoal, somewhat like the process described before. A schematic layout of such a system is shown by means of the flow scheme in Figure 11.

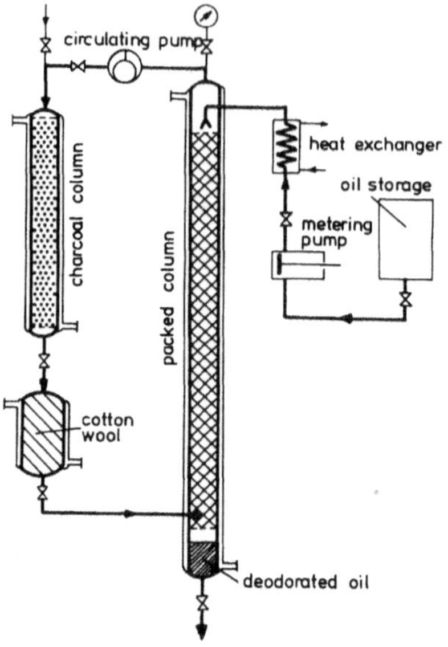

Figure 11 Deodorization of Vegetable Oil: Isobaric Procedure

The equipment for this procedure consists of a 14 m long heatable pressure tube of 50 mm internal diameter filled with glas balls and a similar column of 5 m length filled with the activated charcoal. The supercritical carbon dioxide is circulated at the extraction temperature of 80°C and pressures between 150 and 225 bar. After heating in a heat exchanger the crude oil is applied at the top of the column by means of a metering pump and flows downwards in countercurrent to the supercritical carbon dioxide into a collecting vessel below the column. The loaded supercritical phase leaves the column at the top and flows through the second column which is filled with the activated charcoal on which the dissolved material is adsorbed.

Palm kernel oil was deodorized in this apparatus and the degree of the deodorization was being evaluated on the basis of the acid value of the refined oil. Palm kernel oil with an acid value of 0.3 was fed into the system at 2 l/h and the supercritical carbon dioxide was circulated at approximatly 20 kg/h. If the extraction pressure is varied it could be shown that there is a linear dependence of the acid value of the deodorized oil on the pressure of the carbon dioxide as shown in Figure 12.

Normally such a trial was terminated after an oil throughput of approximately 30 l and then the activated charcoal was removed. The weight of this coal was found to have increased by 2.95 % of the oil inserted. The extraction of the charcoal with chloroform recovered 90 % of this adsorbed material. The acid value of this product was 4.4 whereas the acid value of the refined oil varied between 0.02 and 0.07 depending on the extraction pressure.

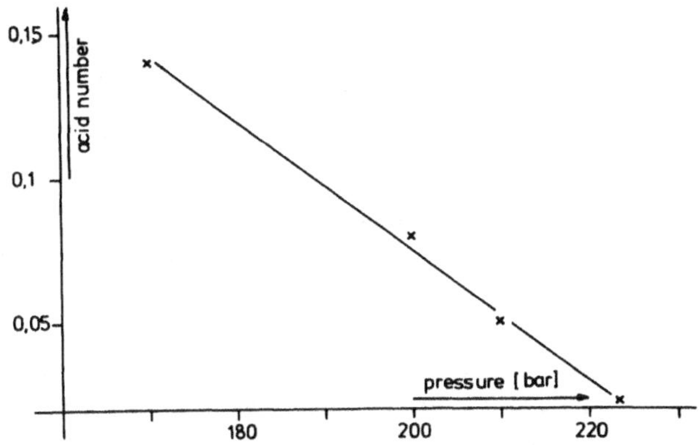

Figure 12 Deodorization of Palmkernel Oil: Dependence on Pressure

This experiments show that the free fatty acids are taken up preferentially into the supercritical carbon dioxide, but also parts of the oil are removed.

In a further study crude soybean oil was treated with supercritical carbon dioxide but this time the material dissolved in the supercritical fluid was isolated in an isothermal manner by pressure release. The resulting product was collected in a series of fractions and analysed by gaschromatography and the content of the free fatty acids, steroids and tocopherols, i.e. the sum of α-, β- γ- and δ-tocopherol, being determined. The predominant steroids were identified by mass spectrometry as β-ergosterol, stigmastenol and β-sitosterol.

Table 3 Deodorization of Soybean Oil with Supercritical Carbon Dioxide

Starting material: 1000 g raw soybean oil containing
0.232 % = 2.32 g free fatty acids
0.058 % = 0.58 g tocopherols
0.242 % = 2.42 g steroids

extract		fatty acids		tocopherol		steroids	
g	% *)	g	% **)	g	% **)	g	% **)
16.3	1.7	0.84	32.6	0.04	6.9	0.19	7.9
16.4	3.6	0.60	62.1	0.04	13.8	0.18	15.3
17.3	3.6	0.49	79.3	0.04	20.7	0.15	21.5
14.4	7.0	0.25	90.1	0.03	25.9	0.14	27.3
27.5	9.9	0.26	101.3	0.05	34.5	0.21	36.0

*) related to the raw oil (sum of all extracts)
**) related to the content in the raw oil

1000 g crude soybean oil were subjected to extraction with supercritical carbon dioxide at 150 bar and 50°C in an autoclave of 2.7 l capacity, the carbon doxide circulating rate was approximatly 20 kg/h. The loaded supercritical phase was depressurized to 15 bar and the extract produced during two hours was collected in fractions. The results of the experiment are summarized in Table 3.

The table of results shows that the free fatty acids were extracted preferentially as supposed; if about 10 % of the initial quantity of oil was separated by extraction this extract contained all the free fatty acids but only one-third of the steroids and tocopherols present in the crude oil.

3.3 Deoiling of Soybeans

In the field of isolation of constituents from natural materials the production of a storage-stable hop extract [10,11] is already in large-scale operation by several companies in West Germany. The recovery of oil from oilseeds will be described using the defatting of soybeans as an example [8]. The curves in Figure 13 represent typical extraction curves, in which the total yield of extract is plotted against the extraction time.

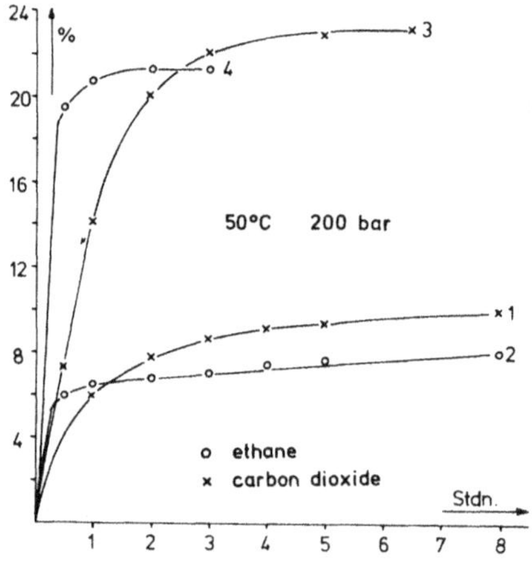

Figure 13 Deoiling of Soybeans: Extraction Curves

All trials were conducted with 500 g raw material in a 2 l autoclave at 50°C and 200 bar with an extraction agent flow rate of 20 l/h. This operation was non-continuous. In this study besides carbon dioxide we used also ethane as a supercritical extraction agent to demonstrate, that an unpolar solvent like ethane is more suitable to extract an unpolar product than the more polar carbon dioxide. These curves show that the extraction rate with ethane is higher than with carbon dioxide, the curves 1 and 3 are initially steeper than the curves 2 and 4. The higher total yield of extract achieved with carbon dioxide is due to the fact that at least part of the water in the beans is extracted with the oil when carbon dioxide is used, whereas no water is extracted using supercritical ethane. In these studies two different raw materials were used, both being deoiled with ethane and carbon dioxide. Coarse soybean grits, particle size 1 - 2 mm, were inserted in the trials 1 and 2, whereas fine soy flour was used in the other two trials. The content of material extractable in 12 hours with boiling hexane (68.9°C) was 12.4 % for the grits and 23.1 % for the soy flour. The difference is attributable to the different origins of the two materials.

Table 4 Deoiling of Soybeans with Ethane and Carbon Dioxide

	grits		flour	
	% oil	% water	% oil	% water
hexane *)	12.0	---	23.1	---
carbon dioxide	7.9	2.0	21.1	2.0
ethane	7.9	---	21.2	---

*) liquid extraction at 68.9°C (12 hours)
% related to the iserted material

The results of such an experiment, summerized in table 4 show no difference in the yield between ethane and carbon dioxide, however, with both supercritical fluids at 50°C and 200 bar only 64 % of the boiling hexane-extractable material could be extracted from the grits but 92 % from the flour. This shows that the relative large particles of the grits could be only partially deoiled at the low extraction temperature of 50°C.

References

[1] Extraction with Supercritical Gases
G.M.SCHNEIDER, E.STAHL, G.WILKE Editors;
Verlag Chemie , Weinheim, West Germnay 1980
[2] K.ZIEGLER, H.G.GELLERT et al., Ang.Chem. **64,** 323-329 (1952)
[3] K.ZIEGLER, DB Patent 10 14 088 (21.07.1960)
[4] G.M.SCHNEIDER, Ang.Chem.Int.Ed.Engl. **17,** 716-727 (1978)
[5] G.M.SCHNEIDER, Ber.Bunsenges.Phys.Chem., **83,** 841-848 (1984)
[6] K.ZOSEL, Studiengesellschaft Kohle mbH,
DB Patent 20 05 293 (26.09.1973), US Patent 4 260 639 (7.04.1981)
[7] O.VITZHUM, P.HUBERT, Sudiengesellschaft Kohle mbH,
DB Patent 21 27 642 (3.06.1971), US Patent 4 167 589 (1979)
[8] K.ZOSEL, Studiengesellschaft Kohle mbH,
DB Patent 14 93 190 (4.01.1973), US Patent 3 969 196 (13.07.1976)
[9] O.VITZHUM, P.HUBERT, Studiengesellschaft Kohle mbH,
DB Patent 21 27 611 (3.06.1971), US Patent 4 123 559 (1978)
[10] O.VITZHUM et al., Studiengesellschaft Kohle mbH,
DB Patent 21 27 618 (3.06.1971), US Patent 4 104 409 (1978)
[11] P.KLÜSTERS, Forum der Brauerei **1,** 17-20 (1983)
[12] W.EISENBACH, Studiengesellschaft Kohle mbH,
DOS 30 21 653 (10.06.1980), Canadian Patent 1 179 546
[13] K.ZOSEL, Studiengesellschaft Kohle mbH,
DAS 23 63 418 (19 08.1976), US Patent 4 331 695 (1982)
[14] W.EISENBACH, K.NIEMANN, Erdöl und Kohle, Erdgas, Petrochemie
34, 296 (1971)

[15] W.EISENBACH, P.GOETTTSCH, R.SONNENTAG Chem.Ing.Tech. **55**, 655 (1983)
[16] W.EISENBACH, K.NIEMANN, P.GOETTSCH, Chemical Engineering on Supercritical Fluid Conditions 419-433 (1983)
[17] H.COENEN, P.RINZA, Chem.Ing.Tech. **54**, 386 (1982)

LIST OF SHORT COMMUNICATIONS PRESENTED AT THE SUMMER SCHOOL ON :

"Carbon Dioxide : Chemical and Biochemical Uses as a Source of Carbon"

Pugnochiuso (Italy) , 22 June - 3 July , 1986.

CHEMISTRY SECTION

M. Aresta *, D. Ballivet- Tkatchenko** and M.C. Bonnet**
CO_2 Interaction with Transition Metal Complexes.
* Dipartimento di Chimica, Via Amendola 173, 70126 Bari (Italy)
**Institut de Recherches sur la Catalyse, CNRS-LCC, 205 Route de Narbonne 31400 Toulose (France)

J. Augustynski
Cathodic Reduction of Bicabornate Anions at Palladium.
Départment de Chimie Minérale, Analytique et Appliquée, 30, Quai E. Ansermet, 1211 Geneva 4, Switzerland

D. Ballivet - Tkatchenko and J. Tanji
CO_2 Electroreduction in the Presence of Unsatured Substrates : Formation of Organic Carboxylates.
Institut de Recherches sur la Catalyse, CNRS-LCC, 205 Route de Narbonne 31400 Toulose (France)

J.C.J. Bart
Copper-Zinc Oxide-Alumina Methanol Catalysts Revisited.
DSM Research BV - PO BOX 18 - Geleen - The Netherlands

A. Belforte*, F. Calderazzo*, P.F. Zanazzi**
Synthesis and Reactivity of N,N-dialkylcarbamato and N,N-dialkylamido Complexes of Manganese (II).
* Dipartimento di Chimica e Chimica Industriale - Sezione Chimica Inorganica - Università di Pisa - Via Risorgimento 35 - 56100 Pisa, Italy
** Dipartimento di Scienze della Terra - Sezione Cristallografia - Università di Perugia - 06100 Perugia - Italy

E. Carmona
Some New Ethylene and Carbon Dioxide Complexes of Molybdenum and Tungsten.
Departamento de Quimica Inorganica - Facultad de Quimica - Universidad de Sevilla - 41012 Sevilla - Espana

K. G. Caulton, G.V. Goeden, T.H. Lennen, E.G. Lundquist, B.L. Sutherland and J.C. Huffman

Carbon Dioxide Conversion Using Heterobimetallic Polyhydrides.

Department of Chemistry - Indiana University - Bloomington, IN 47405 USA

A. Ciccarese, E. Quaranta and M. Aresta

Competition of Carbon Dioxide and Dioxygen towards Styrene in the Presence of Rh(I) Catalysts.

Centro di Biologia - Università degli Studi - Lecce - Italy

G. Filardo*, S. Gambino*, A. Gennaro**, G. Silvestri*, E. Vianello**

Electrochemical Carboxylation of Styrene.

* Istituto di Ingegneria Chimica - Palermo (Italia)

** Dipartimento di Chimica Fisica - Padova (Italia)

U.Kanne

Formation of C-O,C-N,C-C Bonds by Ni(O)-Induced Activation of Carbon Dioxide and Unsaturated Compounds.

Institut fur Technische Chemie und Petrolchemie der Rheinisch-Westfalischen Technischen Hochschule Aachen , Worringer Weg 1, D-5100 Aachen (FGR)

W. Kipshagen

Carboxylation of Malonitrile by Rhodium -and Iridium-Complexes.

Institut fur Technische Chemie und Petrolchemie der Rheinisch-Westfalischen Technischen Hochschule Aachen , Worringen Weg 1, D- 5100 Aachen (FRG)

G. La Monica, G. A. Ardizzoia, F. Cariati, S. Cenini

Further Ivestigations on the Reaction of CO_2 with Tetrahydroborato Copper(I) Complexes.

Dipartimento di Chimica Inorganica e Metallorganica - Centro C. N. R. Via Venezian , 21 - 20133 Milano

S. Lecolier
The "Group Scientific CO_2 (GS-CO_2)": Why and How to Develop Fine Chemistry Based on Carbon Dioxide.
S.N.P.E. - Centre de Recherches du Bouchet - 91710 Vert le Petit - France

R. Mahe, Y. Sasaki and P.M. Dixneuf
Catalytic Synthesis of Vinylic Carbamate Directly from Alkyne, Carbon Dioxide and Amine.
Université de Rennes, Campus de Beaulieu, 35042 Rennes Cedex (Francia)

A. Mackor, A.H.A. Tinnemans and T. P. M. Koster
Reduction of CO_2 at Titanate Powders in Sunlight and at Electrodes in the Light or Dark.
Institute of Applied Chemistry TNO, P.O. Box 5009, 3502 JA Utrecht , The Netherlands.

J. Mascetti and M. Tranquille
Study of CO_2 Transition Metal Atom Complexes in Low Temperature Matrices.
Laboratoire de Spectroscopie Infrarouge - UA 124 - 351 - Cours de la Libération - 33405 TALENCE Cedex (France)

G.K. Parashar
Reaction of Carbon Dioxide with Transition Metals.
Swami Shraddhanand College, Alipur, Delhi-110036 (India)

E. Quaranta and M. Aresta
Interaction of IA Group Metals Naphtalenides with Carbon Dioxide the Role of the Reaction Conditions.
Dipartimento di Chimica - Università degli Studi - Bari - Italy

F. Rivetti
Carbonic Acid Esters: Promising Intermediates for Extensive Use of CO_2 as a Synthetic Building Block in Organic Chemistry.
EniChem Sintesi - Laboratorio di Bolgiano - Via Maritano 26 - 20097 San Donato Milanese - Italy

M. Rosi*, A. Sgamellotti*, F. Tarantelli*, C. Floriani**, M.F. Guest***
A Theoretical Investigation on the Interaction of Formaldehyde with Transition Metal Complexes.
* Dipartimento di Chimica - Università di Perugia - 06100 Perugia
** Department of Chemistry - Columbia University - New York 10027
*** S. E. R. C. Daresbury Laboratory, Warrington

BIOCHEMISTRY SECTION

L.M. Casano, E.M. Valle and R.H. Vallejos
Effect of Photoperiod on CO_2 Assimilation in Soybean.
Centro de Estudios Fotosintéticos y Bioquìmicos - CONICET-Fund. M. Lillo-Universidad Nacional de Rosario - Suìpacha 531 - (2000) Rosario - Rep. Argentina.

N.M. Dixon
Effects of CO_2 on Anaerobic Bacterial Growth and metabolism
Department of Botany and Microbiology - University College of Wales - Aberystwyth - Dyfed SY23 3DA - UK

J. Menaia and J.C. Duarte
Biotechnology of Methanogenic Bacteria.
LINETI - Department of Chemical Indusries - 2745 Queluz de Baxio - Portugal

A. Miceli, D. Pacoda, A. Cardinali and R. Gallerani
Some Peculiar Properties of Ribulose 1-5 Bisphosphate Carboxylase Purified from Artichoke Leaves.
Laboratorio di Chimica Biologica - Centro di Biolologia - Univerità di Lecce - Italia

J.P. Street and R.S. Brown
Tri(2'-pheniyl-5'-methylimidazol-4'-yl-methyl) amine: Zn^{++}. New Ligand-Zn Complexes as Models for Carbonic Anhydrase.
Department of Chemistry - University of Alberta - Canada, T6G 2G2

V. Valenti, S. Scagliarini, P. Pupillo
Replacement of Isoenzymes Involved in Carbon Metabolism During Photosynthetic Differentiation of the Maize Leaf.
Fisiologia Vegetale - Istituto Botanico - Università di Bologna - Via Irnerio 42 - 40126 Bologna - Italy

LIST OF PARTICIPANTS IN THE SUMMER SCHOOL ON :

"Carbon Dioxide : Chemical and Biochemical Uses as a Source of Carbon"

Pugnochiuso (Italy) , 22 June - 3 July , 1986.

LIST OF PARTICIPANTS

Country : CANADA

R.S. BROWN	Department of Chemistry, E3-43 Chemistry Building East, University of Alberta, Canada T6G 2G2.
D.T. CANVIN	School of Graduate Studies and Research, Queen's University, Kingston, Canada K7L 3N6.
J.P. STREET	Department of Chemistry, University of Alberta, Edmonton, Alberta, Canada T6G 2G2.
H. SLEBOCKA TILK	Department of Chemistry, University of Alberta, Edmonton, Alberta, Canada T6G 2G2.

Country : FRANCE

D. BALLIVET	C.N.R.S., Laboratoire de Chimie de Coordination, 205 route de Narbonne, 31400 Toulouse Cedex, France.
M. BONNET	C.N.R.S., Laboratoire de Chimie de Coordination, 205 route de Narbonne, 31400 Toulouse Cedex, France.
P. DIXNEUF	Laboratoire de Chimie de Coordination Organique, Campus de Beaulieu, Rennes Cedex, France.
P. ESCAFFRE	Laboratoire de Catalyse et de Chimie Fine, Ecole Nationale Superieure de Chimie, 118 route de Narbonne, 31077 Toulouse, France.
S. LECOLIER	S.N.P.E., Centre de Recherches du Bouchet, 91710 Vert le Petit, France.
J. MASCETTI	Laboratoire de Spectroscopie Infrarouge, UA 124 , 351 cours de la Libération, 33405 Talence Cedex, France.
I. TKATCHENKO	C.N.R.S., Laboratoire de Chimie de Coordination , 205 route de Narbonne, 31400 Toulouse Cedex ,France.

M. TRANQUILLE Laboratoire de Spectroscopie Infrarouge, UA 124, 351 cours de la Libération, 33405 Talence Cedex,France.

R. ZIESSEL Institut la Bel, Université Louis Pasteur, 4 Rue Blaise Pascal, 67000 Strasbourg, France.

Country : GERMANY

J. BECKER Lehrstuhl und Institut fur Technische Chemie und Petrolchemie, 5100 Aachen, Germany.

A. BEHR Lehrstuhl und Institut fur Technische Chemie und Petrolchemie, 5100 Aachen, Germany.

W. EISENBACH Max Planck Institut fur Kohlenforschung, Kaiser Wilhelm Platz 1, D 4330 Mulheim-Ruhr, Germany.

G. FUCHS Angewandte Mikrobiologie, Oberer Eselsberg M 23, D 7900 Ulm, Germany.

H. HOBERG Max Planck Institut fur Kohlenforschung, Kaiser Wilhelm Platz 1, D 4330 Mulheim-Ruhr, Germany.

U. KANNE Lehrstuhl und Institut fur Technische Chemie und Petrolchemie, 5100 Aachen, Germany.

W. KEIM Lehrstul und Institut fur Technische Chemie und Petrolchemie, 5100 Aachen, Germany.

W. KIPSHAGEN Lehrstuhl und Institut fur Technische Chemie und Petrolchemie, 5100 Aachen, Germany.

J. PERES Max Planck Institut fur Kohlenforschung, Kaiser Wilhelm Platz 1, D 4330 Mulheim-Ruhr, Germany.

Country : GREECE

D. PETRIDIS Ministry of Research and Technology Greek Atomic Commission, P.O. Box 60228, 15130 Aghia Paraskevi, Greece.

Country : ITALY

E. ALESSIO	Dipartimento di Scienze Chimiche, Piazzale Europa n. 1, 34127 Trieste,Italy.
M. ARESTA	Dipartimento di Chimica, Via Amendola n. 173, 70126 Bari,Italy.
A. BELFORTE	Dipartimento di Chimica e Chimica Ind.le, Via Risorgimento n. 35, 56100 Pisa,Italy.
I. BERTINI	Dipartimento di Chimica, Via Gino Capponi n. 7, 50121 Firenze,Italy.
A. CAMUS	Dipartimento di Scienze Chimiche, Piazzale Europa n. 1, 34127 Trieste,Italy.
A. CICCARESE	Centro di Biologia, Università degli Studi, Via Prov.le Lecce-Monteroni, 73100 Lecce,Italy.
A. FLAMINI	CNR ITSE, Area della Ricerca di Roma, Casella Postale 10, 00016 Monterotondo Stazione - Roma,Italy.
G. FORTI	Dipartimento di Biologia,Via Celoria n.26, 20100 Milano,Italy.
A. GENNARO	Dipartimento di Chimica Fisica, Via Loredan n. 2 35131 Padova,Italy.
A. GERVASINI	Dipartimento di Chimica Inorganica e Metallorganica , Via G. Venezian n. 21, 20133 Milano,Italy.
R. GALLERANI	Laboratorio di Chimica Biologica, Centro di Biologia , Univ. Studi , Via Prov.le Lecce - Monteroni, 73100 Lecce, Italy.
G. LA MONICA	Dipartimento di Chimica Inorganica e Metallorganica, Via G. Venezian n. 21, 20133 Milano,Italy.
P. LEONI	Dipartimento di Chimica e Chimica Ind.le, Via Risorgimento n. 35, 56100 Pisa,Italy.
C. LUCHINAT	Dipartimento di Chimica, Via Capponi n.7, 50121 Firenze.
G. MESTRONI	Dipartimento di Scienze Chimiche, Piazzale Europa n.1, 34127 Trieste.
R. MONNANNI	Dipartimento di Chimica, Via Capponi n.7, 50121 Firenze, Italy.

P. MURA — Istituto di Strutturistica Chimica "G. Giacomello", Area della Ricerca di Roma, Casella Postale 10, 00016 Monterotondo Stazione (Roma), Italy.

E. QUARANTA — Dipartimento di Chimica, Via Amendola N. 173, 70126 Bari,Italy.

F. RIVETTI — Enichem Sintesi, Laboratorio di Bolgiano, Via Maritano n. 26, 20097 San Donato Milanese,Italy.

M. ROSI — Dipartimento di Chimica, Università di Perugia, Via Elce di Sotto n. 8, 06100 Perugia,Italy.

G. SILVESTRI — Istituto di Ingegneria Chimica, Facoltà di Ingegneria, Facoltà di Ingegneria, V.le delle Scienze, 90128 Palermo,Italy.

V. VALENTI — Fisiologia Vegetale, Istituto Botanico, Via Irnerio n.42, 40126 Bologna,Italy.

E. VIANELLO — Dipartimento di Chimica Fisica, Via Loredan n. 2, 35131 Padova,Italy.

Country : NETHERLANDS

J.C.J. BART — DSM Research BV, P.O. Box 18, 6160 MD Gelee, The Netherlands.

A. MACKOR — TNO Division of Technology for Society, P.O. Box 108, 3700 AC Zeist, The Netherlands.

Country : NORWAY

R. SIREVÅG — Botanical Laboratory University of Oslo, Box 1045, N 0316, Oslo 3, Norway.

Country : PORTUGAL

A.T. CAETANO — L.N.E.T.I., Departamento de Energias Renovaveis, Estrada do Paço do Lumiar n. 22, 1600 Lisboa, Portugal.

J.L. FIGUEIREDO Faculty of Enegineering, University of Porto, 4099 Porto Codex, Portugal.

J. MENAIA Laboratorio Nacional de Engenharia e Tecnologia Industrial, Estradas das Palmeiras, 2745 Queluz, Portugal.

Country : SPAIN

E. CARMONA Faculdad Quimica, Universidad de Sevilla, Espana.

Country : TURKEY

S. ÇELEBI Dokuz Eylul Univ. Muh-Mim. Fakultesi Çevre Muhendlisligi Bolumu, Bornova - Izmir, Turkey.

D. COBANOGLU Firat Universitesi Fen-Ed.Facultesi Biyoloji Bolumu - Elazig,Turkey.

R. TIPIRDAMAZ Faculty of Science - Hacettepe University - Beytepe - Ankara , Turkey.

Country : UNITED KINGDOM

N.M. DIXON Department of Botany and Microbiology - University College of Wales - Aberystwith, Dyfed SY23 3DA , UK

B.E. MANN Department of Chemistry - University of Sheffield - Sheffield S3 7HF ,UK

M.N. SIVAK Research Institute of Photosynthesis - Department of Botany - Sheffield S10 2TN , UK

Country : UNITED STATES OF AMERICA

C. BAUCH	Department of Chemistry - Texas A&M University - College Station,Texas 77843 , USA
M. BUHL	Chemistry Department - University of Missouri - Rolla - Missouri 65401 , USA
K. CAULTON	Department of Chemistry - Chemistry Building - Bloomington , Indiana 47405 , USA
R. EISENBERG	College of Arts and Science - University of Rochester River Station - Rochester - New York 14627 , USA
F.K.FONG	Department of Chemistry - Purdue University - West Lafayette - IN 47907 , USA
J.A.IBERS	Department of Chemistry - College of Arts and Sciences Northwestern University - Evanston, Illinois 60201 ,USA
E. LUNDQUIST	Department of Chemistry - Chemistry Building - Bloomington - Indiana 47405 , USA.
B.SIMMS	Department of Chemistry - Northwestern University Sheridan Road - Evanston - 60201 Illinois

Country : NON NATO

INDIA

G.K. PARASHAR C-444 Saraswati Vihar, Delhi 110034, India.

JAPAN

T. AKAZAWA School of Agriculture, Nagoya University, Chikusa Nagoya, Japan.

S. INOUE Department of Synthetic Chemistry, Faculty of Engineering, University of Tokyo, Hongo, Bunkyo-Ku, Tokyo,Japan.

SOMALILAND

A.H. ISSE Università Nazionale Somala - Dipartimento di Chimica Mogadiscio-Somaliland.

SWITZERLAND

J. AUGUSTYNSKI Département de Chimie Minérale Analytique et Appliquée, Quai Ernest-Anserment 30, 1211 Genéve 4, Switzerland.

R. NAEGELI Laboratorium fur Anorganische Chemie, Universitatsstr. n. 6, CH-8092 Zurich, Switzerland.

CONCLUSIONS

Conclusions.

M.Aresta

The carbon dioxide accumulation in the atmosphere is a fact. It is strictly connected to the mankind energy needs,to the living standard and,thus, to the utilization ofcarbon based fuels.This problem is of wide concern today. It has been suggested that the only serious way to avoid the carbon dioxide accumulation into the atmosphere is the limitation of fuel combustion. The carbon dioxide recovery from the atmosphere has been told to be quite hopeless .
Plants can use CO_2, but their rate of fixation is very low compared to the rate of production of carbon dioxide. Moreover,the uncontrolled deforestation diminishes the natural capability to fix carbon dioxide.

Can Science help to repair the damage caused by the carbon dioxide accumulation in the atmosphere?

Carbon dioxide can be used as a source of carbon in synthetic chemistry. Does this represent the way to drive it out from the environment?
In some special cases it might be helpful either to recycle some carbon or to control the emission in the atmosphere.
The multiplicity of the aspects of the carbon dioxide fixation and its complexity requires the integrated competences of Scientists working in different areas.In the last ten years some efforts have been made in that direction.
Today at least four major uses of carbon dioxide can be identified. These are the utilization :

- as a building block for "fine chemicals" or for intermediates;
- as the starting material for other C1 molecules to be used as a source of energy;
- in "greenhouses" to grow up flowers and plants;
- as a solvent for industrial extractions.

The first three processes convert carbon dioxide,the last one uses

only it.
It is evident that no one of these processes represents the solution to the problem to avoid carbon dioxide accumulation in the atmosphere. Moreover, even if we fix carbon dioxide in a given product, using that product we shall regenerate carbon dioxide. So,the only real solution to the problem would be to convert carbon dioxide at the same rate we produce it.
Will this be possible in the future ?
What do we know about carbon dioxide chemical reduction ? How easy is it?
Can combined technologies (chemical and biochemical) be more productive than a single one ?

The utilization of carbon dioxide as a building block for chemicals can be useful from an economic point of view and might assure more safe working conditions in industrial plants.
It is useful, then, to investigate the carbon dioxide utilization in those synthetic procedures that today are based on carbonyl chloride, carbamoyl halides,dimethyl carbonate,isocyanates and related molecules.
For these reactions, new catalysts (homogeneous and heterogeneous) must be synthetized and characterized .
The knowledge of the way carbon dioxide binds to a metal center and of its reactivity can be helpful to tailor catalysts. One of the most important goals in this area is to convert into catalytic cycles those reactions that today are known as stoichiometric.
The synthesis of carbamates ,organic carbonates and ureas is an area of great industrial interest.
Electrochemistry can be a way for the selective synthesis of several interesting products.

It is of great interest the utilization of carbon dioxide in order to obtain methane,methanol,carbon monoxide, as this process might involve quite large amounts of carbon dioxide and bring to a "man made" process mimicking photosynthesis . If selective catalysts are discovered operating at a high rate, a man made cycle based on carbon dioxide will be possible. The utilization of visible light is a must in this area. One of the goals is the synthesis of new energy transfer agents and of more active catalysts (from a kinetic and selective point of view). Metal complexes can be used in a homogeneous phase,while binary and ternary compounds can be used in a heterogeneous one. A new area is concerned with the utilization of transition metal complexes supported over semiconductor.

Enzymes could be isolated from bio-systems and used free or supported. Studies on the electron transfer process in the carbon dioxide reduction are outstanding and comparation of biochemical to chemical pathways are of fundamental importance. The exact knowledge of the role of the active metal center in the enzymes can bring to the synthesis of metal complexes mimicking enzymes.

A strict collaboration of biochemists and chemists is required to elucidate the carbon dioxide reduction pathway in order to afford molecules such as $HCOOH$,CO,CH_2O,CH_3OH,CH_4. The chemistry and biochemistry of any of the above C1 molecule demands a great attention.

The use of carbon dioxide in industrial extraction procedures can be interesting from an economic and a technological point of view, but it can not suggest how to eliminate or to convert carbon dioxide. The investigation of systems able to separate carbon dioxide from other gases is of great interest,as this aspect is related to the separation from pollutants.

The carbon dioxide utilization in greenhouses must be also considered. It looks like an expiatory attempt to compensate the negative effect caused by the uncontrolled elimination of forests. It has assumed a restrict economic significance, at least today, but it might assume a different one if it is possible to grow , with an accurate control , algae or plants fixing carbon dioxide at a rate much higher than the usual.
The collaboration among plant physiologists, biochemists,agriculture specialists and chemists can give a contribution to this area.

SPECIALISTS REPORTS AND SUGGESTIONS FOR FUTURE WORK.

Comments: Industrial Applications of Carbon Dioxide

W. Keim and A. Behr

Carbon dioxide is one of the most abundant chemicals in nature. Nearly unlimited amounts are found in the atmosphere, dissolved in water, in gas wells and in solids as carbonates. Due to the widely practised combustion of the various carbonic resources the amount of CO_2 is steadily increasing. Based on its availability carbon dioxide represents an ideal compound for industrial applications. Therefore, it can be anticipated that CO_2's usage will further increase. Especially the food industry can be seen as a major future consumer. Also enhanced oil recovery could raise the consumption.

An other area of substantial growth can be seen in using supercritical carbon dioxide for extraction. W.O. Eisenbach presented convincing data for the extraction of natural products such as decaffination, deodorisation, isolation of flavours, separation of oils and fractionation.

Considering the chemical industry one is wondering, why the share of using CO_2 as building block is so small. Besides of various inorganic carbonates the industry can only show a few organic process.

Often CO_2 is considered to be chemically innert. This is a myth as is evident from the ease with which amins or energy rich compounds react with CO_2. R. Eisenberg presented thermodynamic considerations outlining that the synthesis of many chemicals is feasible.

Especially regarding potential future industrial applications the meeting was quite useful.

I. Tkatchenko outlined the different approaches using heterogeneous catalysts to react CO_2. Hydrocarbons and oxygenates such as methanol can be synthesized from CO_2/H_2.

Especially the role of CO_2 in methanol synthesis found many comments and J.C.J. Bart summarized our current standing in this field.

Special attention was paid to the role of enzymes and bacteria. It became quite obvious that biotechnology using these natural "catalysts" could revolutionize the chemical industry. Here the chemists learned from Reidun Sirevag, which described autotrophic bacteria or from G. Fuchs presenting CO_2 fixation in chemotrophic anaerobes.

F. Rivetti outlined remarkable possibilities of the carbonic acid esters which may be used as intermediates for the use of carbon dioxide as raw material in organic synthesis. High attention should be given to future development of the homogeneous transition metal catalyzed reactions of CO_2! Will it become possible to synthesize acrylic acid from ethene and carbon dioxide as shown in a stoichiometric reaction by E. Carmona? Will the future synthesis of cyclic carbonates occur via a rhodium catalyzed reaction of olefins, oxygen and CO_2 as suggested by A. Ciccarese and M. Aresta? Will unusual ligands such as the CO_2-carrier presented by M.C. Bonnet help to fix carbon dioxide on unsaturated hydrocarbons?

A certain number of homogeneous catalyzed reactions may find industrial application in the near future. The ruthenium catalyzed reaction of alkynes, amines and CO_2 as reported by P.M. Dixneuf appears to be an attractive no-phosgene route to vinylcarbamates. The transition metal catalyzed reaction of dienes and carbon dioxide enables the synthesis of a wide range of lactones in an industrial scale. As stressed by S. Lecolier it can be assumed that CO_2 will become soon a useful feedstock in the production of oxygen-containing fine chemicals.

COMMENTS : CO_2 COORDINATION CHEMISTRY

R. EISENBERG, J.A. IBERS.

The coordination chemistry of CO_2 was discussed from the standpoint of the solid state structures of known CO_2 complexes. Relatively few CO_2 complexes have been stable enough to allow crystallyzation and X-ray structural study. The coordination modes of CO_2 include ^{1}-attachment through carbon, ^{2}-attachment through one of the C=O bonds, and simultaneous ^{1}-bonding of C to M with a stabilizing Lewis acid-CO_2 oxygen interaction. In all cases the CO_2 molecule bound to the metal center is bent, characteristic of some degree of reduction.The significance of solid state structures to the development of catalytic cycles was discussed, and a consensus developed that these structures are useful models for CO_2 activation although they are undoubtedly too stable to participate in catalysis themselves. The great stability of CO_2 works against the formation of isolable CO_2 complexes. However, models for the different bonding modes which can be envisioned have been obtained.

The lack of stable CO_2 complexes does not necessarily indicate the the lack of any interaction between CO_2 and a metal center. In cases where a carbonyl complex forms concomitantly with CO_3^{2-} formation, there is clear evidence that a metal- CO_2 interaction has taken place. One possible application of this frequently uncountered and undesidered reaction is to use it as an assay for metal-CO_2 interaction. This can be done routinely using infrared spectroscopy. The availability of ^{13}C NMR spectroscopy now makes it essential to employ $^{13}CO_2$ for unequivocal spectroscopic identification of cordinated CO_2 and confirmation of its source. Attempts should be made to correlate structure with ^{13}C chemical shifts and other spectroscopic properties.

COMMENTS: ELECTROREDUCTIONS OF CO_2

R. EISENBERG, G. SILVESTRI

The electroreduction of CO_2 has shown significant development during the last five years in the use of metal complexes as electrocatalysts. The thermodynamics of the possible reductions were reviewed, but because the different systems reported operate underdistinctly non-standard state conditions, some care should be exercised in determining G's of various reductions.

The types of complexes which have been used as electrocatalyst include tetraaza macrocycles of Ni and Co, $Re(bipy)X(CO)_3$, $Ph(dppe)_2^+$ and $Fe_4S_4(SR)_4^{2-}$. In every case, the nature of the interaction between CO_2 and the electrocatalyst and the intimate mechanism of reduction are unknown. Selectivity for some of the electrocatalyst are quite good, such as in the electroreduction of CO_2 to CO promoted by $Ni(cyclam)^{2+}$in aqueous solution. Determination of the basis of selectivity will require detailed mechanistic information. The effectiveness of any electrocatalyst will be measured only in part by reduction of the overpotential required for reduction; selectivity and catalyst stability are two others.

The effectiveness of all systems for CO_2 electroreduction should be established by unambiguous identification and quantification of all reduction products, determination of current efficiencies and faradaic yields, use of labelled CO_2 to verify the source of reduced carbon products, and assessment of the lifetime of the electroreduction system including the catalyst.

COMMENTS: CARBONIC ANHYDRASE.

I. BERTINI, S. BROWN

The interconversion between CO_2 and HCO_3^- is a primary process absolutely required by any living system. The process at physiological pH is slow and would not allow life in absence of a specific enzyme. Carbonic anhydrase (CA) is such an enzyme which is therefore ubiquitous. There are several isoenzymes even in a single living being which differ both in structure and activity although all seem to catalyze only CO_2 hydration. The evolutionary process and the reasons for the structural differences are not yet fully understood. In vitro a lot of work has been done which has led to a deep understanding of the mechanism, to the knowledge of the rate limiting steps, to the molecular understanding of inhibition. These studies represent a great advancement in the understanding of the function of the enzyme in vivo. For the complete understanding of the mechanism at the molecular level the identification the identification of the binding site for CO_2, if any, is still required. Application of site directed mutagenesis on the enzyme is expected to prove hypotheses and to provide further insights.

As far as the mimics for the properties of CA are concerned, reasonablemodels for the spectroscopic and structural properties of the active site region are available. These have allowed us to provide small-molecule evidence for the ionization of a M^{2+} $-OH_2$ to form M^{2+} $-OH^-$ which has probable bearing on the origin of the activity and spectroscopic pK_a for the enzyme . More problematic are the catalytic models for the enzyme mediated interconversion of CO_2 and HCO_3^-. At least two major areas require further investigation. Evidence exists that low coordination numbers and a hydrophobic environment for the metal are catalytically important. However none of the present catalysts exceeds OH^- in terms of its ability to hydrate CO_2. Secondly, none of the present models has a sufficient ability to bind M^{2+} at high pH, although the enzyme does. The major problem involves competition of the ligand and OH^- for the metal ion, and invariably insoluble $M^{2+}(OH^-)_x$ precipitates are formed. Further substantial effort in these two areas is required as an aid to understanding of the enzyme catalyzed process.

COMMENTS: BIOLOGICAL REDUCTION OF CO_2 .

D.T. CANVIN, M.N. SIVAK

Several carboxylations occur in biological organisms, but the carboxylations of photosynthesis are the most abundant. All photosynthetic processes use reducing power and ATP generated through the conversion of light energy. In C_3 photosynthesis the initial (and only) carboxylation is the carboxylation of ribulose bisphosphate (RuBP) which on subsequent hydrolysis yields two molecules of 3-phosphoglyceric acid (3- PGH). Their reaction is catalyzed by ribulose bisphosphate carboxylase/oxygenase (RuBisco). Reduction of the fixed carbon dioxide (3-PGA) to the level of carbohydrate,using the reducing power and ATP from the light reaction is catalyzed by another enzyme, triose phosphate dehydrogenase. The remaining reaction of the photosynthetic reduction cycle (PCR) are involved in regenerating the carbon dioxide acceptor, RuBP. Except for some bacteria all photosynthetic organisms contain the PCR cycle. The ubiquitousness of their cycle might, in part, be experted because it is the only carboxylating cycle that is able to regenerate and increase the amount of the CO_2-acceptor.

In addiction to carboxylation Rubisco catalyzes the oxygenation (with molecular O_2) of RuBP to yield a molecule of 2-phosphoglycolate (2-PG) and a molecule of 3-phosphoglyceric acid. The conversion of the 2-PG through the glycolate pathway results via the release of one-quarter of the fixed carbon as CO_2 (a process called photorespiration) and the recovery of three-quarters as 3-PGA. Oxygen and CO_2 are competitive substrates for Rubisco and the competition of oxygen at the catalytic site of the enzyme reduces photosynthesis about 30% under ambient condition, while the release of CO_2 adds a further 15% inhibition.

The mechanism of the carboxylation is reasonably well understood but the mechanism of the oxygenation is only partially known. The reduction of photosynthesis that occurs because of the oxygenation reaction is considered to be a substantial limitation to C_3 plant productivity and efforts to curtail or eliminate their reaction are to be encouraged. The probability of progress to this end is likely low, as evolution has quasi no new examples and C_3 plant photosynthesis is still the most common type of photosynthesis in the plant kingdom.

Some plants, however, have developed processes additional to the PGR cycle, which provide CO_2 concentrations at the enzyme higher than the

CO_2 concentration in air and the elevated CO_2 concentrations suppresses the oxygenase reaction. In C_4 plants the leaf cells have differentiated in the mesophyll cells and bundle sheath cells and the latter contain Rubisco and the PCR cycle. The initial carboxylation occurs in the mesophyl cells where the bicarbonate ion is combined with phosphoenolpyruvate to form oxalacetic acid, a reaction catalyzed by phosphoenolpyruvate carboxylase. After reduction to malic acid or amination to aspartate, the product of the carboxylation is transported to the bundle sheath cells where it is decarboxylated to provide CO_2 for a carboxylation catalyzed by Rubisco. The CO_2 concentration in the bundle sheath cells is increased because the rate of C_4-acid transport and decarboxylation exceeds the rate of Rubisco carboxylation and loss of CO_2 is prevented from the bundle sheat cells because the membrane of the cells have a low permeability to CO_2.

In Crassulacean Acid Metabolism (CAM) plants the formation of malic acid through phosphoenolpyruvate generated from starch as the HCO_3^- acceptor. During the day this malic acid is decarboxylated to provide CO_2 for Rubisco and the PCR cycle. As in C_4 plants the CO_2 concentration can be increased because decarboxylations exceed carboxylations and the escape of CO_2 from the plant is prevented by the epidermic and closure of stomata.

In green algae and cyanobacteria(and perhaps in many algae) inorganic carbon transport systems have been developed. These systems can transport either CO_2 or HCO_3 using energy from light and CO_2 concentrations in the vicinity of Rubisco may be a thousand fold higher than that external to the cell. Carbon dioxide leakage from the cell appears to be reduced by cell membranes with low permeability to CO_2. As a result of the above modifications C_4 plants and algae are able to use low concentration of CO_2 more effectivily and maximum photosynthesis is normally achieved at lower CO_2 concentration than in C_3 plants. Although the CO_2 concentration of the atmosphere is steadily increasing, the present CO_2 concentration is far less than that needed to saturate photosynthesis of C_3 plants. A fuller understanding of the CO_2 concentrating mechanism that have evolved in plants may provide a base for increasing the productivity of our economically important C_3 plants.

Index

A

K

L

M

N

O

Q

R

S

GPSR Compliance

The European Union's (EU) General Product Safety Regulation (GPSR) is a set of rules that requires consumer products to be safe and our obligations to ensure this.

If you have any concerns about our products, you can contact us on ProductSafety@springernature.com

In case Publisher is established outside the EU, the EU authorized representative is:

Springer Nature Customer Service Center GmbH
Europaplatz 3
69115 Heidelberg, Germany

Batch number: 10809439

Printed by Printforce, the Netherlands